Fungal Pigments 2021

Fungal Pigments 2021

Editor

Laurent Dufossé

MDPI • Basel • Beijing • Wuhan • Barcelona • Belgrade • Manchester • Tokyo • Cluj • Tianjin

Editor
Laurent Dufossé
ESIROI Agroalimentaire
Université de La Réunion
Saint-Denis
France

Editorial Office
MDPI
St. Alban-Anlage 66
4052 Basel, Switzerland

This is a reprint of articles from the Special Issue published online in the open access journal *Journal of Fungi* (ISSN 2309-608X) (available at: www.mdpi.com/journal/jof/special_issues/fungal_pigments_2021).

For citation purposes, cite each article independently as indicated on the article page online and as indicated below:

LastName, A.A.; LastName, B.B.; LastName, C.C. Article Title. *Journal Name* **Year**, *Volume Number*, Page Range.

ISBN 978-3-0365-5812-7 (Hbk)
ISBN 978-3-0365-5811-0 (PDF)

Contents

About the Editor

Laurent Dufossé

Laurent Dufossé has held the position of Professor of Food Science and Biotechnology since 2006, at Reunion Island University, which is located on a volcanic island in the Indian Ocean, near Madagascar and Mauritius. The island is one of France's overseas territories, with almost one million inhabitants, and the university has 19,000 students. Previously, Professor Dufossé was a researcher and senior lecturer at the Université de Bretagne Occidentale, Quimper, Brittany, France. He attended the University of Burgundy, Dijon, France, where he received his PhD in Food Science in 1993, and has been involved in the field of the biotechnology of food ingredients for more than 30 years. His main research interests over the last 20 years have mainly been the microbial production of pigments and the study of aryl carotenoids, such as isorenieratene, C50 carotenoids, azaphilones and anthraquinones. These studies have had relevance for applications in food science and technology, in areas such as the cheese industry and sea salt industry.

Preface to "Fungal Pigments 2021"

Following the previous *Journal of Fungi* (ISSN 2309-608X) "Fungal Pigments ", Special Issue edited and published in 2017, with 10 papers and the "Fungal Pigments ", printed edition of 134 pages, the time has come to publish a new edition entitled "Fungal Pigments 2021 ", which present the latest scientific advances in this field.

With the impact of globalization in research trends, the search for healthier lifestyles, the increasing public demand for natural, organic, and 'clean labelled'products, as well as the growing global market for natural colorants in economically fast-growing countries all over the world, filamentous fungi started to be investigated as readily available sources of chemically diverse pigments and colorants. For all of these reasons, this Special Issue of the *Journal of Fungi* highlight exciting findings, which may pave the way for alternative and/or additional biotechnological processes for industrial applications of fungal pigments and colorants. Research papers and reviews about fungal biodiversity from terrestrial and marine origins were welcome, bringing new elements about fungi as potential sources of well-known carotenoid pigments (e.g., beta-carotene, lycopene) and other specific pigmented polyketide molecules, such as Monascus and Monascus-like azaphilones, which are yet not known to be biosynthesized by any other organisms, like higher plants. These polyketide pigments also include promising, and unexplored hydroxy-anthraquinoid colorants from Ascomycetous species. The investigation of biosynthetic pathways of the carotenoids and polyketide-derivative-colored molecules (i.e., azaphilones, hydroxyanthraquinones, and naphthoquinones) in pigment-producing fungal species brought some articles. Contributions about alternative greener extraction processes of fungal colored compounds, along with current industrial applications, description of their limits, and further opportunities for the use of fungal pigments in beverage, food, pharmaceutical, cosmetic, textile, and painting areas are part of this Special Issue.

Laurent Dufossé
Editor

Journal of
Fungi

torial

ıngal Pigments: More Insights from Colorful Fungi

ırent Dufossé

CHEMBIOPRO Laboratoire de Chimie et Biotechnologie des Produits Naturels, ESIROI Agroalimentaire, Université de La Réunion, 15 Avenue René Cassin, F-97400 Saint-Denis, Ile de La Réunion, France; laurent.dufosse@univ-reunion.fr; Tel.: +33-262-692-402-400

Following the previous Journal of Fungi (ISSN 2309-608X) Fungal Pigments Special Issue edited and published in 2017 (weblink https://www.mdpi.com/journal/jof/special_issues/fungal_pigments), with 10 papers and the Fungal Pigments printed book edition of 134 pages (weblink https://www.mdpi.com/books/book/570), the time has come to open a new edition entitled Fungal Pigments 2021 which presents the latest scientific advances in this field from August 2020 to August 2021.

With the impact of globalization in research trends, the search for healthier lifestyles, the increasing public demand for natural, organic, and "clean labelled" products, as well as the growing global market for natural colorants in economically fast-growing countries all over the world, filamentous fungi started to be investigated as readily available sources of chemically diverse pigments and colorants. For all of these reasons, this new Special Issue of the Journal of Fungi intends to highlight exciting findings, which may pave the way for alternative and/or additional biotechnological processes for industrial applications of fungal pigments and colorants. Research papers and reviews on fungal biodiversity from terrestrial and marine origins were welcomed, thus contributing new findings on fungi as potential sources of well-known carotenoid pigments (e.g., beta-carotene, lycopene) and other specific pigmented polyketide molecules, such as *Monascus* and *Monascus*-like azaphilones, which are not yet known to be biosynthesized by any other organisms such as higher plants. These polyketide pigments also include promising and unexplored hydroxy-anthraquinoid colorants from Ascomycetous species. The investigation into the biosynthetic pathways of the carotenoids and polyketide-derivative colored molecules (i.e., azaphilones, hydroxyanthraquinones, and naphthoquinones) in pigment-producing fungal species contribute to the additional articles. Contributions on alternative greener extraction processes of fungal colored compounds, along with current industrial applications, descriptions of their limits, and further opportunities for the use of fungal pigments in beverage, food, pharmaceutical, cosmetic, textile, and painting areas are also projected to be part of this Special Issue.

tion: Dufossé, L. Fungal
ıents: More Insights from
rful Fungi. *J. Fungi* **2022**, *8*, 1109.
s://doi.org/10.3390/jof8101109

ived: 17 October 2022
pted: 19 October 2022
ished: 20 October 2022

lisher's Note: MDPI stays neutral
regard to jurisdictional claims in
ished maps and institutional affilns.

All these subjects and more are covered by the 11 articles published in this Special Issue:

weblink https://www.mdpi.com/journal/jof/special_issues/fungal_pigments_2021.

* *Investigation on various chemical classes of fungal pigments (melanins, azaphilones, quinones, etc.)*

Genomic Analysis and Assessment of Melanin Synthesis in *Amorphotheca resinae* KUC3009 by Jeong-Joo Oh et al.; https://doi.org/10.3390/jof7040289.

Fungal Melanins and Applications in Healthcare, Bioremediation and Industry by Ellie Rose Mattoon et al.; https://doi.org/10.3390/jof7060488.

Recent Findings in Azaphilone Pigments by Lúcia P. S. Pimenta et al.; https://doi.org/10.3390/jof7070541.

Characterization of a Biofilm Bioreactor Designed for the Single-Step Production of Aerial Conidia and Oosporein by *Beauveria bassiana* PQ2 by Héctor Raziel Lara-Juache et al.; https://doi.org/10.3390/jof7080582.

* *Molecular characterization*

Molecular Characterization of Fungal Pigments by Miriam S. Valenzuela-Gloria et https://doi.org/10.3390/jof7050326.

* *Biological properties*

Seven New Cytotoxic and Antimicrobial Xanthoquinodins from *Jugulospora vestita* Lulu Shao et al.; https://doi.org/10.3390/jof6040188.

* *Toxicity assessment and safety evaluation of fungal pigments*

Safety Evaluation of Fungal Pigments for Food Applications by Rajendran Poorniamr et al.; https://doi.org/10.3390/jof7090692.
Preliminary Examination of the Toxicity of Spalting Fungal Pigments: A Comparis between Extraction Methods by Badria H. Almurshidi et al.; https://doi.org/10.3390/j 020155.

* *Use of by-products or waste for industrial production of fungal pigments*

Production of Bio-Based Pigments from Food Processing Industry By-Products (ple, Pomegranate, Black Carrot, Red Beet Pulps) Using *Aspergillus carbonarius* by E Bezirhan Arikan et al.; https://doi.org/10.3390/jof6040240.

* *Prospective aspects and brainstorming*

Does Structural Color Exist in True Fungi? by Juliet Brodie et al.; https://doi.org/10. 0/jof7020141.
Fungal Biomarkers Stability in Mars Regolith Analogues after Simulated Space a Mars-like Conditions by Alessia Cassaro et al.; https://doi.org/10.3390/jof7100859.

I, as Guest Editor, trust all readers of this Special Issue enjoy the contents and I wo like to deeply thank all 62 authors who contributed (sorted by their last names), Prof. David S. Perlin, Editor-in-Chief of the Journal of Fungi, the numerous reviewers, and whole team at MDPI (editing, production, website, etc.):

Aguilar, Cristóbal N.	Dufossé, Laurent	Paul de Vera, Jean-Pierre
Aguilar, Oscar	Gomes, Dhionne C.	Pimenta, Lúcia P.S.
Aguilar-Zárate, Mayra	Harper, Bryan	Poorniammal, Rajendran
Aguilar-Zárate, Pedro	Harper, Stacey	Prabhu, Somasundaram
Almurshidi, Badria H.	Hernández-Almanza, Ayerim	Rabbow, Elke
Arikan, Ezgi Bezirhan	Ingham, Colin J.	Ramona Michel, Mariela
Ascacio-Valdés, Juan Alberto	Kannan, Jegatheesh	Robinson, Seri C.
Ávila-Hernández, José Guadalupe	Kim, Gyu-Hyeok	Rodríguez-Durá, Luis Víctor
Balagurusamy, Nagamani	Kim, Jee Young	Saladino, Raffaele
Baqué, Mickael	Kim, Jung Woo	Shao, Lulu
Botta, Lorenzo	Kim, Young Jun	Stadler, Marc
Böttger, Ute	Kwon, Sun Lul	Stchigel, Alberto M.
Brodie, Juliet	Lara-Juache, Héctor Raziel	Surup, Frank
Canli, Oltan	Lee, Changsu	Takahashi, Jacqueline A.
Cardoso, Patrícia G.	Lee, Myeong-Eun	Valenzuela-Gloria, Miriam S
Caro, Yanis	Marin-Felix, Yasmina	Van Court, R.C.
Casadevall, Arturo	Mattoon, Ellie Rose	Veana, Fabiola
Cassaro, Alessia	Muñiz-Márquez, Diana Beatriz	Vega Gutierrez, Sarath M.
Chávez-González, Mónica L.	Oh, Jeong-Joo	Vignolini, Silvia
Cordero, Radames J.B.	Onofri, Silvano	Wong-Paz, Jorge Enrique
Dizge, Nadir	Pacelli, Claudia	

Conflicts of Interest: The author declares no conflict of interest.

Journal of
Fungi

ticle

;enomic Analysis and Assessment of Melanin Synthesis in *morphotheca resinae* KUC3009

›ng-Joo Oh [1], Young Jun Kim [2], Jee Young Kim [1], Sun Lul Kwon [1], Changsu Lee [3], Myeong-Eun Lee [4], ng Woo Kim [5] and Gyu-Hyeok Kim [1,*]

1. Division of Environmental Science & Ecological Engineering, College of Life Sciences & Biotechnology, Korea University, 145, Anam-ro, Seongbuk-gu, Seoul 02841, Korea; oheuy1027@korea.ac.kr (J.-J.O.); kjy4142@korea.ac.kr (J.Y.K.); sun-lul@korea.ac.kr (S.L.K.)
2. Life Science and Biotechnology Department, Underwood Division, Underwood International College, Yonsei University, Seoul 03722, Korea; youngjn.kim@yonsei.ac.kr
3. Microbiology and Functionality Research Group, World Institute of Kimchi, Gwangju 61755, Korea; lckslick@gmail.com
4. Department of Biotechnology, College of Life Sciences & Biotechnology, Korea University, 145, Anam-ro, Seongbuk-gu, Seoul 02841, Korea; myeongeun88@gmail.com
5. Department of Biomedical Engineering, Sungkyunkwan University, Suwon, 2066 Seobu-ro, Jangan-gu, Suwon 16419, Korea; didch1789@gmail.com

* Correspondence: lovewood@korea.ac.kr; Tel.: +82-2-3290-3014

Abstract: This study reports the draft genome of *Amorphotheca resinae* KUC30009, a fungal isolate with promising industrial-scale melanin production potential. The mechanisms for melanin or melanin-related pigment formation of this strain were examined through bioinformatic and biochemical strategies. The 30.11 Mb genome of *A. resinae* contains 9638 predicted genes. Genomic-based discovery analyses identified 14 biosynthetic gene clusters (BGCs) associated with secondary metabolite production. Moreover, genes encoding a specific type 1 polyketide synthase and 4-hydroxynaphthalene reductase were identified and predicted to produce intermediate metabolites of dihydroxy naphthalene (DHN)-melanin biosynthesis pathway, but not to DHN-melanin. These findings were further supported by the detection of increased flaviolin concentrations in mycelia and almost unchanged morphologies of the culture grown with tricyclazole. Apart from this, the formation of melanin in the culture filtrate appeared to depend on the laccase-like activity of multi-copper oxidases. Simultaneously, concentrations of nitrogen-containing sources decreased when the melanin formed in the media. Interestingly, melanin formation in the culture fluid was proportional to laccase-like activity. Based on these findings, we proposed novel strategies for the enhancement of melanin production in culture filtrates. Therefore, our study established a theoretical and methodological basis for synthesizing pigments from fungal isolates using genomic- and biochemical-based approaches.

Keywords: *Amorphotheca resinae*; fungal melanin; bioinformatics; melanin pigments

ation: Oh, J.-J.; Kim, Y.J.; Kim, J.Y.; on, S.L.; Lee, C.; Lee, M.-E.; Kim, ; Kim, G.-H. Genomic Analysis Assessment of Melanin Synthesis *morphotheca resinae* KUC3009. *J. gi* **2021**, *7*, 289. https://doi.org/390/jof7040289

demic Editor: Laurent Dufossé

eived: 22 March 2021
epted: 9 April 2021
lished: 12 April 2021

1. Introduction

Contrary to their synthetic counterparts, the demand for natural pigments has been steadily increasing in recent years in response to global market shifts and consumer preferences [1,2]. Filamentous fungi have been gaining recognition as a potential microbial source of natural pigments; however, the industrial applications of these microorganisms are not as widespread as those involving algae or bacteria [3]. Fungi can potentially produce a wide range of pigments, such as carotenoids, *Monascus* pigments, and melanins [4–8], and can utilize a wide range of substrates, thus making the medium composition design and fermentation process more flexible [9].

However, fungal pigments are secondary metabolites, and most of their synthesis pathways and optimal production conditions remain largely unknown [10], which limits their optimization and widespread adoption in industrial applications. Fortunately, with the accumulation of high-throughput sequencing data, bioinformatics tools can be us to identify putative genes or gene clusters involved in metabolite production. Afterwa comprehensive prediction of the provisional biosynthetic pathways of pigments would possible based on the identified putative genes or gene clusters [11]. Therefore, genon studies could provide fundamental insights into the pathways associated with seconda metabolite synthesis in fungi, thus paving the way for their adoption in industrial-scale p cesses. We previously reported the promising capacity of the fungus *Amorphotheca resin* KUC3009 to produce melanin at an industrial scale due to its antioxidant activity and hi metal ion adsorption capability [8,12]. However, general information on the mechanisr of melanin or melanin-related pigment biosynthesis in *A. resinae* is not yet available. The fore, understanding the mechanisms involved in melanin biosynthesis using bioinforma tools would provide critical insights to optimize scalable pigment production using fun

Our study sequenced and assembled the whole genome of *A. resinae* KUC3009. Aft gene annotation, the putative genes and gene clusters involved in pigment formati were comprehensively investigated. Afterward, the involvement of putative genes or ge clusters was further corroborated with biochemical and molecular approaches. Based these results, we proposed potential strategies for the optimization of scalable melan production. Therefore, our study establishes a robust foundation for the production secondary metabolites using *A. resinae*, as well as fundamental guidelines for future stud to improve upon.

2. Materials and Methods

2.1. Fungal Culture and DNA/RNA Extraction from Mycelia

A. resinae KUC3009 obtained from the Korea University Culture (KUC) collecti was sub-cultured on sterilized cellophane membrane disks placed on potato dextrose ag for seven days at 25 °C. The mycelia on the cellophane membrane were then harveste and their genomic DNA was extracted using a DNeasy plant mini kit (Qiagen, Valenc CA, USA) according to the manufacturer's instructions [13]. The quantity and quality the extracted DNA were analyzed via the PicoGreen® method and gel electrophoresis. To RNA was also extracted from the same cultures using the RNeasy plant mini kit (Qiag CA, USA) according to the manufacturer's instructions. The quantity and quality of t extracted RNA were assessed using an Agilent 2100 bioanalyzer (Agilent Technologi Santa Clara, CA, USA) coupled with a DNA 1000 chip.

2.2. Library Construction and Whole-Genome Sequencing

Library construction and genome sequencing were carried out by Macrogen Co. L (Seoul, Korea). A DNA library with fragment sizes of approximately 20 kb was prepar with the SMRTbell template prep kit 1.0 (Pacific Biosciences, Menlo Park, CA, USA) and quenced using the PacBio Sequel platform. Additionally, short reads were also construct and sequenced with an Illumina HiSeq 4000 sequencer (San Diego, CA, USA). De no assembly of PacBio sequence reads was performed using the hierarchical genome assemb process (HGAP) v4.0. After assembly, HiSeq reads were implemented to ensure a mo accurate genome sequence using Pilon v1.21 [14]. An RNA-Seq library was also prepar and sequenced with an Illumina HiSeq 4000 system for high-quality genome annotatio

2.3. Genome Analysis, Annotation, and Phylogenetic Analyses

The completeness of the draft genome assembly was evaluated with BUSCO v3.C using 1315 core genes of the Ascomycota dataset [15]. Genomic similarity compariso between *A. resinae* and other strains were performed using orthoANI [16]. After t draft genome was analyzed, the gene locations were predicted using Maker (v2.31.8) [1 tRNA and rRNA genes were predicted using tRNAscan (v1.4) and barrnap (v0.7, htt //github.com/tseemann/barrnap, accessed on 1 November 2018), respectively [18]. T functions of the predicted genes were annotated using Protein BLAST+ (v2.6.0), after whi the annotated genes were classified according to KOG analysis [19]. Gene clusters relat to secondary metabolism were analyzed using antiSMASH Fungi v6.0 and secondary metabolite regions were identified using a "relaxed" strictness [20]. Lastly, Signal P5.0 was used to predict the presence of the signal peptide in translated products from the putative genes [21].

The internal transcribed spacer (ITS) region was selected for phylogenetic analysis of the selected fungus. The ITS region was amplified using the ITS1F (5′-CTT GGT CAT TTA GAG GAA GTA A-3′) and LR3 (5′-CCG TGT TTC AAG ACG GG-3′) primers. Polymerase chain reaction (PCR) was performed on a Bio-Rad MyCycler (Hercules, CA, USA) with the following protocol: initial denaturation at 95 °C for 5 min; 34 cycles at 95 °C (30 s), 55 °C (30 s), and 72 °C (30 s); final 5-min extension at 72 °C. DNA sequencing was carried out by Macrogen (Seoul, Korea) using the Sanger method with a 3730xl DNA analyzer (Life Technologies, Carlsbad, CA, USA). The ITS sequences were deposited in the GenBank database under accession numbers JN033458.2. The obtained ITS sequences were proofread and edited using reference sequences obtained from the GenBank database using MEGA v7.0, after which multiple alignments were conducted using MAFFT v7.130 [22,23]. The sequence alignments were manually modified when necessary. Additionally, a phylogenetic tree was constructed based on the ITS sequences of *A. resinae* and *Cladosporium*-like species using the "randomized accelerated maximum-likelihood" (RAxML) model coupled with the GTR+G evolution model and 1000 bootstrap replicates [24,25].

2.4. Tricyclazole Inhibition Assay and Measurement of Flaviolin in Mycelia

A. resinae was subcultured on potato dextrose agar for 10 days. Spores were then collected in 0.02% Tween-80, and their concentrations were adjusted to 1×10^6 spores/mL. Afterward, cellophane membrane disks were inoculated with 10 μL of spore suspension and placed on potato dextrose agar (PDA) supplemented with different concentrations of tricyclazole ranging from 0 to 100 ug/mL. Colony sizes and colors were evaluated after 10 days via microscope imaging.

The secondary metabolites in the grown mycelia were extracted as described by Lisec et al. [26]. The mycelia on the cellophane membrane were harvested and flash-frozen with liquid nitrogen. The frozen mycelia were then ground, after which 100 mg of sample was transferred to 1.5 mL microcentrifuge tubes containing 1.0 mL of methanol to extract the metabolites. After adding 60 μL of a ribitol solution (0.2 mg/mL), the sample was incubated for 10 min at 70 °C with mild mixing. The incubated sample was centrifuged at 3500 rpm for 20 min, and the supernatant was transferred to a new tube. The supernatant was then mixed with 750 μL of chloroform and 1400 μL of distilled water, and the mixture was thoroughly vortexed. The mixed solution was centrifuged at 3500 rpm for 20 min, after which 150 μL of the supernatant was transferred to a new 1.5 mL tube. The extract was then fully dried and stored at −80 °C. Synthetic flaviolin was also prepared according to a previous study, and stock solutions were fully dried prior to analysis [27].

To derivatize both the synthetic and naturally occurring flaviolin in the mycelia, the samples were mixed with 40 μL methoxyamination reagent (20 mg/mL of methoxyamine hydrochloride dissolved in pyridine) and incubated for 2 h at 37 °C with constant mixing (150 rpm). After incubation, 70 μL of N-methyl-N-(trimethylsilyl)trifluoroacetamide (MSTFA) was added and incubated at 60 °C for 1 h. The concentrations of the resulting products were then determined via gas chromatography-mass spectrometry (GC–MS) (Agilent 6890 N GC equipped with a quadrupole Agilent 5973N MS spectrometer; Santa Clara, CA, USA). The samples were injected in splitless mode at 230 °C. The GC oven was held at an initial temperature of 80 °C for 2 min, which was increased to 325 °C at a 15 °C/min rate, then held at this temperature for 6 min. The separation process was carried out using a 25 m × 0.25 mm (inner diameter) × 0.25 μm DB–5MS UI capillary column with helium as the carrier gas at a 2 mL/min flow rate. Full scan acquisitions were performed over an m/z 50–800 range. Mass spectrometry was conducted at a 70 eV ionization energy, 230 °C source temperature, and 290 °C transfer line temperature. Automatic tuning of the instrument was conducted according to the manufacturer's instructions.

2.5. Characterization of Melanin Production in the Culture Filtrate

The dry weights of the fungal biomass, melanin in the culture filtrate, and resid concentrations of glucose and total nitrogen (TN) were monitored throughout the cu vation process. Spore suspensions (10^6 spores/mL) of the strain were prepared using above-described methods. Afterward, 1 mL of the suspension was inoculated into indiv ual flasks containing 100 mL of sterilized peptone yeast glucose (PYG; peptone: 10 g yeast: 5 g/L; glucose: 20 g/L) media and PYG media supplemented with 1 mM CuS The glucose concentration of the PYG media was adjusted from 5 g/L to 20 g/L. The in ulated culture media were maintained at a constant 150 rpm agitation on a rotary shake 27 °C. At each measurement point, the biomass and cell-free culture media were separa by centrifugation. The cells were weighed after filtering the sample through Whatm 1.2 µm glass fiber filters (Clifton, NJ, USA). The obtained supernatant was then acidifi (pH 2) with 1 M HCl and incubated for 24 h at 21 °C to enable melanin formation. T newly formed melanin pellets were weighed after filtering the sample as described abo The filtered supernatant excluding the melanin pellets was used to measure the resid glucose and TN concentrations. The residual glucose concentrations were monitored usi a high-performance liquid chromatography (HPLC) system (Shimadzu, Tokyo, Jap equipped with a refractive index detector (RID-20A, Shimadzu, Tokyo, Japan) and Aminex HPX-87H ion exchange column (300 mm × 7.8 mm) (Bio-Rad, Hercules, CA, US at a 0.5 mL/min flow rate. The mobile phase was a 5 mM H_2SO_4 solution prepared deionized water. TN concentrations were determined using the HS-TN(CA)-L kit and HS-1000PLUS water analyzer (Humas, Daejeon, Korea).

2.6. Laccase-Like Activity Assay

The laccase-like activity in the culture filtrate was determined as described by previc studies using 2,2′-azino-bis(3-ethylbenzthiazoline-6-sulfonate) (ABTS) as the substrate [28– Briefly, 200 µL of culture filtrate was added to the enzyme assay solutions (100 mM acet buffer, pH 5.0), resulting in a final concentration of 1 mM of ABTS. ABTS oxidation w monitored based on the increase in A_{420} (ε_{420} = 36,000 M^{-1} cm^{-1}). One unit of enzy activity was defined as the amount of enzyme required to oxidize 1 µmol of ABTS p minute at 25 °C.

2.7. Preparation of Synthetic Melanin and Its Characterization

The structural properties of the melanin synthesized using the culture filtrate supp mented with 1 mM $CuSO_4$ and 10 mM L-3,4-dihydroxyphenylalanine (DOPA) soluti were analyzed using Fourier-transform infrared spectroscopy (FT-IR). The synthesiz melanin was precipitated by adjusting solution pH to 2.0 and then washed three times w deionized water. The melanin was air-dried and then mixed with KBr to prepare melan KBr pellets using a pellet press. The characteristic absorption spectrum of the pellet w collected using a Nicolet 6700 spectrometer (Thermo Scientific, Waltham, MA, USA). T measurement was carried out in a wavenumber range of 4000–650 cm^{-1} with a resoluti of 4 cm^{-1}. The number of scans per sample was 32. Automatic background and ba line corrections were applied to the obtained spectrum. The obtained spectrum was r further modified.

3. Results and Discussion

3.1. Taxonomy

A previous study reported that *A. resinae* strain KUC3009 cultures cause discolorati of chromated copper arsenate-treated wood [32]. This strain was formerly classified *Cladosporium* sp.3, as sufficient molecular data were not available at the time. Therefc our study reclassified this strain based on phylogenetic and morphological analyses. BLAST search of the GenBank database using 526 bp ITS sequences of the selected stra indicated that the *Oidiodendron* and *Myxotrichum* genera were highly similar. Based these similarities, we then sought to align 24 taxa, including *Byssoascus striatosporus*, the outgroup taxon, after which a phylogenetic tree was constructed (Figure 1A). The isolate clustered closely with *A. resinae* isolates ATCC200942 and CBS406.68 with genetic similarities of 100% and 93%, respectively. Figure 1B,C shows light and electron microscopic images of the selected strains, respectively. Mycelia cultured on PDA were amorphous and showed strong pigmentation with a dark brown to black color, which was potentially due to the presence of a melanoid membrane. Furthermore, their subglobose or broadly ellipsoid to ovoid morphology with a 210 × 25 μm conidial size was consistent with previous reports [33]. Based on these observations, the isolate was identified as *A. resinae*.

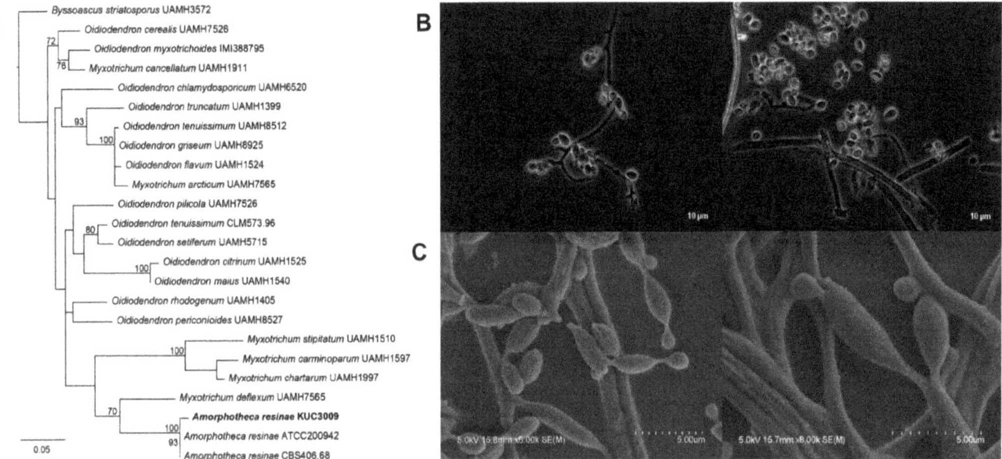

Figure 1. Molecular and morphological characterization of *A. resinae* KUC3009. (**A**) Phylogenetic tree based on internal transcribed spacer (ITS) sequence alignment generated by maximum-likelihood phylogenetic analysis. All bootstrap support values exceeded 70%. (**B**) Fungal culture imaging with light microscopy and (**C**) field emission-scanning electron microscopy.

3.2. Genome Sequencing

The genome of *A. resinae* KUC3009 was sequenced using a combination of PacBio and Illumina HiSeq reads with a 140x coverage. The genome assembly was approximately 30.11 Mb long and included 35 contigs with an average length of 183,235 bp (Table 1). The quality and completeness of the assembly were evaluated via BUSCO analysis. This analysis is routinely used to cross-analyze gene contents based on evolutionarily informed expectations of gene content [15]. Interestingly, 98.9% of the 1315 groups of genes required for the correct assembly of ascomycetes were present in *A. resinae* contigs, indicating that the *A. resinae* genome assembly was highly robust (BUSCO results are available in Supplementary File, Table S1). Only 0.4% and 0.7% of the gene groups were fragmented or missing, respectively.

The genomic features of *A. resinae* KUC3009 were compared with those of its sequenced relatives in the *Myxotrichaeae* family. The genome size of *A. resinae* KUC3009 (30.11 Mb) is similar to that of *A. resinae* ATCC 22711 (28.63 Mb) but smaller than that of *Oidiodendron maius* Zn (46.43 Mb) (Supplementary File, Table S2) [34]. For reference, the genome size of *A. resinae* was below the average genome size of common ascomycetes [35,36]. All species shared very similar genomic G+C content (i.e., approximately 47%). The average shared identity of three strains at the nucleic acid level was obtained with the OrthoANI calculator (Supplementary File, Table S2) [16]. The sequenced *A. resinae* strains were genetically closer to each other but were relatively distant from *O. maius*.

Table 1. Genome assembly statistics for *A. resinae* KUC3009.

Assembly Statistics	Value	Gene Statistics	Value
Number of contigs	35	Number of genes	9638
Length of the largest contig	3,753,173	Number of tRNAs	298
Average contig length	860,288	Number of rRNAs	228
Total contig length	30,110,100	Protein length (amino acids, median)	465
N50	2,338,627	Exon length (bp, median)	452
Genome coverage	140×	Intron length (bp, median)	86
G+C content (%)	47.5	Average exon number per gene	3.0

3.3. Genome Annotation

By utilizing several different gene predictors, we found that the *A. resinae* genor contained 9638 genes, 298 tRNAs, and 228 rRNAs (Table 1). The gene density was 3.20 gen per 10 kilobases (kb), and the predicted average protein size was 465 amino acids. The gen typically exhibited exons and introns with average lengths of 452 and 86 bp, respective Moreover, each gene contained an average of 3 exons. Additionally, the average length the predicted proteins was 465 amino acids.

The predicted proteins from the genes were annotated and then functionally classifi using eukaryotic orthologous group (KOG) analyses (Figure 2). The annotated protei were classified into the following categories: "intracellular processes," "metabolism," " formation storage and processing," and "poorly characterized function" [37]. The resu indicate that the proportion of the genes involved in the intracellular processes catego was the highest, whereas the other categories were relatively insignificant. Among t metabolism category, the number of genes involved in lipid metabolism and the trar port was slightly higher than that of the genes involved in carbohydrate metabolis and transport. *Cladosporium resinae*, the former name of *A. resinae*, is a common jet c deteriorating microorganism, thus highlighting its high capacity to degrade saturat hydrocarbons [38,39]. Consistent with previous studies, gene annotation identified h mologs of alkane degradation enzymes, such as long-chain alcohol oxidase [40]. The results indicated that the genes and enzymes of *A. resinae* make this fungal species pote tially well-suited for bioremediation.

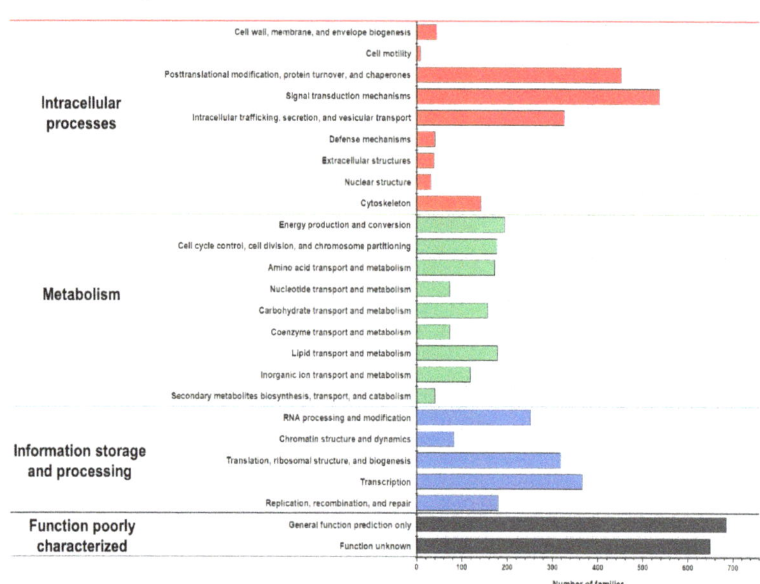

Figure 2. Eukaryotic orthologous group (KOG) distribution of predicted proteins from the *A. resinae* KUC3009 genome.

3.4. Secondary Metabolite Biosynthesis Clusters

Fungi produce numerous secondary metabolites, which have a multitude of roles in cellular processes, such as transcription and development [41]. Many of these compounds have significant applications in the medical field (e.g., antibiotics and antitumor drugs), as well as in the agriculture sector (e.g., insecticides). Based on recent genome sequencing results, the ability of fungi to produce secondary metabolites has been largely underestimated, as the gene clusters associated with secondary metabolite biosynthesis are not expressed under laboratory growth conditions [10].

The biosynthesis gene clusters (BGCs) identified using antiSMASH v6.0 were classified according to their types, and BGCs assigned to the production of a certain product are described in Figure 3. Only 14 BGCs within five classifications were found in the genome of *A. resinae* KUC3009. The number of identified BGCs was significantly lower than that of other common ascomycetes. For reference, the number of predicted BGCs for *Penicillium* and *Aspergillus* ranges between 29 and 85 [36,42,43]. Here, only five clusters were associated with the production of certain metabolites: 1,3,6,8-tetrahydroxynaphthalene (1,3,6,8-THN), neurosporin A, brefeldin, phomopsins, and asperphenamate. One particular BGC in Type 3 polyketide synthase (T3-PKS) is predicted to produce phomopsins, and the gene exhibited a 100% similarity with that of *Phomopsis leptostromiformis*. Notably, phomopsins are a group of hexapeptide mycotoxins with potent antimitotic activity and, therefore, represent promising antitumor agents [44]. Beyond computational analysis, additional studies to prove the gene's function are needed for using the strain as a source of antitumor production.

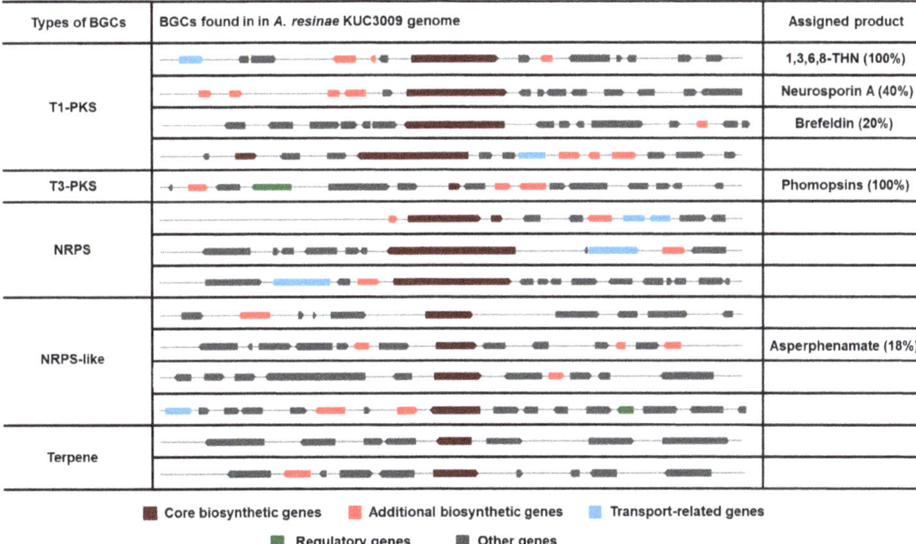

Figure 3. All identified biosynthetic gene clusters (BGCs) in the genome of *A. resinae* KUC3009 and their predicted assigned product. Values in parentheses indicate the similarity with a known cluster.

Our antiSMASH v6.0 analyses indicated that the genome of *A. resinae* retains T1-PKS BGC to produce 1,3,6,8-THN (Figure 3), which is the intermediate metabolite of the dihydroxy naphthalene (DHN)–melanin synthesis pathway (Figure 4A). This T1-PKS BGC of *A. resinae* exhibited a 100% similarity with that of *Bipolaris oryzae* and *Nodulisporium* sp. ATCC74245 (Figure 4B). These modules typically contain conserved domains that are comprised of acyl-carrier protein transacylase (SAT), β-ketoacyl synthase (KS), acyltransferase (AT), two acyl-carrier proteins (ACPs), and thioesterase (TE) (Figure 4C) [45,46].

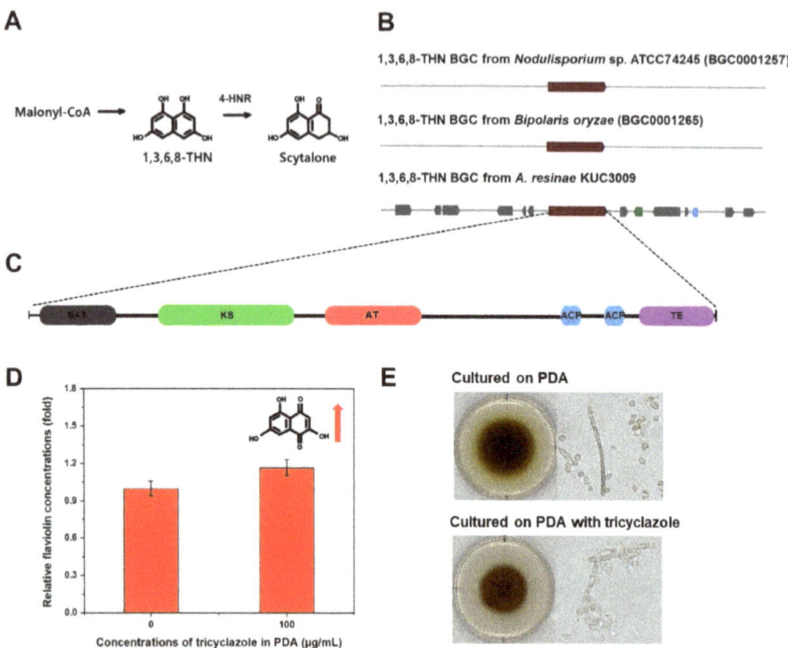

Figure 4. (**A**) Putative scytalone synthesis pathway of *A. resinae* KUC3009 (**B**) 1,3,6,8-THN BGC alignment between *A. resinae* and previously analyzed species. (**C**) Domain structure of the T1-PKS gene, consisting of an acyl-carrier protein transacylase (SAT) domain in gray, a β-ketoacyl synthase (KS) domain shown in green, an acyltransferase (AT) domain in pink, two acyl carrier protein (ACP) domains in blue, and a thioesterase (TE) domain in purple. (**D**) Comparison of flaviolin content in *A. resinae* mycelia cultured on PDA and tricyclazole-supplemented PDA (100 μg/mL). (**E**) Comparison of morphological properties of *A. resinae* cultured on PDA and tricyclazole-supplemented PDA (100 μg/mL).

The T1-PKS gene cluster associated with 1,3,6,8-THN synthesis is predicted to be involv in the scytalone synthesis but not with DHN-melanin synthesis (Figure 3). Through geno annotation analysis, we confirmed that complete putative tetra-hydroxynaphthalene reduct (4-HNR) coding genes were present in the genome, whereas scytalone dehydratase a tri-hydroxynaphthalene reductase (3-HNR) coding genes were not identified. For referer 1,3,6,8-THN molecule goes through a series of catalytic reactions to form the final prod (1,8-DHN) aided by 4-HNR, scytalone dehydratase, and 3-HNR, serially. Finally, the DH molecules are polymerized to form DHN-melanin using oxidation enzymes, such as lacc or other phenol oxidases [47,48].

A series of biochemical evidence supports the existence of the T1-PKS gene associa with 1,3,6,8-THN and 4-HNR genes. Given that flaviolin is the auto-oxidative product 1,3,6,8-THN, its detection in mycelia suggests that T1-PKS genes produce 1,3,6,8-THN [Additionally, the amounts of flaviolin increased when the 4-HNR enzyme is inhibited [We then sought to detect flaviolin and monitor its concentration after tricyclazole treatm to confirm whether our results were consistent with previous literature. GC–MS analy confirmed the existence of flaviolin in mycelia cultured on PDA. Moreover, the flavio concentration in mycelia exhibited a 1.2-fold increase when cultured on PDA supplemen with 100 μg/mL of tricyclazole (Figure 4D).

The study with tricyclazole inhibition assays supported that 1,3,6,8-THN BGC fr *A. resinae* is not involved in the DHN melanin formation. As tricyclazole inhibits DH melanin formation by repressing activity of 3-HNR as well as 4-HNR, fungal pigmentati with DHN-melanin was also inhibited when cultured with tricyclazole [50]. When cultu on tricyclazole-supplemented media, the effect of tricyclazole on pigment formation

A. resinae was poor, suggesting the pigmentation is not related to DHN-melanin synthesis. However, the gray-brown color of *A. resinae* cultures slightly changed to reddish-brown, which was likely due to an accumulation of shunt products (Figure 4E).

3.5. *A. resinae KUC3009 Pigment Production in Culture Filtrate*

We previously reported that *A. resinae* KUC3009 produced considerable amounts of melanin in the culture filtrate. The melanin synthesized by *A. resinae* KUC3009 in PYG media exhibited a high similarity with *Sepia* melanin in terms of their nitrogen content (approximately 7%) and indole-based chemical structures [8]. *Sepia* melanin is a typical type of eumelanin synthesized through the Raper–Mason pathway. L-tyrosine is a precursor, enzymatically oxidized by tyrosinase for melanin formation. Tyrosinase catalyzes the substrate using two consecutive activities of tyrosine hydroxylase and L-DOPA oxidase, resulting in the formation of L-dopaquinone [51–53]. Multi-copper oxidases, such as laccase, also can exhibit the activity of L-DOPA oxidase and be responsible for melanin formation [54].

The melanin derived from *A. resinae* slightly differed from *Sepia* melanin due to the following points; *Sepia* melanin results from the reaction of tyrosinase and L-tyrosine. However, L-tyrosine was not used in the preparation of the PYG media, and this strain cannot utilize the L-tyrosine as a substrate for melanin production (data not shown). Additionally, tyrosinase genes of *A. resinae* searched through gene annotation are predicted to be partial and intracellular (Supplementary File, Table S3), suggesting its activity in the culture filtrate would be insignificant in melanin formation. Instead, we focused on determining whether the pigments in the culture filtrate were formed through the laccase-like activity of multi-copper oxidase. Those enzymes are also known to possess a wide range of substrate specificity for melanin formation, and four putative complete genes predicted to have signal peptides were identified in the *A. resinae* genome (Supplementary File, Table S4).

A. resinae cultures were assessed based on the fungal biomass, melanin production, glucose concentrations, and laccase-like activity in culture filtrates (Figure 5). The fungal biomass increased prior to the stationary phase as the cells consumed the glucose in the culture medium. At a certain point of the stationary phase, both autolysis and melanin formation in the cultivation media occurred (Figure 5A–C). Interestingly, there was a surge in laccase-like activity prior to the melanin formation (Figure 5D–F). Interestingly, laccase-like activity was proportional to fungal biomass and melanin production in the cultivation media, and its surge exhibited a time lag as the initial glucose concentration in the media increased. The former was likely that enzyme production would be proportional to the fungal biomass, whereas the latter was possibly attributed to a glucose-suppressive effect, which has been reported in previous studies [55–58]. These results suggest a potential link between laccase-like activity in the culture filtrate and the melanin production of *A. resinae*.

The secreted enzymes can oxidize the nearby phenolic compounds to form melanins, whereas melanin can be formed via oxidation by enzymes released from the fungus during autolysis [48]. To unveil the exact mechanism of melanin formation in *A. resinae*, we induced melanin formation while keeping the cells alive by supplementing them with copper ions. As a result, both media's fungal biomass increased with the incubation time, suggesting cells in both media were in the growth phase. However, the amount of melanin and laccase-like activity were significantly high only when supplemented with copper after 3 and 6 days of cultivation (Figure 6A,B). The results indicated that melanin formation in the culture filtrate could occur before the autolysis phase, suggesting secretion of the enzyme is a key determinant in melanin formation in media.

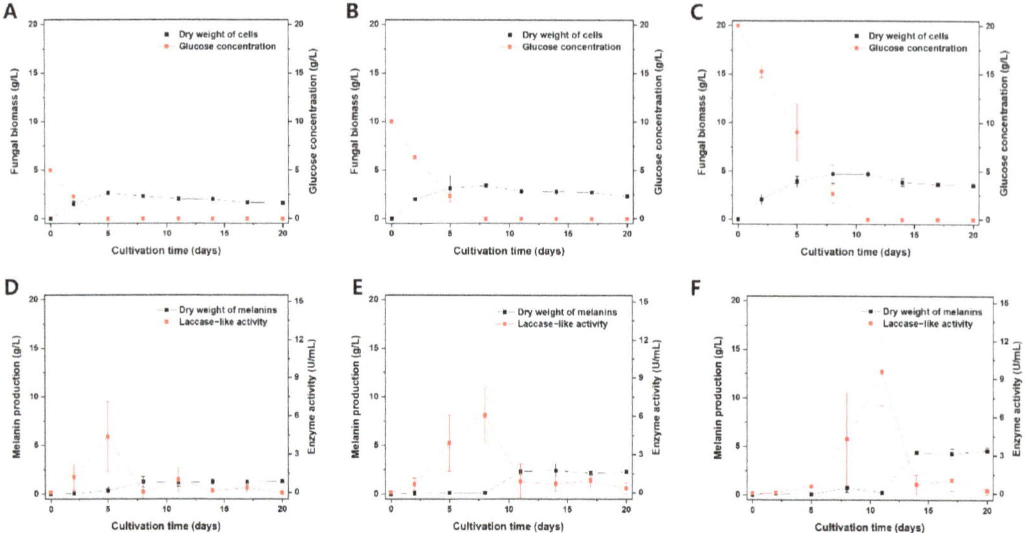

Figure 5. Effect of initial glucose concentration on the *A. resinae* cultivation profile. Fungal biomass and glucose concentration of media when cultivated at an initial glucose concentration of (**A**) 5 g/L, (**B**) 10 g/L, and (**C**) 20 g/L. Melanin production and laccase-like activity in culture filtrate when cultivated at an initial glucose concentration of (**D**) 5 g/L, (**E**) 10 g/L, and (**F**) 20 g/L.

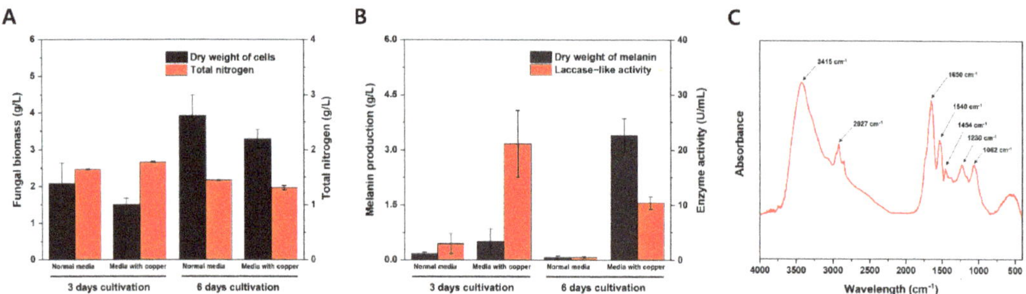

Figure 6. Effect of copper addition on *A. resinae* cultivation profile. (**A**) Fungal biomass and total nitrogen concentration of media. (**B**) Melanin production and laccase-like activity in the culture filtrate. (**C**) FT-IR spectrum of melanin synthesized using the culture filtrate and L-DOPA.

The substrate for melanin formation was not confirmed in this study. However, w discovered that the number of nitrogen sources in the metal-supplemented media w significantly decreased even though fungal biomass in copper-supplemented media w lower than that of normal PYG media (Figure 6A,B). Considering that the nitrogen conte accounted for 7% of purified melanins derived from *A. resinae*, this suggests that *A. resin* used nitrogen-containing substrates to synthesize melanin. Moreover, the purified melan derived from *A. resinae* exhibited features of indolic moieties, which was likely due to t oxidation of indole-containing substrates derived from media components or metabolit derived from the strain. After removing mycelia, the culture filtrate supplemented wi 1 mM $CuSO_4$, exhibiting laccase-like activity, could synthesize the melanin using L-DOP as a substrate. Figure 6C indicates the FT-IR spectrum of melanin synthesized using cultu filtrate and L-DOPA, which exhibited a similar spectrum with that of L-DOPA melan catalyzed by other laccases [59]. They commonly have the band at around 3400 cm and peak at 1650 cm^{-1} due to the stretching vibrations of–OH and–NH_2 groups and t vibrations of aromatic rings, respectively. These properties are commonly found within the melanin consisting of indole-based constituents [60].

Therefore, based on the above-described results, melanin production would likely be favored by enhanced laccase-like activity in the culture filtrate. Moreover, the selection of carbon sources and their concentrations should be prioritized to avoid glucose repression. Additionally, characterizing the effect of environmental factors on laccase-like activity is crucial.

4. Conclusions

Our study reported the draft genome of *A. resinae* and characterized its melanin synthesis mechanisms by combining bioinformatics and biochemical approaches. The genome size of *A. resinae* was relatively small than the average genome size of Ascomycetes. Moreover, only 14 BGCs were identified in the genome assembly, which contrasted with the higher BGC abundance of common ascomycetes. Genes encoding a specific T1-PKS and 4-HNR are predicted to produce intermediate metabolites of DHN-melanin biosynthesis but not proceed to DHN-melanin. These findings were further supported by the detection of increased flaviolin concentrations in mycelia and almost unchanged morphologies of the culture grown with tricyclazole. In the melanin formation in culture filtrates, it was observed that laccase-like activity and nitrogen sources are key determinants. Additionally, melanin formation in culture fluid was proportional to the laccase-like activity in the fluid. Melanin synthesis using L-DOPA and the culture filtrate exhibiting laccase-like activity were observed. Therefore, future studies should focus on enhancing laccase-like activity in culture filtrates to optimize industrial-scale melanin production using *A. resinae*.

Supplementary Materials: The following materials are available online at https://www.mdpi.com/article/10.3390/jof7040289/s1, Table S1: Assessment of genome completeness using the BUSCO software. Table S2: Comparisons of genome features between *A. resinae* KUC3009 and its close relatives belonging to the family Myxotrichaeae. Table S3: Putative tyrosinase genes found in the *A. resinae* genome. Table S4: Putative multi-copper oxidase genes found in the *A. resinae* genome.

Author Contributions: Conceptualization, J.-J.O.; methodology, J.-J.O., Y.J.K., J.Y.K., S.L.K., C.L., M.-E.L., J.W.K.; formal analysis and investigation, J.-J.O., Y.J.K., J.Y.K.; data curation, J.-J.O., Y.J.K., J.W.K.; writing—original draft preparation, J.-J.O.; writing—review and editing. J.-J.O., Y.J.K., J.Y.K., G.-H.K.; supervision, G.-H.K. All authors have read and agreed to the published version of the manuscript.

Funding: This study was supported by a Korea University Research Grant.

Institutional Review Board Statement: Not applicable.

Informed Consent Statement: Not applicable.

Data Availability Statement: Not applicable.

Conflicts of Interest: There are no conflict to declare.

ferences

Lagashetti, A.C.; Dufossé, L.; Singh, S.K.; Singh, P.N. Fungal pigments and their prospects in different industries. *Microorganisms* **2019**, *7*, 604. [CrossRef]
Vinha, A.F.; Rodrigues, F.; Nunes, M.A.; Oliveira, M.B.P. Natural pigments and colorants in foods and beverages. In *Polyphenols: Properties, Recovery, and Applications*; Woodhead Publishing: Cambridge, UK, 2018.
Dufosse, L.; Fouillaud, M.; Caro, Y.; Mapari, S.A.; Sutthiwong, N. Filamentous fungi are large-scale producers of pigments and colorants for the food industry. *Curr. Opin. Biotechnol.* **2014**, *26*, 56–61. [CrossRef]
Dufossé, L. Red colourants from filamentous fungi: Are they ready for the food industry? *J. Food Compos. Anal.* **2018**, *69*, 156–161. [CrossRef]
Xu, F.; Yuan, Q.P.; Zhu, Y. Improved production of lycopene and β-carotene by *Blakeslea trispora* with oxygen-vectors. *Process Biochem.* **2007**, *42*, 289–293. [CrossRef]
Feng, Y.; Shao, Y.; Chen, F. *Monascus* pigments. *Appl. Microbiol. Biotechnol.* **2012**, *96*, 1421–1440. [CrossRef]
Ribera, J.; Panzarasa, G.; Stobbe, A.; Osypova, A.; Rupper, P.; Klose, D.; Schwarze, F.W. Scalable biosynthesis of melanin by the basidiomycete *Armillaria cepistipes*. *J. Agric. Food Chem.* **2018**, *67*, 132–139. [CrossRef]

8. Oh, J.J.; Kim, J.Y.; Kwon, S.L.; Hwang, D.H.; Choi, Y.E.; Kim, G.H. Production and characterization of melanin pigments deriv from *Amorphotheca resinae*. *J. Microbiol.* **2020**, *58*, 648–656. [CrossRef]
9. Hillel, D.; Hatfield, J.L. *Encyclopedia of Soils in the Environment*; Elsevier: Amsterdam, The Netherlands, 2005.
10. Brakhage, A.A.; Schroeckh, V. Fungal secondary metabolites–strategies to activate silent gene clusters. *Fungal Genet. Biol.* **2(** *48*, 15–22. [CrossRef] [PubMed]
11. Yang, Y.; Liu, B.; Du, X.; Li, P.; Liang, B.; Cheng, X.; Du, L.; Huang, D.; Wang, L.; Wang, S. Complete genome sequence a transcriptomics analyses reveal pigment biosynthesis and regulatory mechanisms in an industrial strain, *Monascus purpur* YY-1. *Sci. Rep.* **2015**, *5*, 1–9. [CrossRef] [PubMed]
12. Oh, J.J.; Kim, J.Y.; Kim, Y.J.; Kim, S.; Kim, G.H. Utilization of extracellular fungal melanin as an eco-friendly biosorbent treatment of metal-contaminated effluents. *Chemosphere* **2021**, *272*, 129884. [CrossRef] [PubMed]
13. Kohler, A.; Murat, C.; Costa, M. High quality genomic DNA extraction using CTAB and Qiagen genomic-tip. In *INRA Na Equipe Ecogénomique*; Champenoux: Meurthe-et-Moselle, France, 2011.
14. Walker, B.J.; Abeel, T.; Shea, T.; Priest, M.; Abouelliel, A.; Sakthikumar, S.; Cuomo, C.A.; Zeng, Q.; Wortman, J.; Young, S.K.; e Pilon: An integrated tool for comprehensive microbial variant detection and genome assembly improvement. *PLoS ONE* **201** e112963. [CrossRef]
15. Simão, F.A.; Waterhouse, R.M.; Ioannidis, P.; Kriventseva, E.V.; Zdobnov, E.M. BUSCO: Assessing genome assembly a annotation completeness with single-copy orthologs. *Bioinformatics* **2015**, *31*, 3210–3212. [CrossRef]
16. Yoon, S.H.; Ha, S.M.; Lim, J.; Kwon, S.; Chun, J.A. large-scale evaluation of algorithms to calculate average nucleotide iden *Antonie Van Leeuwenhoek* **2017**, *110*, 1281–1286. [CrossRef] [PubMed]
17. Cantarel, B.L.; Korf, I.; Robb, S.M.; Parra, G.; Ross, E.; Moore, B.; Holt, C.; Alvarado, A.S.; Yandell, M. MAKER: An easy-to- annotation pipeline designed for emerging model organism genomes. *Genome Res.* **2008**, *18*, 188–196. [CrossRef] [PubMed]
18. Lowe, T.M.; Chan, P.P. tRNAscan-SE On-line: Search and Contextual Analysis of Transfer RNA Genes. *Nucleic Acids Res.* **2016**, W54–W57. [CrossRef] [PubMed]
19. Tatusov, R.L.; Fedorova, N.D.; Jackson, J.D.; Jacobs, A.R.; Kiryutin, B.; Koonin, E.V.; Krylov, D.M.; Mazumder, R.; Mekhedov, S Nikolskaya, A.N.; et al. The COG database: An updated version includes eukaryotes. *BMC Bioinform.* **2003**, *4*, 1–14. [CrossR [PubMed]
20. Blin, K.; Shaw, S.; Steinke, K.; Villebro, R.; Ziemert, N.; Lee, S.Y.; Medema, M.H.; Weber, T. antiSMASH 5.0: Updates to secondary metabolite genome mining pipeline. *Nucleic Acids Res.* **2019**, *47*, 81–87. [CrossRef]
21. Armenteros, J.J.A.; Tsirigos, K.D.; Sønderby, C.K.; Petersen, T.N.; Winther, O.; Brunak, S.; Heijne, G.; Nielsen, H. SignalP improves signal peptide predictions using deep neural networks. *Nat. Biotechnol.* **2019**, *37*, 420–423. [CrossRef] [PubMed]
22. Kumar, S.; Stecher, G.; Tamura, K. MEGA7: Molecular evolutionary genetics analysis version 7.0 for bigger datasets. *Mol. E Evol.* **2016**, *33*, 1870–1874. [CrossRef]
23. Katoh, K.; Standley, D.M. MAFFT multiple sequence alignment software version 7: Improvements in performance and usabi *Mol. Biol. Evol.* **2013**, *30*, 772–780. [CrossRef]
24. Stamatakis, A. RAxML-VI-HPC: Maximum likelihood-based phylogenetic analyses with thousands of taxa and mixed mod *Bioinformatics* **2006**, *22*, 2688–2690. [CrossRef]
25. Seifert, K.A.; Hughes, S.J.; Boulay, H.; Louis-Seize, G. Taxonomy, nomenclature and phylogeny of three cladosporium-l hyphomycetes, *Sorocybe resinae*, *Seifertia azaleae* and the *Hormoconis* anamorph of *Amorphotheca resinae*. *Stud. Mycol.* **2007**, 235–245. [CrossRef] [PubMed]
26. Lisec, J.; Schauer, N.; Kopka, J.; Willmitzer, L.; Fernie, A.R. Gas chromatography mass spectrometry–based metabolite profiling plants. *Nat. Protoc.* **2006**, *1*, 387–396. [CrossRef]
27. Snyder, S.A.; Tang, Z.Y.; Gupta, R. Enantioselective total synthesis of (−)-napyradiomycin A1 via asymmetric chlorination of isolated olefin. *J. Am. Chem. Soc.* **2009**, *131*, 5744–5745. [CrossRef] [PubMed]
28. Reiss, R.; Ihssen, J.; Richter, M.; Eichhorn, E.; Schilling, B.; Thöny-Meyer, L. Laccase versus laccase-like multi-copper oxidase comparative study of similar enzymes with diverse substrate spectra. *PLoS ONE* **2013**, *8*, e65633. [CrossRef]
29. Bourbonnais, R.; Paice, M.G. Oxidation of non-phenolic substrates: An expanded role for laccase in lignin biodegradation. *FE Lett.* **1990**, *267*, 99–102. [CrossRef]
30. Kirk, T.K.; Croan, S.; Tien, M.; Murtagh, K.E.; Farrell, R.L. Production of multiple ligninases by Phanerochaete chrysosporiu Effect of selected growth conditions and use of a mutant strain. *Enzym. Microb. Technol.* **1986**, *8*, 27–32. [CrossRef]
31. Johannes, C.; Majcherczyk, A. Laccase activity tests and laccase inhibitors. *J. Biotechnol.* **2000**, *78*, 193–199. [CrossRef]
32. Lee, Y.M.; Jang, Y.; Kim, G.H.; Kim, J.J. Phylogenetic analysis and discoloration characteristics of major molds inhabiting woo Part 3. Genus *Cladosporium*. *Holzforschung* **2012**, *66*, 537–541. [CrossRef]
33. Parbery, D.G. *Amorphotheca resinae*, gen. nov., sp. nov.: The perfect state of *Cladosporium resinae*. *Aust. J. Bot.* **1969**, *17*, 331– [CrossRef]
34. Kohler, A.; Kuo, A.; Nagy, L.G.; Morin, E.; Barry, K.W.; Buscot, F.; Canbäck, B.; Choi, C.; Cichocki, N.; Clum, A.; et al. Converg losses of decay mechanisms and rapid turnover of symbiosis genes in mycorrhizal mutualists. *Nat. Genet.* **2015**, *47*, 410– [CrossRef] [PubMed]
35. Mohanta, T.K.; Bae, H. The diversity of fungal genome. *Biol. Proced. Online* **2015**, *17*, 1–9. [CrossRef]

Choque, E.; Klopp, C.; Valiere, S.; Raynal, J.; Mathieu, F. Whole-genome sequencing of *Aspergillus tubingensis* G131 and overview of its secondary metabolism potential. *BMC Genom.* **2018**, *19*, 1–16. [CrossRef]
Wang, Y.; Coleman-Derr, D.; Chen, G.; Gu, Y.Q. OrthoVenn: A web server for genome wide comparison and annotation of orthologous clusters across multiple species. *Nucleic Acids Res.* **2015**, *43*, 78–84. [CrossRef]
Walker, J.D.; Cooney, J.J. Oxidation of n-alkanes by *Cladosporium resinae*. *Can. J. Microbiol.* **1973**, *19*, 1325–1330. [CrossRef]
Walker, J.D.; Cooney, J.J. Pathway of n-alkane oxidation in *Cladosporium resinae*. *J. Bacteriol.* **1973**, *115*, 635–639. [CrossRef]
Goswami, P.; Cooney, J.J. Subcellular location of enzymes involved in oxidation of n-alkane by *Cladosporium resinae*. *Appl. Microbiol. Biotechnol.* **1999**, *51*, 860–864. [CrossRef]
Keller, N.P. Fungal secondary metabolism: Regulation, function and drug discovery. *Nat. Rev. Microbiol.* **2019**, *17*, 167–180. [CrossRef] [PubMed]
Nielsen, J.C.; Grijseels, S.; Prigent, S.; Ji, B.; Dainat, J.; Nielsen, K.F.; Frisvad, J.C.; Workman, M.; Nielsen, J. Global analysis of biosynthetic gene clusters reveals vast potential of secondary metabolite production in *Penicillium* species. *Nat. Microbiol.* **2017**, *2*, 1–9. [CrossRef]
Inglis, D.O.; Binkley, J.; Skrzypek, M.S.; Arnaud, M.B.; Cerqueira, G.C.; Shah, P.; Wymore, F.; Wortman, J.R.; Sherlock, G. Comprehensive annotation of secondary metabolite biosynthetic genes and gene clusters of *Aspergillus nidulans*, *A. fumigatus*, *A. niger* and *A. oryzae*. *BMC Microbiol.* **2013**, *13*, 1–23. [CrossRef]
Ding, W.; Liu, W.Q.; Jia, Y.; Li, Y.; Van Der Donk, W.A.; Zhang, Q. Biosynthetic investigation of phomopsins reveals a widespread pathway for ribosomal natural products in Ascomycetes. *Proc. Natl. Acad. Sci. USA* **2016**, *113*, 3521–3526. [CrossRef]
Fulton, T.R.; Ibrahim, N.; Losada, M.C.; Grzegorski, D.; Tkacz, J.S. A melanin polyketide synthase (PKS) gene from *Nodulisporium* sp. that shows homology to the pks1 gene of *Colletotrichum lagenarium*. *Mol. Gen. Genet.* **1999**, *262*, 714–720. [CrossRef] [PubMed]
Moriwaki, A.; Kihara, J.; Kobayashi, T.; Tokunaga, T.; Arase, S.; Honda, Y. Insertional mutagenesis and characterization of a polyketide synthase gene (PKS1) required for melanin biosynthesis in *Bipolaris oryzae*. *FEMS Microbiol. Lett.* **2004**, *238*, 1–8. [PubMed]
Langfelder, K.; Streibel, M.; Jahn, B.; Haase, G.; Brakhage, A.A. Biosynthesis of fungal melanins and their importance for human pathogenic fungi. *Fungal Genet. Biol.* **2003**, *38*, 143–158. [CrossRef]
Bell, A.A.; Wheeler, M.H. Biosynthesis and functions of fungal melanins. *Annu. Rev. Phytopathol.* **1986**, *24*, 411–451. [CrossRef]
Romero-Martinez, R.; Wheeler, M.; Guerrero-Plata, A.; Rico, G.; Torres-Guerrero, H. Biosynthesis and Functions of Melanin in *Sporothrix schenckii*. *Infect. Immun.* **2000**, *68*, 3696–3703. [CrossRef] [PubMed]
Pal, A.K.; Gajjar, D.U.; Vasavada, A.R. DOPA and DHN pathway orchestrate melanin synthesis in Aspergillus species. *Med. Mycol.* **2014**, *52*, 10–18. [PubMed]
Solano, F. Melanins: Skin pigments and much more—Types, structural models, biological functions, and formation routes. *New J. Sci.* **2014**, *1*, 1–28. [CrossRef]
Mason, H.S. The chemistry of melanin: III. Mechanism of the oxidation of dihydroxyphenylalanine by tyrosinase. *J. Biol. Chem.* **1948**, *172*, 83–99. [CrossRef]
Raper, H.S. The aerobic oxidases. *Physiol. Rev.* **1928**, *8*, 245–282. [CrossRef]
Janusz, G.; Pawlik, A.; Świderska-Burek, U.; Polak, J.; Sulej, J.; Jarosz-Wilkołazka, A.; Paszczyński, A. Laccase properties, physiological functions, and evolution. *Int. J. Mol. Sci.* **2020**, *21*, 966. [CrossRef] [PubMed]
Castro-Sowinski, S.; Martinez-Drets, G.; Okon, Y. Laccase activity in melanin-producing strains of Sinorhizobium meliloti. *FEMS Microbiol. Lett.* **2002**, *209*, 119–125. [CrossRef] [PubMed]
Butler, M.J.; Day, A.W. Fungal melanins: A review. *Can. J. Microbiol.* **1998**, *44*, 1115–1136. [CrossRef]
Piscitelli, A.; Giardina, P.; Lettera, V.; Pezzella, C.; Sannia, G.; Faraco, V. Induction and transcriptional regulation of laccases in fungi. *Curr. Genom.* **2011**, *12*, 104–112. [CrossRef] [PubMed]
Yang, J.; Wang, G.; Ng, T.B.; Lin, J.; Ye, X. Laccase production and differential transcription of laccase genes in *Cerrena* sp. in response to metal ions, aromatic compounds, and nutrients. *Front. Microbiol.* **2016**, *6*, 1558. [CrossRef] [PubMed]
Al Khatib, M.; Harir, M.; Costa, J.; Baratto, M.C.; Schiavo, I.; Trabalzini, L.; Pollini, S.; Rossolini, G.M.; Basosi, R.; Pogni, R. Spectroscopic characterization of natural melanin from a *Streptomyces cyaneofuscatus* strain and comparison with melanin enzymatically synthesized by tyrosinase and laccase. *Molecules* **2018**, *23*, 1916. [CrossRef]
Pralea, I.-E.; Moldovan, R.C.; Petrache, A.M.; Ilieș, M.; Hegheș, S.C.; Ielciu, I.; Nicoară, R.; Moldovan, M.; Ene, M.; Radu, M.; et al. From extraction to advanced analytical methods: The challenges of melanin analysis. *Int. J. Mol. Sci.* **2019**, *20*, 3943. [CrossRef]

Journal of
Fungi

iew

ıngal Melanins and Applications in Healthcare, ioremediation and Industry

ie Rose Mattoon [1], Radames J. B. Cordero [2,*] and Arturo Casadevall [2]

1 Krieger School of Arts and Sciences, Johns Hopkins University, Baltimore, MD 21218, USA; emattoo1@jhu.edu
2 Department of Molecular Microbiology and Immunology, Johns Hopkins Bloomberg School of Public Health, 615 North Wolfe Street, Baltimore, MD 21205, USA; acasade1@jhu.edu
* Correspondence: rcorder4@jhu.edu

Abstract: Melanin is a complex multifunctional pigment found in all kingdoms of life, including fungi. The complex chemical structure of fungal melanins, yet to be fully elucidated, lends them multiple unique functions ranging from radioprotection and antioxidant activity to heavy metal chelation and organic compound absorption. Given their many biological functions, fungal melanins present many possibilities as natural compounds that could be exploited for human use. This review summarizes the current discourse and attempts to apply fungal melanin to enhance human health, remove pollutants from ecosystems, and streamline industrial processes. While the potential applications of fungal melanins are often discussed in the scientific community, they are successfully executed less often. Some of the challenges in the applications of fungal melanin to technology include the knowledge gap about their detailed structure, difficulties in isolating melanotic fungi, challenges in extracting melanin from isolated species, and the pathogenicity concerns that accompany working with live melanotic fungi. With proper acknowledgment of these challenges, fungal melanin holds great potential for societal benefit in the coming years.

Keywords: industrial microbiology; melanin; fungi; radioprotection; biotechnology; fungal pigments

tion: Mattoon, E.R.; Cordero, .; Casadevall, A. Fungal nins and Applications in thcare, Bioremediation and stry. *J. Fungi* **2021**, *7*, 488. s://doi.org/10.3390/jof7060488

lemic Editor: Laurent Dufossé

ived: 27 May 2021
pted: 13 June 2021
ished: 18 June 2021

isher's Note: MDPI stays neutral regard to jurisdictional claims in ished maps and institutional affil- ns.

1. Introduction

The term melanin refers to a diverse set of dark polymeric pigments found in all kingdoms of life. In fungi, melanin plays a panoply of protective roles against stress (Table 1; Figure 1) [1]. For example, fungal melanin can protect against ionizing radiation, including ultraviolet, X-ray, gamma-ray, and particulate radiation [2–4]. In addition, melanin can play a role in thermoregulation and protection against both heat and cold shock [5,6]. For example, melanized *Cryptococcus neoformans* cells were more likely to survive both heat shock and cold shock compared to non-melanized species [6]. Melanized endophytes in a mutualistic relationship with plants will often help their symbionts thermoregulate by dissipating heat and absorbing ROS [7]. Darkly pigmented yeasts and mushrooms tend to be more common at higher absolute latitudes and colder climates, suggesting that the pigment's ability to capture radiation energy and dissipate it as heat provides an advantage in generating thermal energy [5,8]. Part of melanin's role in thermotolerance may be attributable to its ability to react with and neutralize Reactive Oxygen Species (ROS), helping fungal organisms withstand the oxidative stress that often accompanies higher temperatures [9].

In addition, melanin can help fungi withstand chemical stressors. In halotolerant black yeast, melanin synthesis inhibitors diminished the yeast's ability to grow in hypersaline environments [10]. Researchers posited that this could be due to the stabilizing effect that melanin has on the fungal cell wall, which would have permitted a more effective response to osmotic stress [10]. In addition, melanization has also been shown to protect fungal cells from heavy metal stress and hydrolytic enzymes [11,12]. In areas with low water content, melanin is associated with stress response to dry conditions. For example, heavy melanization was associated with survival in microcolonial rock fungi exposed long periods of desiccation [13]. Some fungal species in spalted woods will melanize response to periods of low water content, creating black zones in the wood [14]. These merous protective functions enable melanized fungi to reside in some of civilization's m extreme environments, from deep-sea vents to the International Space Station [15,16].

Figure 1. Panel of four images of different melanized fungi. *Xylaria polymorpha* (top left), a mushro also referred to as 'Dead Man's Fingers.' The stromata of *X. polymorpha* have an average total len including rooting bases, of 5 to 8 cm by an average 2 cm diameter [27]. "Black fungi—*Xyl polymorpha* (dead man's fingers)" by ohi007 is licensed under CC BY-NC-SA 2.0; *Cladophialosph bantiana* (top right), a pathogenic mold. *C. bantiana* conidia are approximately 5 to 10 µm in length ["File:Cladophialophora bantiana UAMH10767.jpg" by Medmyco is licensed under CC BY-SA *Inonotus obliquus* (bottom left), also known as the Chaga mushroom. Chaga appears as a sclerc ranging from 5 to 40 cm in diameter [29]. "Chaga mushroom (*Inonotus obliquus*)" by Distant F Gardens is licensed under CC BY-NC-SA 2.0; *Cryptococcus neoformans* (bottom right), a pathoge yeast. Cryptococcal cells range from 5 to 10 µm in diameter [30]. "Cryptococcosis—GMS stain" Pulmonary Pathology is licensed under CC BY-SA 2.0. Images presented are not to scale.

Table 1. Examples of the functions of fungal melanin in living fungal organisms.

Function	Source
Photoprotection	[2–4]
Thermoregulation	[5–8]
Energy Harvesting	[5,22–24]
Metal Binding	[17–19]
Chemical Stress Response	[10–12,25]
Antioxidant	[7,9,26]
Anti-Desiccation	[14,26]
Virulence	[20,21]

Given the numerous functional groups present in melanotic pigments, melanin c bind and interact with many different organic and inorganic molecules [1,17]. One table example is melanin's affinity for chelating metal ions, which can be toxic to fungal cells [17–19]. Fungal melanin also plays a role in protecting the fungus against the human immune response and is often associated with pathogenicity in *Aspergillus fumigatus*, *Cryptococcus neoformans*, and *Talaromyces marneffei*, among others [20].

Lastly, melanin is correlated with fungal virulence and has been implicated as a possible antifungal target [21]. While fungal melanin's role in virulence will be limited in this review, please see [20] for a more thorough overview of this topic.

Some of the functions that melanin holds in biology can be utilized for societal benefit. These biological functions of radioprotection, stress response, and substrate binding, among others, often stem from melanin's unique physical and chemical properties. When properly understood, scientists may be able to use melanin to carry out parallel purposes in the fields of industry, healthcare, and bioremediation.

2. Fungal Melanins

Investigators wishing to work with melanin have several different options for sourcing the material, whether from animals such as cuttlefish, bacteria, fungi, or even synthetic reservoirs. Each of these options has different chemical properties and extraction protocols, conferring various advantages and disadvantages depending on the desired application [31]. For example, there are significant differences between fungal eumelanin and synthetic eumelanin, likely due to the fact that in vivo melanogenesis is vesicle-associated and enzymatically catalyzed, while synthetic melanogenesis often relies on spontaneous autopolymerization in solution [32]. In addition, synthetic melanins can be costly to produce [33]. Melanin from animal and plant sources can be less expensive to produce than the synthetic route, but it can be difficult to purify, as melanin is often associated with other biomolecules. Thus, microbial melanin is often touted as a low-cost and high-yield alternative to synthetic, animal, or plant melanin [33]. More specifically, fungal melanin is attractive due to the well-known protective functions in fungal organisms adapted to survive extreme environments [1].

Fungal organisms can produce different kinds of melanins (Figure 2). Most fungal melanins are allomelanins, which do not contain nitrogen [34]. DHN melanin is a common type of allomelanin derived from the polymerization of 1,8-dihydroxynaphthalene (DHN) [35]. The polyketide pathway that produces allomelanins begins with an endogenously produced molecule of acetyl coA or malonyl coA, which undergoes several reductions and dehydrations to produce 1,8-DHN [36]. The complete literature for scientific studies of fungal melanins dates back to the 1960s [37].

Some basidiomycetous fungi, such as *Cryptococcus neoformans*, produce eumelanin [38]. Eumelanins are derived from the amino acid tyrosine, and unlike allomelanins, they contain nitrogen [39]. Fungi synthesize eumelanins via the L-3,4-dihydroxyphenylalanine (L-dopa) pathway, which begins by using laccase to oxidize L-dopa into dopaquinone [40]. Eventually, the pathway produces dihydroxyindoles that can polymerize into eumelanins [40]. *Cryptococcus neoformans* cannot carry out this pathway without an exogenous substrate [40]. Pyomelanins are also derived from tyrosine, but they are produced by fungi such as *Aspergillus fumigatus* via the tyrosine degradation pathway, which involves the oxidative phosphorylation of homogentisate (HGA) [41].

Although not heavily discussed in recent reviews, some mushrooms and basidiomycetes have been shown to produce melanin from the precursor glutaminyl-hydroxybenzene (GHB) [42]. The presence of GHB on mushroom caps of *Agaricus biosporus* has been correlated with their susceptibility to browning, and the transformed product is commonly dubbed GHB-melanin [43]. Although a synthesis pathway has not been fully outlined, it is thought to occur as GHB is transformed by a polyphenol oxidase (typically tyrosinase) into 2-hydroxy-*p*-iminobenzoquinone (2-HpIBQ) and polymerized [42,44]. The diversity of pathways that organisms use to synthesize melanin explains in part why these polymers are not as well understood as other biopolymers such as proteins or nucleic acids [45].

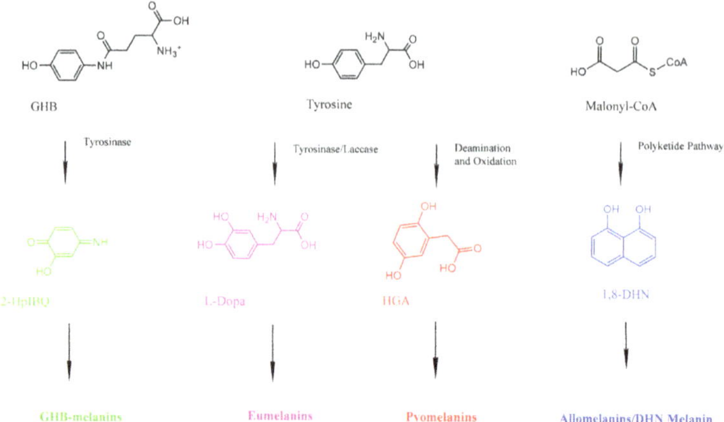

Figure 2. A simplified diagram showing the precursors for the three different kinds of fung melanin. GHB-melanins (far left, green) are synthesized through a series of reactions with tyrosina Eumelanins (center left, pink) are synthesized via the L-Dopa pathway, which uses tyrosine as precursor. Pyomelanins (center right, red) also use tyrosine as a precursor but are ultimately deriv from HGA. Allomelanins (far right, blue) are derived from 1,8-DHN.

When extracting fungal melanins for human applications, eumelanins are often p ferred as allomelanins are attached to the inner side of the fungal cell walls [46]. In contra eumelanins are used by fungi to neutralize toxic environmental compounds, making the melanins easier to extract extracellularly [46]. Currently, there are several different metho used to culture melanotic fungi, but most methods use a combination of tyrosine and me ions [46]. Recently, fungal strains were genetically modified to become more melanotic f extraction purposes, principally by overexpressing genes for tyrosinases [33].

Prior to melanin extraction, isolating melanotic yeast can be a challenge depending the yeast used. Some melanized yeasts are polyextremophiles, which can be difficult to isola due to their slow growth rates and low competitive ability [47]. However, researchers ha proposed the method of enriching black extremophile yeast of the order Chaetothyriales aromatic hydrocarbon, which can inhibit the growth of other microbial species while allowi black fungi to be cultured [47].

Another challenge associated with the industrial use of fungal melanin is the potent pathogenicity of fungal species. This requires scientists to identify non-pathogenic speci of melanized fungi, attenuate pathogenic melanotic fungi, or to introduce melanization a non-pathogenic species [48]. Alternatively, when using pathogenic species for melan extraction, operations must be carried out in laboratories certified for a given biosafety lev

While the challenges of fungal melanin extraction have made industrial-scale i plementation difficult in the past, newer extraction techniques are simpler and produ higher yields of the compound [49]. For example, a 2015 paper recorded a 10% yield fungal melanin from *Auricularia auricula* after several steps, including treatment with ly enzymes, guanidinium thiocyanate, chloroform, and HCl [50]. In contrast, a 2019 pap recorded that an optimized strain of *Armillaria cepistipes* was able to produce a 99% yield melanin following a simpler extraction procedure [49].

This review provides an overview of the various developments attempting to u specifically fungi-derived melanin for human application in fields including healthca bioremediation, and industry. The outlined projects are in varying stages of developme mostly due to the complicated nature of melanin extraction.

3. Health

Melanin's unique properties allow it to have diverse applications in the field of human health, whether in a pharmaceutical, medical device, or antimicrobial. One of the most heavily studied health applications of melanin involves protection from radiation, given that fungal melanin is known for its radioabsorptive properties.

For example, in an experiment where mice were fed black mushrooms *Auricularia auricila-judae* and soon after irradiated, they tended to exhibit improved survival than the control over the course of 45 days [51]. Researchers postulated that melanin's ability to dissipate Compton electron energy and scavenge free radicals would shield the mice's gastrointestinal (GI) tract, preventing cellular apoptosis. When mouse GI tissue was examined 24 h post-irradiation, researchers found fewer apoptotic cells in the tissue of mice that had been fed black mushrooms. In addition, mice fed white mushrooms supplemented with melanin had the same outcome as those who were given melanized mushrooms to begin with, pointing to the role of melanin in the increased survival rates. However, because melanin is insoluble, its protective effects remained mostly limited to the gastrointestinal tract [51]. Melanin extract from *Auricularia auricula* also showed promise in reducing oxidative stress and enhancing survival in liver cells exposed to high doses of ethanol, providing a theoretical basis for the substance's ability to treat alcoholic liver disease. The investigators associated their results with the liver cell's activation of the antioxidase Nrf2 and the inhibition of the cytochrome CYP2E1, which produces ROS as it metabolizes ethanol. Additional investigations would further elucidate the pathway producing this phenomenon [52].

In a separate mice study, those given melanin from the fungus *Gliocephalotrichum simplex* not only experienced better survival from irradiation, but also showed improvement in spleen parameters, reduced oxidative stress in the liver, and reduced production of inflammatory cytokines [53]. The authors suggested that a key mechanism of melanin's protective property in the study was its ability to reverse the decrease in phosphorylation of the transducing protein ERK that is commonly seen upon radiation exposure [53]. Radioprotective technologies are needed for the protection of multiple vulnerable demographics. For example, radiation can have a harmful effect on patients receiving it for diagnostic or therapeutic purposes [51]. Cardiac diagnostic procedures alone account for about one-fifth of the radiation exposure per person per year in the US [54]. In addition, some occupations receive high levels of radiation exposure, including healthcare professionals [54] and military personnel [55].

Currently, melanin has been proposed as a potential vehicle to protect astronauts from space radiation [56]. Particulate radiation is of particular concern in space travel, especially as it can generate secondary radiation upon interaction with spacesuits or spacecraft components [56]. The idea of fungal melanins as a potential material to protect against dangerous radiation in space originated from a series of studies demonstrating the ability of fungal melanin to attenuate and protect against different types of ionizing radiation [4,57–59]. A comparison of melanized and non-melanized forms of *Cryptococcus neoformans* and *Cryomyces antarcticus* found in both cases that the melanized cells were more resistant to a deuteron dose nearly 300,000 times higher than the dose lethal for humans [4]. In addition, samples of *C. antarcticus* were exposed for over a year on the International Space Station (ISS) to the radiation conditions in low Earth orbit. The samples not only survived, but also maintained a mutational load below 5% and sustained metabolic activity [60]. The organism's thick layers of melanin have been implicated as a possible explanation for its survival [60]. Given melanin's ability to protect microorganisms from both ionizing and particulate radiation, it may be a favorable material for protection in manned space travel [56]. For example, a recent preprint concluded that a melanotic *Cladosporium sphaerospermum* plate on the International Space Station produced attenuated radiation levels compared to a control plate [61]. However, more research about the radioprotective properties of melanin is needed to confirm that the application of melanin to space materials is possible and beneficial.

Allomelanin has recently been investigated for both its porosity and its ability absorb nerve gas stimulants in solution [62]. Given fungal melanin's natural ability absorb harmful materials while allowing essential cellular materials such as water a nutrients to pass through, the material's intrinsic porosity is a valid possibility. In additi melanin's extensive binding capacity can permit it to bind to many possibly harm materials. Researchers worked with both synthetic allomelanin derived from 1,8-DI and natural melanin ghosts derived from fungal cells. While the melanin ghosts provid a higher binding capacity for paraoxon and diazinon than some synthetic allomelani they were not as porous as the synthetic analogues [62]. In addition, investigators no allomelanin's potential in protective, breathable fabrics by applying synthetic melanin nylon-cotton fabric. Researchers also noted that the obtained melanin ghosts contain other polysaccharides such as chitin and glucan, making it more difficult to examine properties of natural allomelanin [62].

Although not specifically for radioprotection, the melanized medicinal mushro *Inonotus obliquus* has shown in vitro effects against tumor growth and diabetes mellitus [In one investigation, B16-F10 melanoma cells that were exposed to aqueous extracts c *obliquus* for 48 h exhibited decreasing, dose-dependent viability [64]. The study also no that the antitumor properties performed in vivo as well; mice that received the extr through an intraperitoneal route showed inhibited tumor mass growth. These effe were associated with G0/G1 arrest through the down-regulation of p53, pRb, and p proteins, although the investigators acknowledged the need for further research to elucid a complete mechanism of action [64]. Thus, it is possible that melanin may not b contributor to the extract's antitumor properties. However, melanin's presence in *I. obliqu* may still have health benefits. Water-soluble melanin complexes from the organism ha shown insulin-sensitizing activity [65]. When the melanin complex was extracted filtration and rotary evaporation, it reduced adiposity in high-fat obese mice [65].

Fungal melanin has also been implicated to have potential use in the functio biointerfaces used for stem cell manipulation [66]. Eumelanins were proposed for t purpose due to their antioxidant and electrical properties. However, eumelanin can vulnerable to degradation from alkaline or oxidative stress. Manini et al. propose alternative use of fungal allomelanins, citing the material's resistance to degradation a smoothness [66]. Such a mycomelanin film was able to encourage the differentiation stem cells towards an endodermal lineage [66].

4. Bioremediation

Melanized fungi and fungal melanin have both been proposed as mechanisms removing various toxins from a polluted environment. Fungi can degrade volatile ganic compounds (VOCs), and the ability of melanotic fungi to survive acidic or c conditions makes them great candidates for VOC absorption [67]. In western countr where individuals spend much of their time indoors, VOCs emitted from constructi materials, appliances, and cleaning chemicals have been implicated as a cause of s building syndrome (SBS) by the United States Environmental Protection Agency (EPA) [Researchers found that melanized fungal cultures placed in an indoor air model were a to reduce the VOC content by more than 96% after 48 h [69]. However, researchers no that which melanotic fungal species were used was an important factor for the feasibi of using fungi to eliminate VOCs in indoor spaces [69]. For example, some species p duced putative volatile metabolites in low concentrations [69]. In addition, fungal spec prone to rapid growth and/or sporulation may become prone to aerosolization, presenti challenges both as an allergen and as a pathogen to immunocompromised individuals [The researchers ultimately concluded that slow-growing melanotic fungi may be optin for future applications [69].

Melanin is also known to have properties that facilitate the absorption of heavy met making it a good candidate for addressing heavy metal pollution in waterways and environment. One natural example of this is that of fungi isolated from a uranium mine

Brazil [70]. Some of these species showed changes in melanin production, and most of these species were deemed to have a high potential for uranium absorption from water upon a biosorption test [70]. In a different setting, nanofiber membranes made with fungal melanin from *Armillaria cepistipes* were able to absorb heavy metals from water [71]. With these particular membranes, the fungal melanin was better able to absorb heavy metals over essential metals [71]. Fungal melanin derived from *Amorphotheca resinae* for the purposes of heavy metal absorption was able to be reused, pointing to the economic advantages of fungal melanin for removing metals from the environment [72]. After metal ions were absorbed in a pH of 5, exposing the melanin to acidic conditions allowed the metal ions to dissociate. For up to five cycles of binding and eluting these metals, the melanin's binding capacity was maintained [72].

Fungi can also play a major role in the absorption of nuclear pollution, given their resistance against and sometimes affinity for radiation. Fungal growth has been detected in and around the Chernobyl Nuclear Power Plant [73] and some fungi have been documented growing towards sources of radiation in a sort of radiotropism [74]. The extensive surface area of hyphae in saprotrophic fungi makes them excellent absorbers of radionuclides in the environment [75]. In one experiment, *Rhizopus arrhizua* and *Aspergillus niger* were able to remove 90–95% of radiothorium from solution [76]. One issue with using fungi for radioabsorption involves the concern that radioactive compounds could be transferred up the food chain from wild mushrooms to animals and people [77]. This was the case following the 1986 Chernobyl accident, in which cesium-137 remained highly concentrated in the fruiting bodies of edible fungi. In parts of Eastern Europe and even Great Britain, wild fungi consumption contributed to human intake of radioactive cesium [77]. However, this does not rule out the possibility of using extracted fungal melanin for radionuclide absorption applications.

5. Industry

Fungal melanin has found one potential application in the packaging of pork lard. Most commonly, fatty products such as lard go rancid due to the oxidation of lipids [78]. Luposiewicz et al. added fungal melanin to the gelatin coatings of pork lard and found that lard with the modified coatings tended to have lower oxidative rancidity [78]. This may be due to melanin's antioxidant properties counteracting the effect of oxygen free radicals in the environment [78].

Polylactic acid is a promising bioplastic, but its use is currently limited in food packaging applications due to its low thermal stability and solvent resistance [79]. When modified with fungal melanin, polylactic acid demonstrated improved barrier properties [79]. However, these properties decreased when too much fungal melanin was added [79].

Given its distinctive black color, microbial melanin has clear potential in industrial dyes. The study of natural dyes is growing in popularity, as they have been seen as safer and more environmentally friendly than synthetic dyes for food, textiles, and materials [80]. Researchers were recently able to dye poplar veneer using melanin secreted by *Lasiodiplodia theobromae* [81]. Beyond aesthetic effects, this dyeing of wood may be able to mimic the color of more expensive tree species while utilizing more efficient, faster-growing trees [82]. While the experiment shows that such a method is feasible, more data are needed to determine the fastness of the dye, or the material's ability to maintain its color over time [81].

Although the literature on natural fungal melanin's use in textiles is limited, investigators recently examined the ability of synthetic allomelanin to adhere to nylon-cotton fabric swatches [62]. Melanin derived from other microbial species, such as *Streptomyces spp.*, has successfully dyed wool without mordanting [83]. Barriers to natural fungal melanin's widespread use in textiles may include the size of the polymer and the need for large-scale extraction mechanisms.

6. Conclusions

Melanin has played a function in stress responses in biological organisms long befc humans took notice of it, and the scientific community has yet to discover many importa properties of microbial melanin, including its structural characterization and large-sca extraction. The unique properties fungal melanin exhibits in nature can be applied human use, and such applications are numerous and imminent. However, the scienti community must also focus on confronting the challenges associated with adopting fung melanin by identifying or creating non-pathogenic melanotic species, refining metho of isolating cultures, and improving methods of melanin extraction. The potential f microbial melanin's use to improve human health, environments, and industries furth emphasizes the importance of further study into these pigments from multiple fiel of interest.

Author Contributions: Conceptualization, E.R.M. and R.J.B.C.; writing—original draft preparati E.R.M. and R.J.B.C.; writing—review and editing, A.C. All authors have read and agreed to t published version of the manuscript.

Funding: This research received no external funding.

Conflicts of Interest: Radames J.B. Cordero and Arturo Casadevall have a business interest MelaTech, LLC, a biotech company that may be affected by the content presented in this artic MelaTech, LLC had no role in the article.

References

1. Cordero, R.J.; Casadevall, A. Functions of fungal melanin beyond virulence. *Fungal Biol. Rev.* **2017**, *31*, 99–112. [CrossR [PubMed]
2. Wang, Y.; Casadevall, A. Decreased susceptibility of melanized Cryptococcus neoformans to UV light. *Appl. Environ. Microb* **1994**, *60*, 3864–3866. [CrossRef] [PubMed]
3. Khajo, A.; Bryan, R.A.; Friedman, M.; Burger, R.M.; Levitsky, Y.; Casadevall, A.; Magliozzo, R.S.; Dadachova, E. Protection melanized Cryptococcus neoformans from lethal dose gamma irradiation involves changes in melanin's chemical structure a paramagnetism. *PLoS ONE* **2011**, *6*, e25092. [CrossRef] [PubMed]
4. Pacelli, C.; Bryan, R.A.; Onofri, S.; Selbmann, L.; Shuryak, I.; Dadachova, E. Melanin is effective in protecting fast and slc growing fungi from various types of ionizing radiation. *Environ. Microbiol.* **2017**, *19*, 1612–1624. [CrossRef]
5. Cordero, R.J.B.; Robert, V.; Cardinali, G.; Arinze, E.S.; Thon, S.M.; Casadevall, A. Impact of yeast pigmentation on heat captu and latitudinal distribution. *Curr. Biol.* **2018**, *28*, 2657–2664.e3. [CrossRef]
6. Rosas, Á.L.; Casadevall, A. Melanization affects susceptibility of Cryptococcus neoformans to heat and cold1. *FEMS Microbiol. L* **2006**, *153*, 265–272. [CrossRef]
7. Redman, R.S.; Sheehan, K.B.; Stout, R.G.; Rodriguez, R.J.; Henson, J.M. Thermotolerance generated by plant/fungal symbio *Science* **2002**, *298*, 1581. [CrossRef]
8. Krah, F.-S.; Büntgen, U.; Schaefer, H.; Müller, J.; Andrew, C.; Boddy, L.; Diez, J.; Egli, S.; Freckleton, R.; Gange, A.C.; et European mushroom assemblages are darker in cold climates. *Nat. Commun.* **2019**, *10*, 2890. [CrossRef]
9. Jacobson, E.S.; Tinnell, S.B. Antioxidant function of fungal melanin. *J. Bacteriol.* **1993**, *175*, 7102–7104. [CrossRef]
10. Kejžar, A.; Gobec, S.; Plemenitaš, A.; Lenassi, M. Melanin is crucial for growth of the black yeast Hortaea werneckii in its natu hypersaline environment. *Fungal Biol* **2013**, *117*, 368–379. [CrossRef]
11. García-Rivera, J.; Casadevall, A. Melanization of Cryptococcus neoformans reduces its susceptibility to the antimicrobial effe of silver nitrate. *Med. Mycol.* **2001**, *39*, 353–357. [CrossRef]
12. Bloomfield, B.J.; Alexander, M. Melanins and resistance of fungi to lysis. *J. Bacteriol.* **1967**, *93*, 1276–1280. [CrossRef]
13. Gorbushina, A.A.; Kotlova, E.R.; Sherstneva, O.A. Cellular responses of microcolonial rock fungi to long-term desiccation a subsequent rehydration. *Stud. Mycol.* **2008**, *61*, 91–97. [CrossRef]
14. Tudor, D.; Robinson, S.C.; Cooper, P.A. The influence of moisture content variation on fungal pigment formation in spalted wo *AMB Express* **2012**, *2*, 69. [CrossRef]
15. Le Calvez, T.; Burgaud, G.; Mahé, S.; Barbier, G.; Vandenkoornhuyse, P. Fungal diversity in deep-sea hydrothermal ecosyster *Appl. Environ. Microbiol.* **2009**, *75*, 6415–6421. [CrossRef]
16. Novikova, N.; De Boever, P.; Poddubko, S.; Deshevaya, E.; Polikarpov, N.; Rakova, N.; Coninx, I.; Mergeay, M. Survey environmental biocontamination on board the International Space Station. *Res. Microbiol.* **2006**, *157*, 5–12. [CrossRef]
17. Buszman, E.; Pilawa, B.; Zdybel, M.; Wilczyński, S.; Gondzik, A.; Witoszyńska, T.; Wilczok, T. EPR examination of Zn^{2+} and Cu binding by pigmented soil fungi Cladosporium cladosporioides. *Sci. Total Environ.* **2006**, *363*, 195–205. [CrossRef]
18. Hong, L.; Simon, J.D. Current understanding of the binding sites, capacity, affinity, and biological significance of metals melanin. *J. Phys. Chem. B* **2007**, *111*, 7938–7947. [CrossRef]

Fogarty, R.V.; Tobin, J.M. Fungal melanins and their interactions with metals. *Enzym. Microb. Technol.* **1996**, *19*, 311–317. [CrossRef]
Smith, D.F.Q.; Casadevall, A. The role of melanin in fungal pathogenesis for animal hosts. *Curr. Top. Microbiol. Immunol.* **2019**, *422*, 1–30. [CrossRef]
Nosanchuk, J.D.; Ovalle, R.; Casadevall, A. Glyphosate inhibits melanization of Cryptococcus neoformans and prolongs survival of mice after systemic infection. *J. Infect. Dis.* **2001**, *183*, 1093–1099. [CrossRef] [PubMed]
Dadachova, E.; Bryan, R.A.; Huang, X.; Moadel, T.; Schweitzer, A.D.; Aisen, P.; Nosanchuk, J.D.; Casadevall, A. Ionizing radiation changes the electronic properties of melanin and enhances the growth of melanized fungi. *PLoS ONE* **2007**, *2*, e457. [CrossRef] [PubMed]
Robertson, K.L.; Mostaghim, A.; Cuomo, C.A.; Soto, C.M.; Lebedev, N.; Bailey, R.F.; Wang, Z. Adaptation of the black yeast Wangiella dermatitidis to ionizing radiation: Molecular and cellular mechanisms. *PLoS ONE* **2012**, *7*, e48674. [CrossRef] [PubMed]
Pinkert, S.; Zeuss, D. Thermal Biology: Melanin-Based Energy Harvesting across the Tree of Life. *Curr. Biol.* **2018**, *28*, R887–R889. [CrossRef] [PubMed]
Larsson, B.S. Interaction between chemicals and melanin. *Pigment. Cell Melanoma Res.* **1993**, *6*, 127–133. [CrossRef] [PubMed]
Fernandez, C.W.; Koide, R.T. The function of melanin in the ectomycorrhizal fungus Cenococcum geophilum under water stress. *Fungal Ecol.* **2013**, *6*, 479–486. [CrossRef]
Rogers, J.D.; Callan, B.E. *Xylaria polymorpha* and its allies in continental united states. *Mycologia* **1986**, *78*, 391–400. [CrossRef]
Emmons, C.W.; Binford, C.H.; Utz, J.; Kwon-Chung, K. Medical Mycology. In *Philadelphia: Lea & Febiger*, 3rd ed.; 1977; Available online: https://www.cabdirect.org/cabdirect/abstract/19772703192 (accessed on 17 June 2021).
Zheng, W.; Miao, K.; Liu, Y.; Zhao, Y.; Zhang, M.; Pan, S.; Dai, Y. Chemical diversity of biologically active metabolites in the sclerotia of *Inonotus* obliquus and submerged culture strategies for up-regulating their production. *Appl. Microbiol. Biotechnol.* **2010**, *87*, 1237–1254. [CrossRef]
Okagaki, L.H.; Strain, A.K.; Nielsen, J.N.; Charlier, C.; Baltes, N.J.; Chrétien, F.; Heitman, J.; Dromer, F.; Nielsen, K. Cryptococcal cell morphology affects host cell interactions and pathogenicity. *PLoS Pathog* **2010**, *17*, e1000953.
Pralea, I.-E.; Moldovan, R.-C.; Petrache, A.-M.; Ilieș, M.; Hegheș, S.-C.; Ielciu, I.; Nicoară, R.; Moldovan, M.; Ene, M.; Radu, M.; et al. From extraction to advanced analytical methods: The challenges of melanin analysis. *Int. J. Mol. Sci.* **2019**, *20*, 3943. [CrossRef]
Camacho, E.; Vij, R.; Chrissian, C.; Prados-Rosales, R.; Gil, D.; O'Meally, R.N.; Cordero, R.J.B.; Cole, R.N.; McCaffery, J.M.; Stark, R.E.; et al. The structural unit of melanin in the cell wall of the fungal pathogen Cryptococcus neoformans. *J. Biol. Chem.* **2019**, *294*, 10471–10489. [CrossRef]
Martínez, L.M.; Martinez, A.; Gosset, G. Production of melanins with recombinant microorganisms. *Front. Bioeng. Biotechnol.* **2019**, *7*, 285. [CrossRef] [PubMed]
Britton, G. *The Biochemistry of Natural Pigments*; Cambridge University Press: Cambridge, UK, 1983.
Tsai, H.F.; Wheeler, M.H.; Chang, Y.C.; Kwon-Chung, K.J. A developmentally regulated gene cluster involved in conidial pigment biosynthesis in Aspergillus fumigatus. *J. Bacteriol.* **1999**, *181*, 6469–6477. [CrossRef] [PubMed]
Bell, A.A.; Wheeler, M.H. Biosynthesis and functions of fungal melanins. *Annu. Rev. Phytopathol.* **1986**, *24*, 411–451. [CrossRef]
Langfelder, K.; Streibel, M.; Jahn, B.; Haase, G.; Brakhage, A.A. Biosynthesis of fungal melanins and their importance for human pathogenic fungi. *Fungal Genet. Biol.* **2003**, *38*, 143–158. [CrossRef]
Williamson, P.R. Biochemical and molecular characterization of the diphenol oxidase of Cryptococcus neoformans: Identification as a laccase. *J. Bacteriol.* **1994**, *176*, 656–664. [CrossRef]
Millington, K.R. Improving the whiteness and photostability of wool. In *Advances in Wool Technology*; Woodhead Publishing Limit: Cambridge, UK, 2009; pp. 217–247.
Eisenman, H.C.; Casadevall, A. Synthesis and assembly of fungal melanin. *Appl. Microbiol. Biotechnol.* **2012**, *93*, 931–940. [CrossRef]
Heinekamp, T.; Thywißen, A.; Macheleidt, J.; Keller, S.; Valiante, V.; Brakhage, A.A. Aspergillus fumigatus melanins: Interference with the host endocytosis pathway and impact on virulence. *Front. Microbiol.* **2012**, *3*, 440. [CrossRef]
Solano, F. Melanins: Skin pigments and much more?types, structural models, biological functions, and formation routes. *New J. Sci.* **2014**, *2014*, 1–28. [CrossRef]
Weijn, A.; van den Berg-Somhorst, D.B.P.M.; Slootweg, J.C.; Vincken, J.-P.; Gruppen, H.; Wichers, H.J.; Mes, J.J. Main phenolic compounds of the melanin biosynthesis pathway in bruising-tolerant and bruising-sensitive button mushroom (Agaricus bisporus) strains. *J. Agric. Food Chem.* **2013**, *61*, 8224–8231. [CrossRef]
Weijn, A.; Bastiaan-Net, S.; Wichers, H.J.; Mes, J.J. Melanin biosynthesis pathway in Agaricus bisporus mushrooms. *Fungal Genet. Biol.* **2013**, *55*, 42–53. [CrossRef] [PubMed]
Cao, W.; Zhou, X.; McCallum, N.C.; Hu, Z.; Ni, Q.Z.; Kapoor, U.; Heil, C.M.; Cay, K.S.; Zand, T.; Mantanona, A.J.; et al. Unraveling the Structure and Function of Melanin through Synthesis. *J. Am. Chem. Soc.* **2021**, *143*, 2622–2637. [CrossRef] [PubMed]
Tran-Ly, A.N.; Reyes, C.; Schwarze, F.W.M.R.; Ribera, J. Microbial production of melanin and its various applications. *World J. Microbiol. Biotechnol.* **2020**, *36*, 170. [CrossRef] [PubMed]
Quan, Y.; van den Ende, B.G.; Shi, D.; Prenafeta-Boldú, F.X.; Liu, Z.; Al-Hatmi, A.M.S.; Ahmed, S.A.; Verweij, P.E.; Kang, Y.; de Hoog, S. A comparison of isolation methods for black fungi degrading aromatic toxins. *Mycopathologia* **2019**, *184*, 653–660. [CrossRef]

48. Cordero, R.J.B.; Vij, R.; Casadevall, A. Microbial melanins for radioprotection and bioremediation. *Microb. Biotechnol.* 2(*10*, 1186–1190. [CrossRef]
49. Ribera, J.; Panzarasa, G.; Stobbe, A.; Osypova, A.; Rupper, P.; Klose, D.; Schwarze, F.W.M.R. Scalable Biosynthesis of Melanin the Basidiomycete Armillaria cepistipes. *J. Agric. Food Chem.* **2019**, *67*, 132–139. [CrossRef]
50. Prados-Rosales, R.; Toriola, S.; Nakouzi, A.; Chatterjee, S.; Stark, R.; Gerfen, G.; Tumpowsky, P.; Dadachova, E.; Casadevall Structural Characterization of Melanin Pigments from Commercial Preparations of the Edible Mushroom Auricularia auric *J. Agric. Food Chem.* **2015**, *63*, 7326–7332. [CrossRef]
51. Revskaya, E.; Chu, P.; Howell, R.C.; Schweitzer, A.D.; Bryan, R.A.; Harris, M.; Gerfen, G.; Jiang, Z.; Jandl, T.; Kim, K.; e Compton scattering by internal shields based on melanin-containing mushrooms provides protection of gastrointestinal tr from ionizing radiation. *Cancer Biother. Radiopharm.* **2012**, *27*, 570–576. [CrossRef]
52. Hou, R.; Liu, X.; Wu, X.; Zheng, M.; Fu, J. Therapeutic effect of natural melanin from edible fungus Auricularia auricula alcohol-induced liver damage in vitro and in vivo. *Food Sci. Hum. Wellness* **2021**, *10*, 514–522. [CrossRef]
53. Kunwar, A.; Adhikary, B.; Jayakumar, S.; Barik, A.; Chattopadhyay, S.; Raghukumar, S.; Priyadarsini, K.I. Melanin, a promis radioprotector: Mechanisms of actions in a mice model. *Toxicol. Appl. Pharmacol.* **2012**, *264*, 202–211. [CrossRef]
54. Massalha, S.; Almufleh, A.; Small, G.; Marvin, B.; Keidar, Z.; Israel, O.; Kennedy, J.A. Strategies for minimizing occupatio radiation exposure in cardiac imaging. *Curr. Cardiol. Rep.* **2019**, *21*, 71. [CrossRef]
55. Blake, P.K.; Komp, G.R. Radiation exposure of U.S. military individuals. *Health Phys.* **2014**, *106*, 272–278. [CrossRef]
56. Cordero, R.J.B. Melanin for space travel radioprotection. *Environ. Microbiol.* **2017**, *19*, 2529–2532. [CrossRef]
57. Malo, M.E.; Bryan, R.A.; Shuryak, I.; Dadachova, E. Morphological changes in melanized and non-melanized Cryptococ neoformans cells post exposure to sparsely and densely ionizing radiation demonstrate protective effect of melanin. *Fungal B* **2018**, *122*, 449–456. [CrossRef]
58. Dadachova, E.; Bryan, R.A.; Howell, R.C.; Schweitzer, A.D.; Aisen, P.; Nosanchuk, J.D.; Casadevall, A. The radioprotect properties of fungal melanin are a function of its chemical composition, stable radical presence and spatial arrangem *Pigment Cell Melanoma Res.* **2008**, *21*, 192–199. [CrossRef]
59. Schweitzer, A.D.; Howell, R.C.; Jiang, Z.; Bryan, R.A.; Gerfen, G.; Chen, C.-C.; Mah, D.; Cahill, S.; Casadevall, A.; Dadachova Physico-chemical evaluation of rationally designed melanins as novel nature-inspired radioprotectors. *PLoS ONE* **2009**, *4*, e7 [CrossRef]
60. Onofri, S.; Pacelli, C.; Selbmann, L.; Zucconi, L. The Amazing Journey of Cryomyces antarcticus from Antarctica to Sp In *Extremophiles as Astrobiological Models*; Seckbach, J., Stan-Lotter, H., Eds.; Wiley: New York, NY, USA, 2020; pp. 237–254.
61. Shunk, G.K.; Gomez, X.R.; Averesch, N.J.H. A Self-Replicating Radiation-Shield for Human Deep-Space Exploration: Radiotrop Fungi can Attenuate Ionizing Radiation aboard the International Space Station. *BioRxiv* **2020**. [CrossRef]
62. McCallum, N.C.; Son, F.A.; Clemons, T.D.; Weigand, S.J.; Gnanasekaran, K.; Battistella, C.; Barnes, B.E.; Abeyratne-Perera, Siwicka, Z.E.; Forman, C.J.; et al. Allomelanin: A biopolymer of intrinsic microporosity. *J. Am. Chem. Soc.* **2021**. [CrossRef]
63. Lin, L.; Xu, J. Fungal Pigments and Their Roles Associated with Human Health. *J. Fungi* **2020**, *6*, 280. [CrossRef]
64. Youn, M.-J.; Kim, J.-K.; Park, S.-Y.; Kim, Y.; Park, C.; Kim, E.S.; Park, K.-I.; So, H.S.; Park, R. Potential anticancer properties the water extract of Inonotus [corrected] obliquus by induction of apoptosis in melanoma B16-F10 cells. *J. Ethnopharmacol.* 2(*121*, 221–228. [CrossRef]
65. Lee, J.-H.; Hyun, C.-K. Insulin-sensitizing and beneficial lipid-metabolic effects of the water-soluble melanin complex extrac from Inonotus obliquus. *Phytother. Res.* **2014**, *28*, 1320–1328. [CrossRef] [PubMed]
66. Cavallini, C.; Vitiello, G.; Adinolfi, B.; Silvestri, B.; Armanetti, P.; Manini, P.; Pezzella, A.; d Ischia, M.; Luciani, G.; Menichetti Melanin and Melanin-Like Hybrid Materials in Regenerative Medicine. *Nanomaterials* **2020**, *10*, 1518. [CrossRef] [PubMed]
67. Blasi, B.; Poyntner, C.; Rudavsky, T.; Prenafeta-Boldú, F.X.; Hoog, S.D.; Tafer, H.; Sterflinger, K. Pathogenic yet environmenta friendly? black fungal candidates for bioremediation of pollutants. *Geomicrobiol. J.* **2016**, *33*, 308–317. [CrossRef] [PubMed]
68. United States Environmental Protection Agency. *Indoor Air Facts No. 4 (Revised) Sick Building Syndrome*; United States Envir mental Protection Agency: Washington, DC, USA, 1991.
69. Prenafeta-Boldú, F.X.; Roca, N.; Villatoro, C.; Vera, L.; de Hoog, G.S. Prospective application of melanized fungi for the biofiltrat of indoor air in closed bioregenerative systems. *J. Hazard. Mater.* **2019**, *361*, 1–9. [CrossRef] [PubMed]
70. Coelho, E.; Reis, T.A.; Cotrim, M.; Mullan, T.K.; Corrêa, B. Resistant fungi isolated from contaminated uranium mine in Bra shows a high capacity to uptake uranium from water. *Chemosphere* **2020**, *248*, 126068. [CrossRef] [PubMed]
71. Tran-Ly, A.N.; Ribera, J.; Schwarze, F.W.M.R.; Brunelli, M.; Fortunato, G. Fungal melanin-based electrospun membranes for hea metal detoxification of water. *Sustain. Mater. Technol.* **2020**, *23*, e00146. [CrossRef]
72. Oh, J.-J.; Kim, J.Y.; Kim, Y.J.; Kim, S.; Kim, G.-H. Utilization of extracellular fungal melanin as an eco-friendly biosorbent treatment of metal-contaminated effluents. *Chemosphere* **2021**, *272*, 129884. [CrossRef]
73. Zhdanova, N.N.; Zakharchenko, V.A.; Vember, V.V.; Nakonechnaya, L.T. Fungi from Chernobyl: Mycobiota of the inner regi of the containment structures of the damaged nuclear reactor. *Mycol. Res.* **2000**, *104*, 1421–1426. [CrossRef]
74. Zhdanova, N.N.; Tugay, T.; Dighton, J.; Zheltonozhsky, V.; McDermott, P. Ionizing radiation attracts soil fungi. *Mycol. Res.* 2(*108*, 1089–1096. [CrossRef]
75. Dighton, J.; Tugay, T.; Zhdanova, N. Fungi and ionizing radiation from radionuclides. *FEMS Microbiol. Lett.* **2008**, *281*, 109– [CrossRef]

White, C.; Gadd, G.M. Biosorption of radionuclides by fungal biomass. *J. Chem. Technol. Biotechnol.* **2007**, *49*, 331–343. [CrossRef] [PubMed]
Barnett, C.L.; Beresford, N.A.; Frankland, J.C.; Self, P.L.; Howard, B.J.; Marriott, J.V.R. Radiocaesium intake in Great Britain as a consequence of the consumption of wild fungi. *Mycologist* **2001**, *15*, 98–104. [CrossRef]
Łopusiewicz, Ł.; Jędra, F.; Bartkowiak, A. The application of melanin modified gelatin coatings for packaging and the oxidative stability of pork lard. *World Sci. News* **2018**, *101*, 108–119.
Łopusiewicz, Ł.; Jędra, F.; Mizielińska, M. New Poly (lactic acid) Active Packaging Composite Films Incorporated with Fungal Melanin. *Polymers* **2018**, *10*, 386. [CrossRef]
Shahid, M.; Shahid-ul-Islam; Mohammad, F. Recent advancements in natural dye applications: A review. *J. Clean. Prod.* **2013**, *53*, 310–331. [CrossRef]
Liu, Y.; Zhang, Y.; Yu, Z.; Qi, C.; Tang, R.; Zhao, B.; Wang, H.; Han, Y. Microbial dyes: Dyeing of poplar veneer with melanin secreted by Lasiodiplodia theobromae isolated from wood. *Appl. Microbiol. Biotechnol.* **2020**, *104*, 3367–3377. [CrossRef]
Sun, D.; Sun, D.; Yu, X. Ultrasonic-Assisted Dyeing of Poplar Veneer. *Wood Fiber Sci.* **2011**, *43*, 442–448.
Amal, A.; Keera, A.; Abeer, H.; Samia, A.; El-Hasser, H.; Nadia, A.K.A. Selection of Pigment (Melanin) production in Streptomyces and their application in Printing and Dyeing of Wool Fabrics. *Res. J. Chem. Sci.* **2011**, *1*, 22–28.

Journal of
Fungi

iew

ecent Findings in Azaphilone Pigments

:ia P. S. Pimenta [1], Dhionne C. Gomes [2], Patrícia G. Cardoso [3] and Jacqueline A. Takahashi [1,*]

1 Department of Chemistry, Universidade Federal de Minas Gerais (UFMG), Av. Antonio Carlos, 6627, Belo Horizonte CEP 31270-901, MG, Brazil; lpimenta@qui.ufmg.br
2 Department of Food Science, Universidade Federal de Minas Gerais (UFMG), Av. Antonio Carlos, 6627, Belo Horizonte CEP 31270-901, MG, Brazil; dhionne@gmail.com
3 Department of Biology, Universidade Federal de Lavras, Av. Dr. Sylvio Menicucci, 1001, Lavras CEP 37200-900, MG, Brazil; patricia@ufla.br
* Correspondence: jat@qui.ufmg.br

Abstract: Filamentous fungi are known to biosynthesize an extraordinary range of azaphilones pigments with structural diversity and advantages over vegetal-derived colored natural products such agile and simple cultivation in the lab, acceptance of low-cost substrates, speed yield improvement, and ease of downstream processing. Modern genetic engineering allows industrial production, providing pigments with higher thermostability, water-solubility, and promising bioactivities combined with ecological functions. This review, covering the literature from 2020 onwards, focuses on the state-of-the-art of azaphilone dyes, the global market scenario, new compounds isolated in the period with respective biological activities, and biosynthetic pathways. Furthermore, we discussed the innovations of azaphilone cultivation and extraction techniques, as well as in yield improvement and scale-up. Potential applications in the food, cosmetic, pharmaceutical, and textile industries were also explored.

Keywords: natural pigments; filamentous fungi; azaphilones; production; biotechnological tools; non-mycotoxigenic strains; regulatory issues

tion: Pimenta, L.P.S.; Gomes, D.C.; oso, P.G.; Takahashi, J.A. Recent ings in Azaphilone Pigments. *J. i* **2021**, *7*, 541. https://doi.org/ 90/jof7070541

lemic Editor: Laurent Dufossé

ived: 30 May 2021
pted: 4 July 2021
ished: 7 July 2021

1. Introduction

Color has been used by mankind since the Neolithic period and has been associated to different people such as purple to the Phoenicians, yellow (annatto) to the Mayans, and to different purposes as henna pigments for body and hair coloring in India. In human history, color gained a powerful status in many daily experiences and key decisions. Some studies show, for example, that preference for blues and reds (at the expense of yellowish and greenish hues) influenced auction prices, as reported for Mark Rothko's rectangular paintings [1].

Color is also naturally associated with chemosensory perceptions regarding flavor, quality and freshness, highly interfering in product choice [2]. In this way, consumers expect some foods to have specific colors. However, variation and heterogeneousness of natural color in foods initiated the process of adding pigments to maintain color uniformity while granting high coloring power, as well as stability in aqueous phase and in different pH [3].

Vegetal-derived natural products are source of pigments very important to the food industry. However, the production is limited by yield issues since the gross amounts of vegetal pigments recovered, even from improved cultivars is not sufficiently competitive to fulfill modern industrial demand. Yield improvement is surely the major problem which have been addressed by developing and breeding modified cultivars and new large-scale processes were developed to the production of natural pigments [4]. Insect-derived coloring compounds such as carmine have been introduced in the market, but despite their natural origin, they are not accepted by many countries' regulatory agencies due to ethical issues. In addition, vegetarian, vegan and kosher diet adepts do not accept colorants animal origin [5].

To date, industry still has not overcome the low availability of natural pigments successful alternative was found in the synthesis of coloring agents structurally identica the natural ones to produce compounds like beta carotene, riboflavin and cantaxantl xanthophylls (yellow, orange and red palette) [6]. However, the synthesis of natu pigments did not bring enough economic competitiveness and there are issues related the classification (natural or synthetic) of compounds naturally occurring produced synthetic means.

With the growing demand for industrialized food, the high per-unit cost of natu colorants boosting the cost of final food products, without the benefit of significant co content, led industries to adopt synthetic substitutes, which feature vast color spectru and colorfastness. Azo dyes are some of the most utilized synthetic compounds in t area, offering reproducible stable color. They can be easily synthesized by diazotation aromatic amines and became the first-choice colorants in food industry for decades However, azo compounds have been associated with several diseases, including cancer Moreover, although controversial, meta-analysis studies found evidence on the relations between intake of artificial food coloring agents with allergic response and behavio problems such as hyperactivity in children [9]. These facts led regulatory agencies ban some synthetic colorants and, consequently, food industry is facing the challer of developing novel formulations containing natural food coloring agents to provide complement the color palette of foods.

The replacement of artificially colored products by natural ones is also demanded b new generation of green-minded consumers seeking for "clean label" and safe ingredie The boom of groups opting for environmentally friendly consumption and healthy lifesty led to a big change in food consumer behavior, especially by individuals from the so-cal Generation Z (Gen-Z). This group was pointed to account for about 40% of all consume the largest consumer market share in 2020 [10].

In this scenario, fungi are highly quoted as alternative sources of naturally deriv healthy, safe, stable and low-cost pigments for food industry applications [11]. Fun bio-pigments have the advantage of being produced using inexpensive sources of carb and nitrogen, that can even be obtained from food by-products or from agro-industr residues [12]. One of the most promising classes of fungal pigments in research as industr pigments are azaphilones, compounds that stand out for their yellow, orange and r colors [13]. This class of fungal secondary metabolites encompasses a large number compounds of polyketide origin, containing a pyrone-quinone core, a chiral quaterna center and hydroxyl groups as substituents. Orange-colored azaphilones usually posses heterocycle containing a pyranyl oxygen that is susceptible to aminophilic reactions wh the pyran oxygen atom is exchanged for a nitrogen atom derived from peptides, nucl acids, proteins and others [14]. This exchange alters the absorption of the pigment t goes from orange to red, frequently also altering the biological properties.

Azaphilones research is extremely important and literature reporting new azaphilc derivatives described in the last decades, different fungi sources, and a wide scope biological activities is comprehensive. However, many issues on industrial scaleup wet bench fermentative conditions, optimized production, efficient extraction protocols maximize industrial production and certification of generally recognized as safe (GR strains are areas that still demand research and technological development. An expressi number of works have been addressing the challenge to find a safe, low cost azaphilo source to fit the contemporary demand for edible natural pigments that meet regulat guidelines. The readiness of fungi-derived red colorants for use in food industry w discussed on an interesting paper by Dufossè [15], while production of yellow pigments *Monascus* sp. was addressed by Yang et al. [16].

This review, covering literature from January 2020 to April 2021, focuses on the sta of-the-art of azaphilone research, comprising market scenario, fungi sources reported the period, main cultivation, extraction, and purification techniques, chemistry, scope of biological activity, and potential applications in the food industry. Strategies for yield improvement and scale up, associated with market possibilities for cosmetic, pharmaceutical, and textile industries among other applications will also be discussed.

2. Global Market Size for Yellow, Orange and Red Colored Pigments

The economic crisis imposed by COVID-19 pandemic in 2020 profoundly influenced human life in several ways. Industrial sector suffered huge losses and long-term effects are expected in all sectors that had to adapt marketing policies to minimize economic breakdown [17]. Nevertheless, the market of additives for food industry may achieve 43.3 billion USD by 2021 [18] with the food colorant market alone being responsible for 3.75 billion USD [19]. The contribution of global food pigment (comprising carotenoids, caramel, curcumin and spirulina among others) market is projected to reach 1271.4 million USD by 2025 [20].

Although the good scenario and ascending numbers, prohibition or scrutiny of some artificial red colorant compounds and legal restrictions to their use in the food industry either by the Food and Drug Administration in the USA or by the European Food Safety Agency (EFSA) have been affecting the availability of food pigments and the color palette. A major issue related to synthetic compounds regards the azo-aromatic group present in the chemical structure of many red and yellow synthetic colorants. In the same way, allergy and other allegations have also been affecting the employment of yellow pigments such as tartrazine in foodstuff [21]. This becomes a problem, as the range of color yellow-red is essential in the food industry. Red, yellow, and orange, along with "clear" and white colors were associated with refreshing foods and beverages [22]. Even though the role of color in the market "is still in infancy", high color saturation captures consumer attention. For example, red color is protagonist in avoidance or approach motivations related to fresh fruits preference, as red color is associated with fruit ripening [23]. Color perception, together with visible fat and origin were reported as the main intrinsic attributes that drive choice of pork products by consumers from some emerging markets [24].

Innumerous compounds biosynthesized by plants have yellow, orange and red natural color, as determined by structural features. Some of them meet the requirements for using in food products such as anthocyanins (cyanidin), carotenoids (bixin), indole-derived glycosides (betanin), anthraquinones (carminic acid) and polyphenols (curcumin). Some information and market size for these pigments can be found in Figure 1.

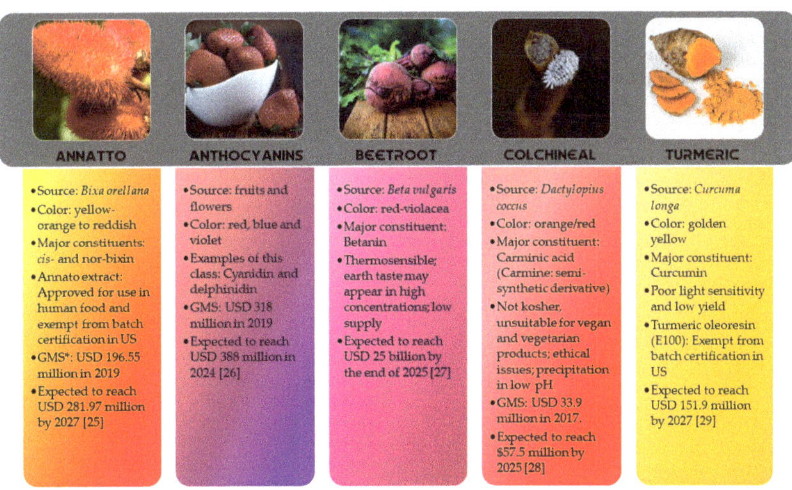

Figure 1. Source, color, constitution and global market size (GMS) data for some vegetal-originated colored compounds [25–29].

3. Chemistry, Biological Activities and Biosynthetic Pathways of Recently (2020–2021 Reported Azaphilones

3.1. New Azaphilone Compounds

Two complementary reviews cover a good part of literature about azaphilones fro 1932 to 2019. Gao et al. [30] reviewed literature from end of 1932 to September 20 reporting data on 373 azaphilones of 18 categories and Chen et al. [31] published data on t chemistry and biology of azaphilones, covering 252 compounds predominantly originat from 32 genera of fungi reported between October 2012 to December 2019 [31]. Natural derived azaphilones reported by Chen et al. [31] were classified in 13 types: nitrogenate citrinins, austdiols, deflectins, bulgariolactones, spiro-azaphilones, O-substituted, lacto hydrogenated, chaetovirins, pulvilloric acid, sclerotiorins, and cohaerins. Azaphilo pigments of atrorosin class produced by *Talaromyces atroroseus* were reviewed by Isbran et al. [32], and Morales-Oyervides et al. [11] reviewed natural colorants produced by fun from *Talaromyces/Penicillium* genus [11].

In this section, it is presented a summary of the new compounds reported aft December 2019 classified according to the fungal genera source. Despite the great numb of 100 new compounds reported from January 2020 to March 2021, the azaphilones we isolated only from nine fungal genera (*Aspergillus, Chaetomium, Hypoxylon, Monasc Muycopron, Penicillium, Phomopsis, Pleosporales,* and *Talaromyces*). The genus *Phomopsis* w not cited in the latest review and now appeared as fungal endophytic sources of chlorinat azaphilone pigments. At this time, it will be presented the new compounds isolated fro each genus (Figures 2–9) displayed according to the species (Table 1).

3.1.1. Azaphilones from *Aspergillus* Genus

Aspergillus genus is one of the three largest genera where azaphilones can be four Recently, 23 azaphilones (**1–23**) were isolated from three species (Figure 2 and Table Sassafrin E (**1**), sassafrin F (**2**), and sassafrinamine A (**3**) were isolated from the filamento fungus *Aspergillus neoglaber* 3020 [33]. Two pigments (Sassafrin E (**1**) and Sassafrin F (were yellow and display the azaphilone core fused to the same angular lactone ring wi different substituents. The third pigment (sassafrinamine A) (**3**) is purple and displays nitrogen into the isochromene system substituted with ethyl-1-ol group (Figure 2). The fu gus *Aspergillus cavenicola* afforded the nitrogenated azaphilones *trans*-cavernamine (**4**), c cavernamine (**5**), amino acid derivatives of *cis*-cavernamines (**6–10**), hydroxy-cavernami (**11**), amino acid derivatives of hydroxy-cavernamines (**12–16**), and two oxygenated deriv tives *cis* and *trans*-cavernines (**17–18**) [34]. The marine-derived fungus *Aspergillus falconen* yielded five mitorubrins derivative azaphilones with different benzoyl moieties: two ne chlorinated azaphilones, falconensins O and P (**19** and **20**) when the fungus was cultivat in a solid rice medium containing 3.5% NaCl and three additional new azaphilone deriv tives (**21–23**) when NaCl was replaced by 3.5% NaBr [35]. From the endophytic *Aspergill terreus* of *Pinellia ternate*, the undescribed dimer of citrinin penicitrinol Q (**24**) was isolat displaying accentuated Gram-positive antibacterial activity [36].

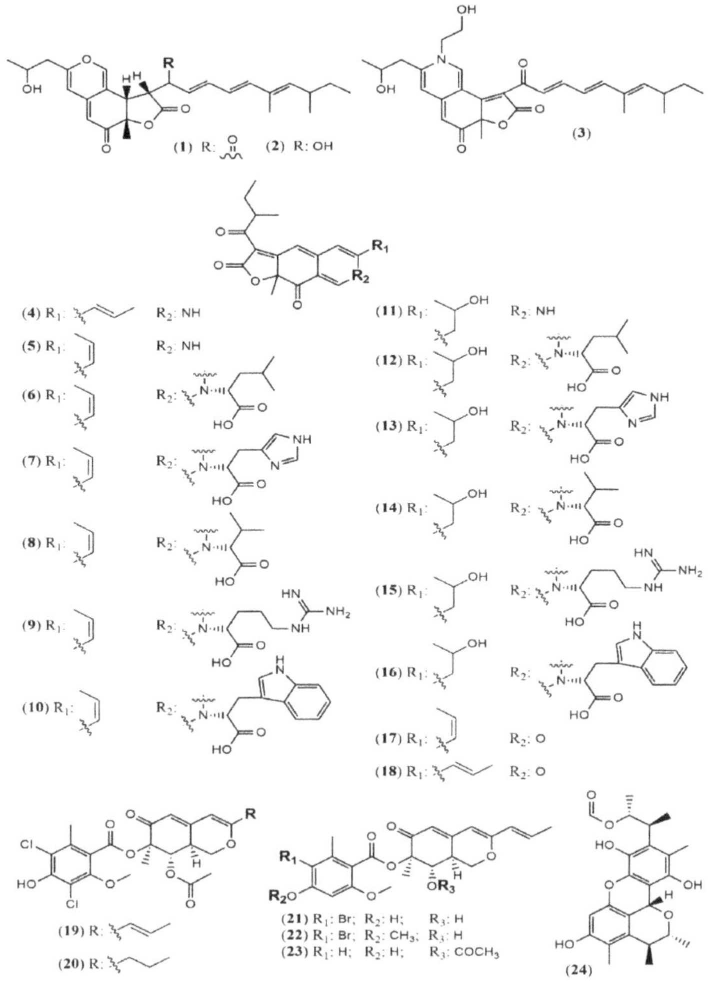

Figure 2. Chemical structures of *Aspergillus* azaphilones **1–2**: sassafrin E-F; **3**: sassafrinamine A; **4**: *trans*-cavernamine; **5**: *cis*-cavernamine; **6–10**: Leu, His, Val, Arg, Trp-cavernamine derivatives; **11**: hydroxy-cavernamines; **12–16**: Leu, His, Val, Arg and Trp-hydroxy-cavernamines.; **17**: *cis*-cavernine; **18**: *trans*-cavernine; **19–20**: falconensins O and P; **21–23**: falconensins Q, R, and S; **24**: penicitrinol Q [33–36].

3.1.2. Azaphilones from *Chaetomium* Genus

Chaetomium is a large genus presenting more than 300 species worldwide. *Chaetomium globosum* represents one of the most studied species and is known as a rich source of azaphilones. Since the last two years, this species has still been contributing with new metabolites. The arthropod-associated endophytic fungus *C. globosum* TW1–1 was investigated considering whether the presence of 1-methyl-l-tryptophan into the growth medium would activate a biosynthetic pathway to produce novel alkaloids [37]. However, instead of nitrogenated metabolites, the authors isolated and identified two chlorinated azaphilones, chaephilone C and D (**25–26**) with anti-inflammatory activity. Their stereostructures were unequivocally confirmed by X-ray analyses. Nevertheless, chaephilone C was also previously reported from the deep sea-derived fungus *Chaetomium* sp. NA-S01-R1 with the same planar structure of **25** but with different stereochemistry, suggesting that its structure should be revised [38]. Two months after the report of chaephilone C (25 new chlorinated azaphilone from *C. globosum*, endophytic of *Polygonatum sibiricum*, w reported and also called chaephilone C (27) [39]. However, this latter compound display a chemical structure similar to (26), but completely different from the former (25).

From the wild-type strain *C. globosum*, a new dimeric azaphilone called cochliodc J (28) was identified in the same medium which cochliodone A had been isolated fore [40]. The deep-sea *C. globosum* MP4-S01–7 provided eight new structurally correla nitrogenated azaphilones **29–36** (Figure 3 and Table 1) [41]. The azaphilone core is same in all compounds with differences only in the lactone acyl substituents and N-alquil groups. Seco-chaetomugilin (37) was isolated for the first time from the et acetate extract of *Chaetomium cupreum* in a bio-guided fractionation for activities agai human breast adenocarcinoma cell lines [42]. Although the authors named the compou isolated as seco-chaetomugilin, it presented the same structure of seco-chaetomugilin previously isolated from *C. globosum* [43]. A screening by LC-MS/MS-GNPS data b of a strain of an endophytic plant fungus *Chaetomium* sp. g1 resulted in the isolation chaetolactam A (38), a unique 9-oxa-7-azabicyclo[4.2.1]octan-8-onering system with t new compounds chaetoviridins derivatives, 11-epi-chaetomugilide B (39), and chae mugilide D (40) [44]. Another plant endophytic fungus *C. globosum* isolated from the des Asteraceae species, *Artemisia desterorum*, yielded globosumone (41), a new stereoisomer the known chaetoviridin E [45].

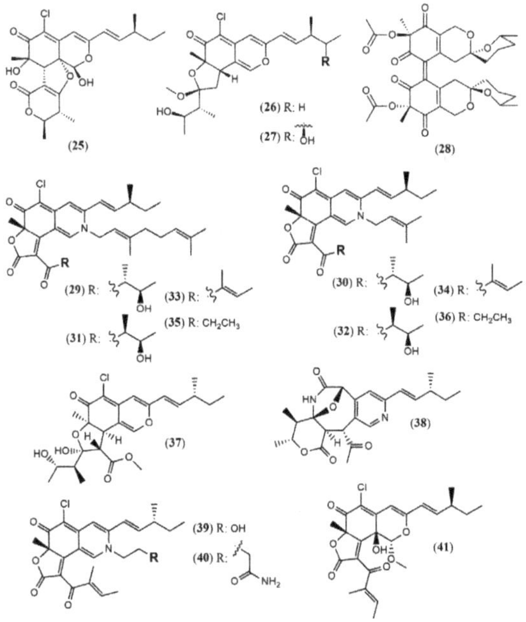

Figure 3. Chemical structures of *Chaetomium* azaphilones: **25**: Chaephilone C (1R,7S,8R,8aR,9E,11S,4′R,5 **26**: chaephilone D; **27**: chaephilone C*; **28**: cochliodone J; **29**: N-(3,7-Dimethyl-2,6-octadienyl aza-2-deoxychaetoviridin A; **30**: 4′-epi-N-(3,7-Dimethyl-2,6-octadienyl)-2-aza-2-deoxychaetoviri A; **31**: N-(3-Methyl-2-butenyl)-2-aza-2-deoxychaetoviridin A, **32**: 4′-epi-N-(3-Methyl-2-buten 2-aza-2-deoxychaetoviridin A; **33**: N-(3,7-Dimethyl-2,6-octadienyl)-2-aza-2-deoxychaetoviridir **34**: N-(3-Methyl-2-butenyl)-2-aza-2-deoxychaetoviridin E; **35**: 4′,5′-dinor-5′-Deoxy-N-(3,7-dimet 2,6-octadienyl)-2-aza-2- deoxychaetoviridin A; **36**: 4′,5′-dinor-5′-Deoxy-N-(3-methyl-2-butenyl aza-2-deoxy- chaetoviridin A; **37**: seco-chaetomugilin; **38**: chaetolactam A; **39**: 11-epi-chaetomugil B; **40**: chaetomugilide D; **41**: globusumone [37–45].

3.1.3. Azaphilones from *Hypoxylon* Genus

Four unprecedented bisazaphilones hybridorubrins A–D (**42–45**) were isolated together with two new mitorubrin-type azaphilones, fragirubrins F–G (**46–47**) [46] from *Hypoxylon fragiforme*. The main differences among them are the acyl substituents in the lenormandin/fragirubrin-type moiety. In this study, the authors determined the azaphilones stereochemistry by electronic circular dichroism (ECD) spectroscopy in a comparative study between isolated and synthetic compounds. The acquired data suggest that the previous stereochemistry reported for rutilins C (**48**), D (**49**) and the mitorubrins [47] must be revised to be (S)-configured at C-8 and C-8a (Figure 4). Another species *Hypoxylon fuscum* complex yielded a new daldinin F derivative possessing a 3′-malonyl group (**50**) [48].

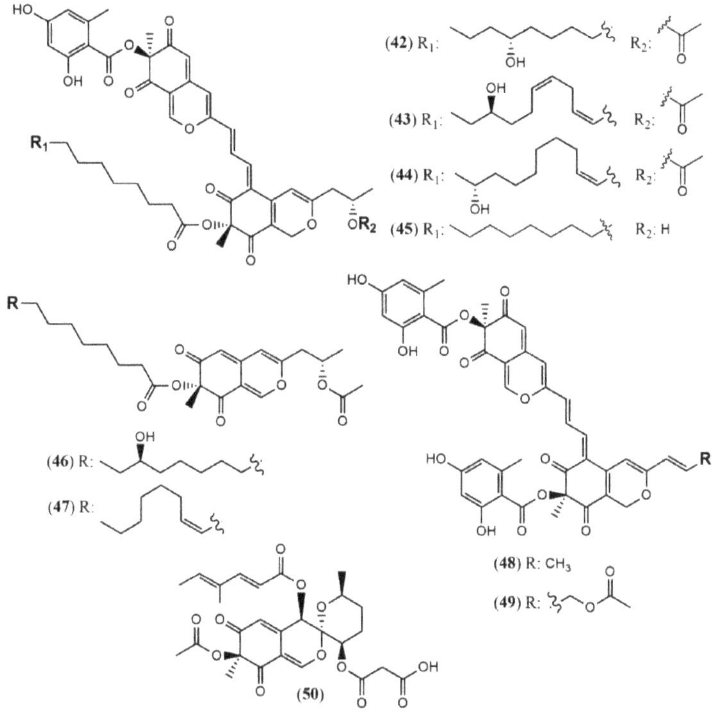

Figure 4. Chemical structures of *Hypoxylon* azaphilones: **42–45**: hybridorubrins A–D; **46–47**: fragirubrins F and G; **48–49**: rutilins C–D; **50**: 3′-malonyl-daldinin F [47,48].

3.1.4. Azaphilones from *Monascus* Genus

Monascus pilosus BCRC 38072, a citrinin-free strain, was able to produce several azaphilone pigments including three new *Monascus* red pigments without citrinin presence: monapilonitrile A (**51**), monapilosine (**52**), and *N*-ethanolic monapilosine (**53**) [49] (Figure 5). Metabolites (**52**) and (**53**) are nitrogenated azaphilones lacking or bearing the N-hydroxyethyl group, respectively.

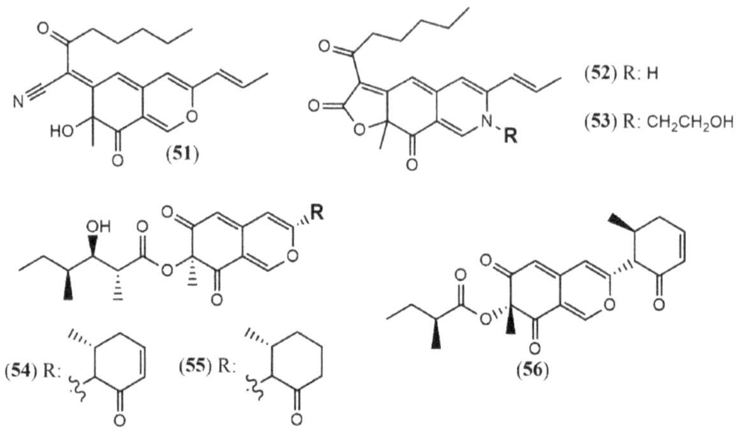

Figure 5. Chemical structures of *Monascus* and *Muyocopron* azaphilones: **51**: monapilonitri 52: monapilosine; **53**: N-ethanolic monapilosine; and *Muyocopron* azaphilones: **54–55**: muyo prones A and B; **56**: lijiquinone 1 [50,51].

3.1.5. Azaphilones from *Muyocopron* Genus

The chemical investigation of the endophyte *Muyocopron laterale* ECN279 isolat from a health leaf of *Conavalia lineata* led to the isolation of the two new azaphilon muyocopronones A and B (**54–55**) [50]. An endophyte fungus F53 from the tradition Chinese medicine plant *Taxus yunnanensis* had its genome sequenced and mined, a the multi-locus phylogeny of F53 allowed its placement within the genus *Muyocopr* with its closest relative being *Muyocopron atromaculans* (MUCL 34983) [51]. Moreover new azaphilone lijiquinone 1 (**56**) with activities against human myeloma cells and t yeast *Candida albicans* and *Cryptococcus albidus* was isolated from its ethyl acetate extra (Figure 5).

3.1.6. Azaphilones from *Penicillium* Genus

The *Penicillium* genus produces a great number of azaphilone metabolites [31]. *Pe cillium citrinum* WK-P9 was isolated as an associated fungus from the sponge *Suber* sp., displaying antibacterial activity. The bio-guided chemical investigation of its eth acetate extract led to the isolation of a new citrinin derivative called penicitrinone (**57**) [52]. Genome mining, epigenetic regulation, optimization of culture conditions, a one-strain-many-compounds (OSMAC) were investigated as a possible way to prioriti the production of other polyketide metabolites different than the rubratoxins in *Peni lium dangeardii* [53]. Only the metabolic shunting strategy, based on the deletion of t key gene *rbtJ* encoding PKS for rubratoxins biosynthesis, and the optimization of cultu conditions successfully led to the production of 35 azaphilones, from which 23 were ne ones. They were identified as nine monomers named dangelones A–G (**58–64**), dangelosi A–B (**65–66**), eight dimers, didangelones A–G (**67–74**), and five trimers, tridangelones A- (**75–79**) [53] (Figure 6). Dangelones A–G (**58–64**) have the same planar structure and t distinctions among them lay on the side chains at C-3. The differences at C-3 side cha are also present in the dimers. Still regarding *Penicillium* endophytic fungi, a strain *Penicillium* sp. T2–11 isolated from the rhizomes of the underground portion of *Gastroc elata* produced a citrinin dimer, named penctrimertone (**80**) [54].

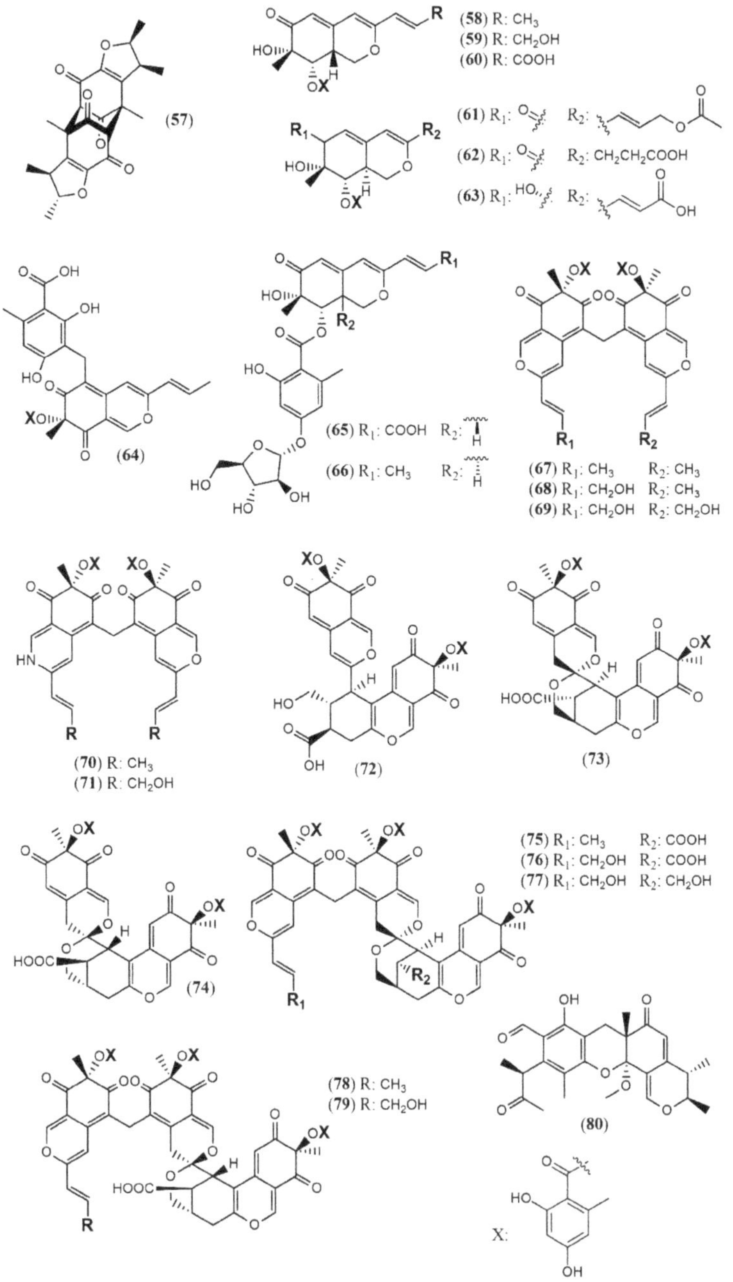

Figure 6. Chemical structures of *Penicillium azaphilones*: **57**: penicitrinone G; **58–64**: Dangelones A–G; 65–66: dangelosides A and B; **67–74**: didangelones A–H; **75–79**: tridangelones A–E; **80**: penctrimertone [52–54].

3.1.7. Azaphilones from *Phomopsis* Genus

Culture of the endophyte fungus *Phomopsis* sp. CGMCC No.5416 yielded the three azaphilones phomopsones A–C (**81–83**), presenting anti-HIV and cytotoxic activity [55]. From the deep-sea-derived fungus *Phomopsis tersa* FS441, five chlorinated azaphilones nar tersaphilones A–E (**84–88**) presenting unique structures were isolated [56] (Figure 7).

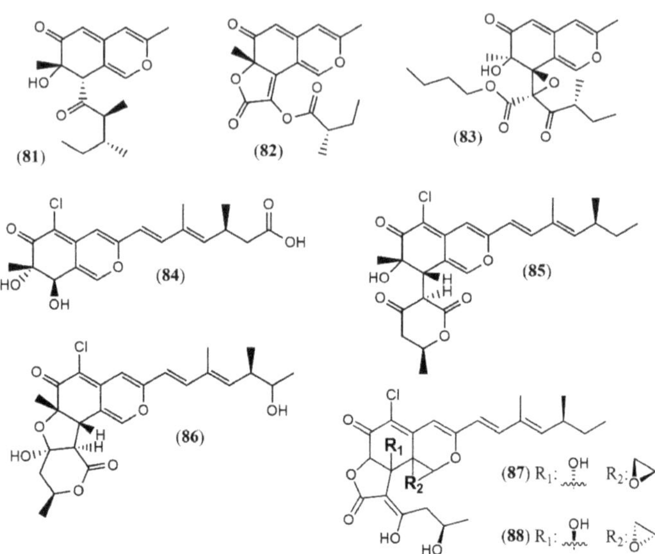

Figure 7. Chemical structures of azaphilones from *Phomopsis*: **81–83**: phomopsones A–C; **84-** tersaphilones A–E [55,56].

3.1.8. Azaphilones from *Pleosporales* Genus

The marine-derived fungus *Pleosporales* sp. CF09-1 produced the uncommon bisa philones dipleosporalones A and B (**89–90**) (Figure 8) [57]. These compounds own a 6/4 ring system that might come from a [2 + 2] cycloaddition reaction between two pinophi B-type monomers and represents the first example of this coupling.

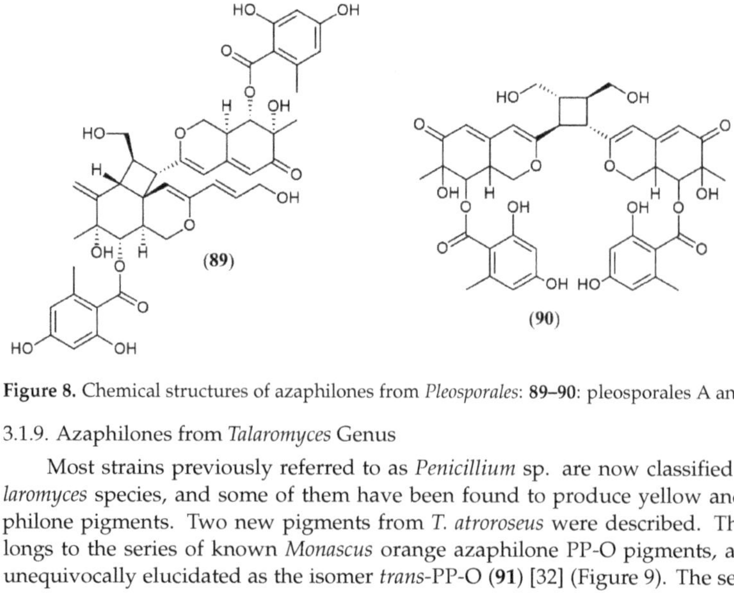

Figure 8. Chemical structures of azaphilones from *Pleosporales*: **89–90**: pleosporales A and B [57].

3.1.9. Azaphilones from *Talaromyces* Genus

Most strains previously referred to as *Penicillium* sp. are now classified in the *laromyces* species, and some of them have been found to produce yellow and red a philone pigments. Two new pigments from *T. atroroseus* were described. The first longs to the series of known *Monascus* orange azaphilone PP-O pigments, and it w unequivocally elucidated as the isomer *trans*-PP-O (**91**) [32] (Figure 9). The second w the unique azaphilone atrosin S, which presented the incorporation of a serine moiety into the isochromene/isoquinoline system. The fungus cultivation in medium enriched with a specific amino acid as sole source of nitrogen could allow seven atrorosin derivatives (atrorosin D, E, H, L, M, Q, and T, depending on the amino acid incorporated) (**92–99**), which were identified by dereplication using HPLC-DAD-MS/HRMS analysis [32]. From the fungus *Talaromyces albobiverticillius* associated with the isopod *Armadillidium vulgare,* two interesting azaphilone pigments talaralbols A and B (**100–101**) was reported [58]. However, talaralbol B presents the same planar structure of trans-PP-O, early described in *T. atroroseus* [32], in which the C-9 stereochemistry was not reported.

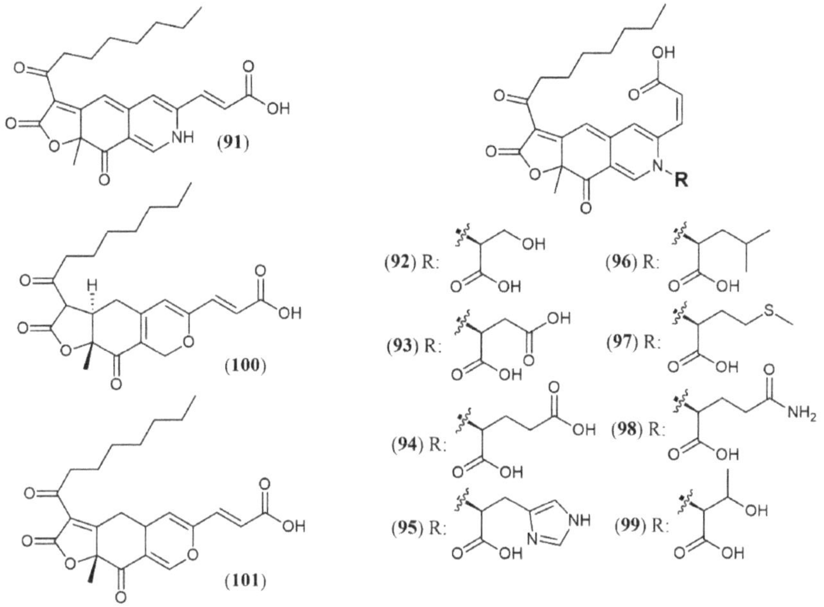

Figure 9. Chemical structures of *Talaromyces* azaphilones: **91**: trans-PP-O; **92–99**: atrosins S (Ser),D (Asp), E (Glu), H (His), L (Leu), M (Met), Q (Gln), and T (Trp); **100–101**: talaralbols A and B [32,58].

3.2. Biological Activities of Azaphilones

Azaphilones, besides being good compounds to replace synthetic pigments, aggregate valuable pharmacological properties. The wide broad range of biological activities that has been reported for azaphilones such as cytotoxic, anti-inflammatory, antimicrobial, antitumoral, antiviral and antioxidant is exemplified in Table 1.

Concerning the activities regarded to the new 101 azaphilones reported, the cytotoxic and antitumor potential are the most evaluated. Remarkably, compounds (**29**), (**30**), and (**33**) showed the most effective anti-gastric cancer activities (MGC803 and AGS cell lines) with IC_{50} values less than 1 μM, being more active than the positive control paclitaxel (3.8 μM) [41]. Additionally, (**29**) and (**30**) induced apoptosis in a concentration-dependent manner and (**30**) inhibited cell cycle progression. The authors also claim that 3,7-dimethyl-2,6-octadienyl group attached to N-2 contributed to the potent cytotoxic activities against MGC803 and AGS gastric cancer cell lines what can induce new investigations with semi-synthetic azaphilone derivatives possessing this group [11]. The azaphilones (**39**) and (**40**) showed moderate activity against leukemia HL-60 and human breast cancer. However, (**39**) exhibited potent apoptosis induction activity by mediating caspase-3 activation and PARP degradation at 3 μM in leukemic cells HL-60 [44]. Another interesting result was the potent cytotoxic activity showed by the dimeric azaphilones (**89**) and (**90**) against five different human cell lines. (**89**) showed more potent cytotoxicity against MGC-803 th cisplatin and possessed a unique 6/4/6 ring system suggesting the new ring may play important role in cytotoxicity [57].

A great number of azaphilones present anti-inflammatory activity [35,37,38,42,49, 52,53,58,59]. The compounds (**21**), (**51**), (**52**) and (**100**) exhibited anti-inflammatory act ities due to potent anti-NO production activity, with IC_{50} values of 11.9, 2.6, 12.5, ar 10.0 µM, respectively, compared to the known iNOS inhibitor quercetin (34.6 ± 1.4 µM) lipopolysaccharide (LPS) -induced nitric oxide (NO) production [35,46,57]. The antimic bial activity of azaphilones also must be highlighted. Two dimeric azaphilones, penicitrir Q (**24**) and penctrimertone (**80**), showed both excellent inhibitory activities against *B. st tilis* with MIC of 6.2 and 4.0 µg/mL, respectively. Moreover, (**24**) also presented inhibito activity against bacteria *Staphylococcus aureus* (4.3 µg/mL) and *Pseudomonas aerugino* (11.2 µg/mL), and the yeast *C. albicans* (4.0 µg/mL) [36].

In vitro antiviral activity against HIV-1 was detected for phomopsones B and C (**82–** (7.6 and 0.5 µM, respectively [52]). Research in antiviral potential of azaphilones may strengthened as they have been focused as possible drug leads for the development of eff tive antiviral agents against SARS-CoV-2 [60,61]. This worldwide impact-generated vir draws attention to the difficulty in developing new non-toxic antiviral drugs, as viruses u cell host metabolism for replication. This is corroborated by previous reports of antivir activity of azaphilone metabolites, such as chermisinone B, isolated from the endophy fungus *Nigrospora* sp. YE3033, and active against A/Puerto Rico/8/34 (H1N1) in CPE say (IC_{50} 0.80 µg/mL) with low cellular toxicity on MDCK cells (CC_{50} 184.75 µg/mL) [6 In vitro HIV-1 replication inhibitory effects in C8166 cells were demonstrated for Helotiali A and B (EC_{50} 8.01 and 27.9 nM, respectively) [63]. In 2019, comazaphilone D was report as a non-competitive inhibitor of neuraminidase from recombinant rvH1N1 (IC_{50} 30.9 µ while rubiginosin A was active against H5N1 (IC_{50} 29.9 µM) [64]. The previous knowled of the antiviral potential of azaphilone derivatives is an advantageous background for t development of new drugs to inhibit SARS-CoV-2.

Table 1. Azaphilones fungal sources and reported biological activities.

Name (No).	**Producing Strains**	**Activity**
Aspergillus		
Sassafrin E-F (**1–2**) Sassafrinamine A (**3**)	*A. neogabler* IBT3020 [33]	Data not reported
Trans-cavernamine(**4**) *Cis*-cavernamine (**5**) *Cis*-cavernamines-Leu, His, Val, Arg, Trp (**6–10**) Hydroxy-cavernamine (**11**) Hydroxy-cavernamines-Leu, His, Val, Arg, Trp (**12–16**) *Cis*-cavernines (**17**) *Trans*-cavernines (**18**)	*A. cavernicola* [34]	Data not reported
Falconensins O (**19**)	*A. falconensis* [35]	Anti-inflammatory (MDA-MB-231 cells line for NF-κB inhibition: 15.7 µM)
Falconensins P (**20**)		Not tested
Falconensins Q (**21**)		Anti-inflammatory (MDA-MB-231 cells line for NF-κB inhibition: 11.9 µM)
Falconensins R (**22**)		Anti-inflammatory (MDA-MB-231 cells line for NF-κB inhibition: 14.6 µM)
Falconensins S = 8-*O*-Acetil-falconensin I (**23**)		Anti-inflammatory (MDA-MB-231 cells line for NF-κB inhibition: 20.1 µM)

Table 1. *Cont.*

ame (No).	Producing Strains	Activity
nicitrinol Q (**24**)	*A. terreus* [36]	Antimicrobial (*S. aureus*: 4.3 mg/mL; *B. subtilis*: 6.2 mg/mL)
Chaetomium		
naephilone C (1R,7S,8R,8aR,9E, S,40R,50R) (25)	*C. globosum* TW1–1 [37]	Anti-inflammatory (inhibit NO production: 15.12 µM)
naephilone D (**26**)		Anti-inflammatory (inhibit NO production: 20.65 µM)
naephilone C * (27)	*C. globosum* [39]	Cytotoxic (HepG-2: 38.6 µM); BST (68.6% of letality at 10 mg/mL)
ochliodone J (28)	*C. globosum* [40]	Cytotoxic (HeLa: 17.3 µM)
R,5′R,7S,11S)-N-(3,7-methyl-2,6- octadienyl)-2-aza-deoxychaetoviridin A (**29**)	*C. globosum* MP4-S01–7 [41]	Antitumor (MGC803 and AGS gastric cells lines: 0.78 and 0.12 µM, induced apoptosis)
-epi-N-(3,7-dimethyl-2,6-octadienyl)-2-aza-2-oxychaetoviridin A (**30**)		Antitumor (MGC803 and AGS gastric cells lines: 0.46 and 0.62 µM, induced apoptosis an altered the cell cycle distribution)
-(3- methyl-2-butenyl)-2-aza-2-oxychaetoviridin A (**31**)		Antitumor (MGC803 and AGS gastric cells lines: 2.7 and 6.5 µM)
- epi-N-(3-methyl-2-butenyl)-2-aza-2-oxychaetoviridin (**32**)		Antitumor (MGC803 and AGS gastric cells lines: 3.0 and 2.9 µM)
-(3,7-dimethyl-2,6-tadienyl)-2-aza-2-deoxychaetoviridin E (**33**)		Antitumor (MGC803 and AGS gastric cells lines: 0.72 and 0.12 µM)
-(3-methyl-2-butenyl)-2-aza-2-oxychaetoviridin E (**34**)		Antitumor (MGC803 and AGS gastric cells lines: 6.8 and 2.0 µM)
,5′-dinor-5′-deoxy-N-(3,7- dimethyl-2,6-tadienyl)-2-aza-2-deoxychaetoviridin A 5)		Antitumor (MGC803 and AGS gastric cells lines: 2.2 and 1.2 µM)
,5′-dinor-5′- deoxy-N-(3-methyl-2-butenyl) -aza-2-deoxychaetoviridin A (**36**)		Antitumor (MGC803 and AGS gastric cells lines: 5. 8 and >10 µM)
co-chaetomugilin (**37**)	*C. cupreum* [42]	Anticancer (MCF-7: 75.25% at 50 mg/mL) Increased ROS production: 19.6% at 5 mg/mL
naetolactam A (**38**)	*Chaetomium* sp. g1 [44]	Cytotoxic (Not detected)
-epi-chaetomugilide B (**39**)		Cytotoxic (HL-60: .3.19 µM; A549: 8.37 µM; MCF-7: 4.65 µM; SW480: 4.21 µM; apoptosis induction mediated by caspase 3 in HL-60 cell: 3 µM)
naetomugilide D (**40**)		Cytotoxic (HL-60: .15.92 µM; MCF-7: 17.97 µM; SW480: 14.09 µM; apoptosis induction mediated by caspase 3 in HL-60 cell: 15 µM)
obosumone (**41**)	*C. globosum* [45]	Cytotoxic (Not detected)
Hypoxylon		
ybridorubrin A (**42**)	*H. fragiforme* [47]	Antimicrobial (% biofilm inhibition of *S. aureus*: 81% at 250 mg/mL)
ybridorubrin B (**43**)		No antimicrobial or cytotoxic activity
ybridorubrin C (**44**)		Antimicrobial (% biofilm inhibition of *S. aureus*: 82% at 250 mg/mL)
ybridorubrin D (**45**)		Antimicrobial (% biofilm inhibition of *S. aureus*: 71% at 250 mg/mL)
agirubrin F (**46**)		Not tested
agirubrin G (**47**)		Not tested
ıtilin C (**48**)		Antimicrobial (% biofilm inhibition of *S. aureus*: 58% at 250 mg/mL)
ıtilin D (**49**)		Not tested

Table 1. *Cont.*

Name (No).	Producing Strains	Activity
3′-Malonyl-daldinin F (**50**)	*H. fuscum* [48]	Cytotoxic (L929 murine fibroblast: weak; KB 3. cervix-cancer cells: weak)
Monascus		
Monapilonitrile (**51**)	*M. pilosus* BCRC 38072 [49]	Anti-inflammatory (inhibit NO production: 2.6 μM)
Monapilosine (**52**)		Anti-inflammatory (inhibit NO production: 12.5 μM)
N-Ethanolic monapilosine (**53**)		Anti-inflammatory (inhibit NO production: 27.5 μM); cytotoxic (LPS-induced RAW264.7: ce viability< 65% at 50 μM)
Muyocopron		
Muyocopronone A (**54**)	*M. laterale* ECN279 [50]	Antimicrobial (Not detected)
Muyocopronone B (**55**)		Antimicrobial (methicillin-resistant *S. aureus* ar vancomycin-resistant *E. faecalis*: MIC at 128 mg/mL)
Lijiquinone 1 (**56**)	*Muyocopron* sp. ** [51]	Antifungal (*C. albicans*: 79 μM; *C. albidus*: 141 μM); Cytotoxic (RPMI-8226: 129 μM)
Penicillium		
Penicitrinone G (**57**)	*P. citrinum* WK-P9 [52]	Antimicrobial (Not detected)
Dangelone A (**58**)	*P. dangeardii* [53]	Cytotoxic (Inactive: IC > 20 mmol)
Dangelone B (**59**)		Cytotoxic (HepG2: 6.82 mmol; MCF-7: 14.98 mmol)
Dangelone C-G (**60–64**)		Cytotoxic (Inactive: IC > 20 μM)
Dangeloside A and B (**65 and 66**)		Cytotoxic (Inactive: IC > 20 μM)
Didangelone A-H (**67–74**)		Cytotoxic (Inactive: IC > 20 μM)
Tridangelone A-E (**75–79**)		Cytotoxic (Inactive: IC > 20 μM)
Penctrimertone (**80**)	*Penicillium* sp. T2–11 [54]	Antimicrobial (*C. albicans*: 4mg/mL; *B. subtilis*: 4mg/mL); cytotoxic (HL-60: 16.77 μM; SMMC-7721: 23.03 μM; A-549: 28.62 μM; MCF- 21.53 μM)
Phomopsis		
Phomopsone A (**81**)	*Phomopsis* sp. CGMCC No.5416 [55]	Antiviral (Not detected); cytotoxic (Not detecte
Phomopsone B (**82**)		Antiviral (HIV-1: 7.6 μM); cytotoxic (A549: 176 μM; MDA-MB-231: 303.0 μM);
Phomopsone C (**83**)		Antiviral (HIV-1: 0.5 μM); cytotoxic (A549: 8.9 μM; MDA-MB-231: 3.2 μM); apoptosis (PANC- cancer cells: 28.54% at 17.3 μM
Tersaphilone A-C (**84–86**)	*P. tersa* FS441 [56]	Cytotoxic (Not detected)
Tersaphilone D (**87**)		Cytotoxic (SF-268: 7.5 μM; MCF-7: 7.8 μM; HepG-2: 14.0 μM; A549: 8.3 μM)
Tersaphilone E (**88**)		Cytotoxic (SF-268: 5.6 μM; MCF-7: 5.4 μM; HepG-2: 9.8 μM; A549: 6.7 μM)
Pleosporales		
Dipleosporalone A (**89**)	*Pleosporales* sp. *CF09-1* [57]	Cytotoxic (MDA-MB-231: 1.9 μM; HeLa: 2.5 μM MGC-803: 1.3 μM; MCF-7: 2.1 μM; A549: 1.0 μM
Dipleosporalone B (**90**)		Cytotoxic (MDA-MB-231: 3.8 μM; HeLa: 3.0 μM MGC-803: 2.0 μM; MCF-7: >10 μM; A549: 3.5 μM)

Table 1. *Cont.*

ame (No).	Producing Strains	Activity
Talaromyces		
ans-PP-O (**91**) trosins S (**92**), D (**93**), E (**94**), (**95**), L (**96**), M (**97**), Q (**98**) nd T (**99**)	*T. atroroseus* [32]	Not tested
alaralbol A (**100**)	*T. albobiverticillius* [58]	Anti-inflammatory (LPS-induced NO production in RAW264.7 cell: 10.0 μM); 31.0% of inhibitory rate)
alaralbol B (**101**)		Not detected

* isolated as endophytic of *Polygonatum sibiricum*; ** closest relative being *Muyocopron atromaculans* (MUCL 34983); SF-268 (human glioblastoma carcinoma), MCF-7 (breast cancer), HepG-2 (liver cancer), HeLa (human cervix carcinoma), and A549 (lung cancer), BST = Brine Shrimp test.

3.3. Recent Insights in the Biosynthesis of Azaphilones

The biosynthesis of azaphilones has been reviewed by Pavesi et al. [65] and was also considered in the two latest reviews [31]. Five biosynthetic pathways were exhaustively discussed, which highlighted the comprehensive study of *Monascus* and *Aspergillus* pathways [65]. Furthermore, a thorough study performed about the precise role of ammonium nitrate in the production of *Monascus* pigments showed that some biosynthetic pathways can present changes due to the regulation and expression of several key genes involved [66]. The expression of the gene mppG (MrPigF), responsible for orange pigments, was significantly downregulated with ammonium nitrate addition, and an improvement in yellow pigment production was followed by an upregulated *mppE* expression. Additionally, ammonium nitrate increased the NH_3 content in the fermentation broth resulting in the increased red pigments yield [66].

Dimeric azaphilones have been described in *the Chaetonium* genus, and the fungal laccase-like multi-copper oxidase gene encoded by CcdJ (CHGG_10025) is believed to dimerize the cochliodones [65]. Cochliodone J (**28**), a new dimeric azaphilone containing a spirotetrahydropyran moiety, was reported, but the mechanism of the spiro ring formation still remains to be determined [40]. Moreover, the unusual fusion between an eight-membered lactam and a six-membered lactone, presented in the structure of chaetolactama A (**38**), has not been investigated yet.

The biosynthetic gene cluster responsible for the sequential and convergent production of azaphilones in *Chaetonium* sp. might count with a hidden gene allegedly responsible for the epimerization of the 7-OH group in chaetoviridin E as well as the oxidation/epoxidation leading to OH groups in C-8a and C-1 positions, followed by methylation of the latter, as in (**41**) [45]. Based on studies with *Monascus*, *Aspergillus*, and *Talaromyces*, two biosynthetic gene clusters were postulated to drive the diverse azaphilones in *H. fragiforme*. However, the biosynthetic dimerizations which led to the compounds (**42**)–(**49**) demand more investigations. This represents a challenge because *Hypoxylaceae* azaphilones are exclusively formed during stromata development, which cannot be induced under laboratory conditions [46]. A reasonable proposal consists on a spontaneous aldol condensation responsible for the heterodimerization of different azaphilones derivatives [46].

The biosynthesis of three different azaphilone skeletons was reported for *P. tersa* FS441. The tersaphilone B (**85**) showed the unique 6/6-6 carbon skeleton with a cleaved tetrahydrofuranyl ring, and the diastereomers tersaphilones D and E (**87–88**) displayed a unique five-membered furan ring open and an epoxide ring in C-8a and C-1 positions [56]. A remarkably biosynthetic proposal was provided to penctrimertone (**80**), which presented a 6/6/6/6 tetracyclic ring system with an unusual aldehyde group in one of the rings [54]. It is supposed to be a citrinin dimer furnished by a citrinin monomer that suffered hydration, oxidation, and reduction affording an orthoquinone methide susceptible to an unusual intermolecular hetero-Diels-Alder reaction with another citrinin molecule [57].

Another interesting observation is the presence of a six-membered ring at the C position of the azaphilones core reported in the *Muyocopron* genus, which is present less than 10% of the hundreds of azaphilones isolated to date. Regarding the compoun **(54–56)**, the gene cluster *lij.* was proposed to control a convergent biosynthetic pathw The LijE would be responsible for the formation of the aromatic ring with a carbon cha attached to the cyclohexanone ring. Reduction of the acyl ester followed by cyclizati and dehydration afforded the azaphilone core. This core would be attached by the C-7 C group to the acyl derivative formed by previous condensation of acetyl-CoA/malonyl-Co and C-methylation controlled by the LijA gene. The compounds (**54–55**) also presentec 2,4-dimethyl-3-hydroxyhexanoate moiety that was reported in only eight compounds this genus. The cyclohexanone ring and 2,4-dimethyl-3-hydroxyhexanoate moiety mig be biomarkers of the Dothideomycetes class and constitute a noteworthy point to be mo investigated [50].

4. Processing and Innovations in Azaphilones Production

Over the last decade, many studies have focused attention on optimizing producti of pigments and growth of different fungal species. Many variables that affect the prod tion, as fermentation process (submerged fermentation, solid-state fermentation, larg scale), culture media composition (carbon and nitrogen source, C/N ratio, co-factors, su factants, tricarboxylic acid intermediates), inoculum type and age (spores and myceliur temperature, pH, oxygen level and agitation; light, humidity, pigment recovery, extractic and isolation have been critically discussed by recent reviews [11,31,67–69]. Some relat aspects of production, processing and innovations in azaphilones production published 2020 and up to March 2021 are highlighted below.

4.1. Overcoming Mycotoxin Issues

The consensual approval of color additives for food industry by international regu tory bodies is of great importance for commercial transactions, so that in-house produc can be exported to other markets without alterations to remove or replace pigments regul ized only in the exporting country. US and EU are good examples. Sixteen color additiv allowed in the EU are not accepted by US regulatory agency, while four color additiv allowed in the US are not permitted in the EU [70]. The ancient knowledge about *Monasc* pigments and utilization of *Monascus* by Asian people for hundreds of years has motivat the search for beneficial and healthy metabolites of *Monascus* azaphilones. Despite t isolation of many *Monascus* metabolites, these pigments were not approved by regulato agencies in the US and UE so far, due to concerns over co-production of the hepatoneph toxic mycotoxin citrinin (**102**, Figure 10). Co-production of azaphilones and citrinin is major issue on this point and optimization of azaphilones production on industrial sca must assure no production of toxic metabolites [71]. For this purpose, genetic techniqu have been used, such as depletion of *ctnE* gen, responsible for the production of citrin (**102**), successfully performed in *Monascus aurantiacus* Li AS3.4384 [72]. The medicin properties reported for azaphilones are a catalyst in the search for fermentative process suitable for the production of these pigments from safe biosynthetic routes, obtained l deletion of citrinin gene.

M. purpureus has also been studied with the aim of inhibiting citrinin (**102**) productic without negative change in pigments biosynthesis. Hong et al. [71] used transcripton sequencing to explore citrinin gene expression in experiments comparing the effect inorganic (ammonium chloride and ammonium nitrate) with organic nitrogen (pepto group) sources in *M. purpureus* M3103 metabolism. It was found that biosynthesis of ami acids was up-regulated by ammonium chloride and ammonium nitrate, enhancing t producing of biosynthetic precursors of pigments while essential genes and transcriptic factors involved in the biosynthesis pathway of citrinin (**102**) were down-regulated by the inorganic nitrogen sources. Therefore, inorganic nitrogen proved to be more favorab for the biosynthesis of citrinin-free pigments (especially orange and red pigments) by *M. purpureus* M3103.

Industry Research and Development Institute in Taiwan is dedicated to investigating new ways to obtain azaphilone pigments using genetic manipulation and optimization of a fermentative process, aiming to avoid the production of citrinin (**102**) (Figure 10). They successfully developed some citrinin-free *Monascus* strains, including the strain *M. pilosus* BCRC 38072, previously mentioned for its production of azaphilones **51–53** [49].

Other mycotoxins are also of concern. *Talaromyces* genus have species reported to produce both, red colorants and mycotoxins (*T. atroroseus* [32], *Talaromyces purpureogenus* [73] and *T. albobiverticillius* [58]) while other species of this genus are not reported to produce known mycotoxins [11,74,75]. Mycotoxins reported from *T. purpureogenus* are rubratoxins A (**103**) and B (**104**), rugulovasins (**105**) and luteoskyrin (**106**), (Figure 10) therefore limiting the use of this species for biotechnological production of food pigments [73]. *T. purpureogenus* CFRM0 produces higher yield of pigments in Potato Dextrose Agar (PDA) and Charcoal Yeast Extract (CYE) rather than in Malt Extract Agar (MEA) and Yeast Extract with Supplements (YES) media (30 °C, 3–4 days), although the growth rate was similar in all conditions [73]. The pigments produced by *T. purpureogenus* CFRM0 were not toxic to female Wistar rats. No alterations related to toxicity were found, including no biochemical, hematological and histological modifications, indicating the safety of this pigment even when administrated in successive days [73].

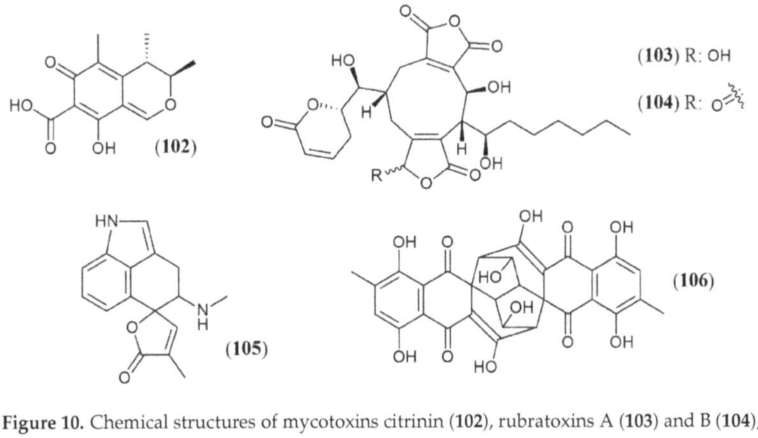

Figure 10. Chemical structures of mycotoxins citrinin (**102**), rubratoxins A (**103**) and B (**104**), rugulovasins (**105**) and luteoskyrin (**106**) [74].

4.2. Color-Directed Production of Pigments

Fungi from *Monascus* genus are the oldest source of azaphilone pigments and this genus is still considered as one of the most proliferous sources of pigments nowadays [76]. Azaphilones produced by *Monascus* species are usually refered as MonAzPs (*Monascus* azaphilone pigments) and are incorporated in many food products as a natural colorant in China, where MonAzPs exceed 20 thousand tons per year. It is estimated that the number of consumers that eat food containing MonAzPs daily is over one billion people [77]. *Monascus* pigments have predominantly three colors, yellow (monascin (**107**) and ankaflavin (**108**)), orange (rubropunctatin (**109**) and monascorubrin (**110**)) and red (rubropunctamine (**111**) and monascorubramine (**112**)) [78]. The structures of the mentioned substances and their chromophores, the part of the molecule responsible for their color, are shown in Figure 11. Several works focus *M. purpureus* metabolism [66,79,80]. Literature is also rich in reports presenting conditions to drive the metabolism of other fungal species to biosynthesize or to improve the production of pigments.

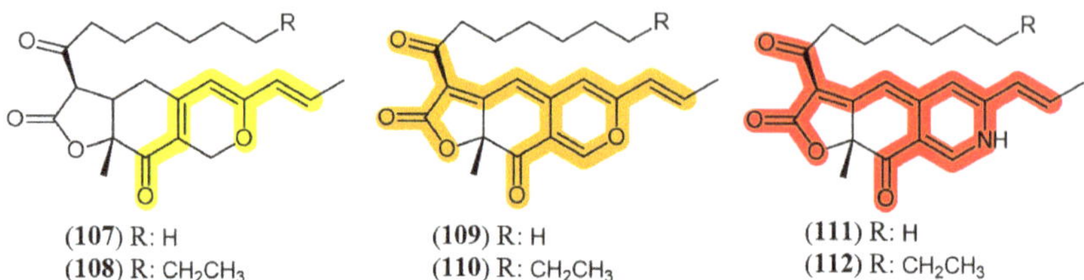

Figure 11. *Monascus* pigments and their chromophores (highlighted in color). Yellow: **107**: monascin; **108**: ankaflavin; Orange: **109**: rubropunctatin; **110**: monascorubrin; Red: **111**: rubropunctamine; **112**: monascorubramine [79].

Color-directed production of pigments is advantageous as this approach would eli nate purification steps slowing down the processing by adding a separation step, to pu or concentrate pigments of the desired color. Therefore, a big challenge in pigments prod tion is to obtain pure extracts, containing fewer substances and, preferably, with only c color [19]. Figure 12 presents some fungal species and associated fermentative paramet that resulted in the production of yellow [66,80–84], orange [14,85] or red [14,79,85– pigments. However, in most of the works, yellow, orange and red azaphilones are produc simultaneously (cocktail pigments phenomenon) in different proportions.

Regarding *Monascus* species, *M. ruber* CCT 3802 has been studied in terms of colo morphology and biomass production during pigments production utilizing cheese wh as substrate [90]. Strain *M. ruber* M7 showed different response to the addition of ace acid, sodium acetate and ammonium acetate to PDA culture medium. The original b orange fleecy colony morphology turned into small compact reddish or tightly-pack orange colony upon increase of acetic acid or acetate. Pigment production, in turn, w enhanced by addition of acetate to the culture medium [91]. Yang et al. [16] reported that expression of key genes for *Monascus* pigment biosynthesis was significantly up regula in the presence of sodium nitrate. Increase in total pigment production and yellow pigm proportion was reported for a *M. purpureus* strain (LQ-6), after adding exogenous cofac methyl viologen and rotenone (1.0 mg/L) to the submerged batch-fermentation [84].

The color of pigments produced by *Talaromyces amestolkiae* DPUA 1275 was sho to be pH-dependent. Low pH (2.59 and 3) directed to small production of yellow p ments while red ones were not detected [86]. On a further study, *T. amestolkiae* DPU 1275 was grown in MSG-glucose medium supplemented with three individual comp nitrogen sources (yeast extract, meat extract and meat peptone), six individual amino ac (glutamic acid, threonine, tyrosine, glycine, cysteine and tryptophan), and two vitami (biotin and thiamine) [92]. Complex nitrogen and amino acid supplementation did n favor red pigments production but small improvement (1.3 times) was detected af thiamine supplementation.

On the other side, the production of yellow and orange colorants was increas adding yeast extract as nitrogen source in the medium in pH above 5.0. In this conditi conidiation and biomass production were enhanced. The higher yield of colorants the monosodium glutamic acid (MSG) glucose medium was attributed to the metabo stress caused by poor nutrition provided by this medium [92]. The production proc was scaled-up to a 4 L stirred-tank bioreactor. In another study, the same group [evaluated the effect of pH and agitation (100 to 600 rpm) in the improvement of pigme production. They reported near 4-fold increase in orange and red pigments production 500 rpm, under the pH-shift strategy from 4.5 to 8.0, after 96 h of cultivation at 2.0 vvm 30 °C. Moreover, the aforementionated work also demonstrated the possibility of usi *T. amestolkiae* colorants in the preparation of cassava starch-based biodegradable films food packaging, resulting in enhancement of protection against butter oxidation, reduci peroxide amount.

YELLOW AZAPHILONES

- Mutant strains of *Monascus anka* produces yellow azaphilones in aerated submerged fermentation bioreactor containing corn steep liquor, ammonium chloride, potassium dihydrogen phosphate, glucose and starch. Agitation at 300 rpm/min improved yellow pigment production [82].
- High osmotic concentration (NaCl 3.5%) increase the production of yellow intracellular metabolites by *M. ruber* CGMCC 10910 by 40%, due to the up-regulation of specific genes [83].
- *T. purpureogenus* was reported to have a pH-related biosynthesis of yellow piments in Czapeck yeast medium [84].
- Increase in total pigment production and yellow pigment proportion by *M. purpureus* strain (LQ-6) can be achieved adding the exogenous cofactor methyl viologen and rotenone (1.0 mg/L), in submerged batch-fermentation [85].
- Expression level of key genes of *M. purpureus* ZH106-E metabolic pathway can be significantly improved by addition of 3% ethanol in submerged culture medium after 168 h, increasing the production of yellow pigments [81]
- *M. purpureus* M9, grown in rice supplemented with ammonium nitrate (10 g/L), improves production of yellow (monascin and ankafavin) and red (rubropunctamine and monascorubramine) azaphilones and reduce the yields of rubropunctatin or monascorubrin (both orange) [67].

ORANGE AZAPHILONES

- Ammonium nitrate successfully increases the yield of orange pigments and inhibit production of red and yellow metabolites by *Monascus* sp. KCCM 10093 [14].
- Choe et al. [14] also report the possibility to manipulate the metabolism in suboptimal conditions.
- Acidic condictions are effective for accumulation of orange pigments by *Monascus ruber* M7 [86].
- Stimulation of the synthesis of orange metabolites in low pH (2.5) is also effective for *Monascus* sp. KCCM 10093, since, under acidic conditions, conversion of orange to red metabolites is low due to greater expression of the genes MrpigA, B, F, and K, responsible for the production of orange pigments [86].
- Improvement of O_2 dispersion in the culture medium by aeration stimulates the production of orange pigments in *Monascus* sp. KCCM 10093, since O_2 is a substrate for the enzyme flavin adenine dinucleotide-dependent monooxygenase, essential in the production of these pigments [14].

RED AZAPHILONES

- High yield of water-soluble red colorants was achieved culturing *T. amestolkiae* DPUA 1275 using glucose, magnesium and iron sulfate, sodium chloride and monosodium glutamate for 14 days at pH 5.0 [87].
- Despite the low fungal biomass production, improvement of 31% in the red colorant production using MSG-glucose medium and high oxygen supply (8.0 $Lmin^{-1}$) after 48 h (30 °C) in comparison to control conditions was reported for *T. amestolkiae* DPUA [88].
- Solid state fermentation using palm oil of *M. purpureus* FTC 5357, followed by ethanol extraction provided red pigments in expressive yield (60%) [80].
- Optimum biomass production, in addition to increasing the presence red pigments can be achieved growing *T. purpureogenus* strain F in Czapek yeast medium at pH 5 [84].
- Production of red pigments by *Monascus* sp. KCCM 10093 was achieved at pH 6.5 [14].
- Rice and peptone are optimum carbon and nitrogen sources, respectively, to produce citrinin-free red pigments by *M. ruber* OMNRC45 under submerged fermentation conditions [89].
- Using response surface methodology, an improvement by 38-fold in the yield of an antioxidant red pigment was related culturing *T. purpurogenus* KKP in a medium containing dextrose and peptone (2.25 and 1.10% respectively), pH 6.0 at 27 °C [90].

Figure 12. Conditions reported for color-directed production of yellow, orange and red azaphilones [14,67,80–90].

4.3. Yield Improvement

Yield is another key bottle neck in the way to produce fungal pigments to supply dustrial demand. Yield improvement can start early in wet bench step, selecting promisi species from under-studied niches. Marine environment has gained prominence in th area in recent decades. In terms of chemical structures, marine metabolites are frequen halogenated in comparison to metabolites biosynthesized by non-marine microorganisr Halogenated fungal metabolites reach 59.2% of metabolites isolated from marine fungi ar among these metabolites, several halogenated pigments of the azaphyllone class have be reported, as penicilazaphilones D (**113**) and E (**114**) isolated from *Penicillium sclerotiori* (Figure 13) [38,93]. It is noteworthy that fungal species isolated from marine environme can also be isolated from terrestrial sources, such as *P. sclerotiorum*, that, despite bei isolated from soil, was also reported of being capable of producing halogenated derivativ (**115** and **116**) (Figure 13) [94,95].

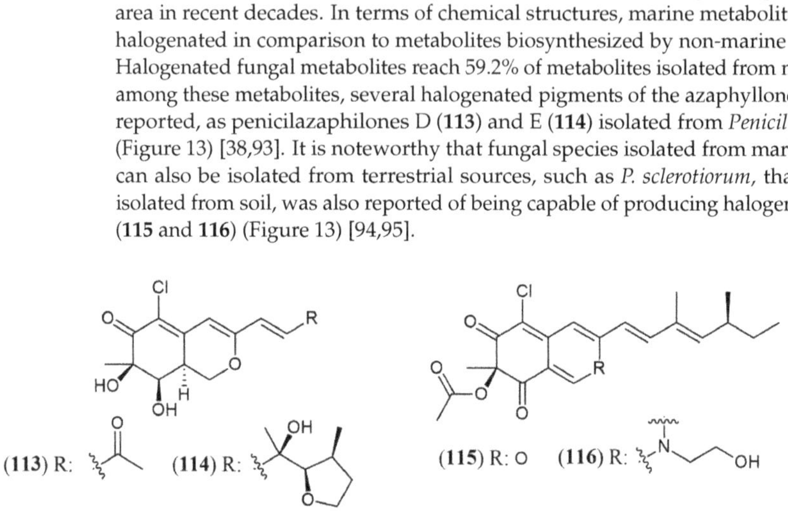

Figure 13. Halogenated azaphilones produced by marine and terrestrial fungi. **113–114**: penicilazaphilones D and E; **115**: sclerotiorin; **116**: N-ethylbenzene-sclerotioramine [95,96].

Enhancement of metabolites yield can be achieved applying stressing conditions d ing fungal development, aiming at activating unconventional metabolic routes related the production of substances linked to defense (biotic stress) or adaptation (abiotic stres This technique is particularly interesting for the production of fungal pigments, since the metabolites are associated with defense against various types of abiotic stress [96]. Abio stress is usually caused by altering nutrients (carbon, nitrogen, minerals) and conditio (temperature, length, oxygen supply) in the culture medium, improving pigments produ tion, although independently of directing to a single pigment color. Increase in bioma development is not a must to enhance pigments production, as optimized conditio for development of fungal biomass not necessarily guarantee maximum production metabolite [82]. In general, in the search for better yields, both, biomass and metaboli yield should increase [97].

The relationship between fungal development and pigments secretion was report for *T. albobiverticillius* (IBT31667). When cultured on Czapek Yeast Agar (CYA), a malt-fr extract, this species produced atrorosins, pigments already reported as metabolites of *atroroseus* IBT 11181 [32]. Production of atrorosins by *T. atroroseus* was accomplished or complex culture medium containing metals solution supplemented with single amino aci as the sole nitrogen source in the range of pH 4–5. In sequence, Tolborg et al. (2019) [9 demonstrated that individual amino acids as the sole nitrogen source led to high bioma production but not necessarily to high amounts of red pigment in *T. atroroseus*. Tolborg group also reported that some amino acids can avoid the cocktail pigments phenomen directing *T. atroroseus* to produce single atrorosins. Corroborating their work, only atroros S (**92**) was detected in the fermentation broth when serine was used as the sole nitrog source. Addition of glutamic acid as a second nitrogen source induced the producti of atrorosin E (**94**). Interestingly, only some aminoacids induced atrorosins biosynthes since individual supplementation of proline, lysine, asparagine and tryptophan as t sole nitrogen source did not result in atrorosins production by *T. atroroseus* [32]. Th strain produced two new azaphilone pigments, talaralbols A and B, along with five know azaphilone metabolites, when subjected to growth under submerged fermentation in malt extract medium (ME) (28 °C, 120 rpm) during 14 days [58].

Pigments production by *T. atroroseus* strain GH2 was studied in two different culture media (pH 5.0, 30 ± 2 °C, 200 rpm, 8 days) [98]. The first one was composed by synthetic Czapek-dox modified medium containing high levels of xylose, with and without nutrients supplementation and the second medium was composed by hydrolyzed corncob, a lignocellulosic waste. *T. atroroseus* GH2 demonstrated a significantly different response to the carbon and nitrogen composition of the culture media, with improved growth and enhanced pigments production in the hydrolyzed corncob medium without any nutrient supplementation. Therefore, *T. atroroseus* was pointed by the authors as a promising pigment-producing microorganism for economically competitive large-scale fermentation at lower cost [98].

Carbon source in the fermentation is a very major parameter to direct fungal metabolism. Parul et al. [83] demonstrated that mannitol is the best carbon source for reproduction and growth of *T. purpureogenus* strain F, but the growth is accompanied by low yield of pigment production, while sucrose causes the opposite effect. The authors correlate this fact to species and strain-specific capacity to produce specific enzymes that will dictate the fungus priorities. Under no stressing conditions and abundant carbon availability, primary metabolism is prioritized and the metabolism will be directed to biomass productions instead of secondary metabolites production [83]. In addition, the rate of carbon source depletion is also important. In large-scale industrial production, the rapid growth of the fungus occurs together with rapid decrease in the carbon source concentration. To avoid decrease of metabolite production rate, the carbon source must be constantly added to the fed-batch fermentation to guarantee a constant concentration of this substrate and, consequently, uninterrupted production of pigments [99]. In the same way, culture medium agitation and aeration ensure better distribution of nutrients and better growth, but at the expense of faster depletion of carbon sources. Therefore, agitation and aeration are factors that must be strictly controlled in industrial production [83].

Another tool to improve the yield of fungi metabolites is to create stress conditions during fungal development, thar results in activation and/or suppression of gene clusters to allow fungal adaptation and survival. Co-cultivation two fungal species is an example of stressing condition that generates metabolic responses to allow survival in multispecies environment. Oppong-Danquah et al. [100] described a specific co-cultivation gene cluster, when studying the co-culture of pigment producer fungus *Plenodomus influorescens* with *Pyrenochaeta nobilis*, where five polyketides were produced, including the yellow azaphilones spiciferinone (**117**) and 8a-hydroxy-spiciferinone (**118**) (Figure 14). The cultivation of *Trichoderma guizhouense* NJAU 4742 in the presence of *Fusarium oxysporum* cells also resulted in increase in azaphilone production, which was demonstrated experimentally by the increased activity of the gene cluster responsible for pigment production. This fungal response was drove to neutralize the high concentration of H_2O_2, produced as a defense mechanism during co-cultivation, since azaphilones are capable of neutralizing free radicals, especially the superoxide anion [101]. The same effect is observed in other oxidative stress conditions related to H_2O_2, such as fungal cultivation in the presence of the fungicides amphotericin B, miconazole and ciclopirox. The production of azaphilones increases as a survival mechanism directed to the neutralization of fungicide effects rather than a decrease in antifungal concentration [101].

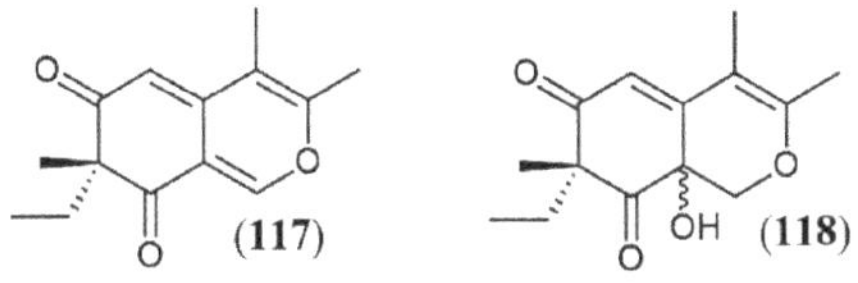

(117) **(118)**

Figure 14. Azaphilones produced by *Plenodomus influorescens* in co-cultivation with *Pyrenochaeta nobilis*: **117**: spiciferinone; **118**: 8a-hydroxy-spiciferinone [102].

Cost minimization for industrial production of azaphilones can be reduced by us agro-industrial waste as material for fungal growth, which also helps to solve the pr lem of pollution associated with the disposal of residues in the environment [19,98]. et al. [102] used rice straw hydrolysate for pigment production by *M. purpureus* M630 reported that this substrate and does not have the ideal carbon content required by fungus. Although supplementation may be necessary in some cases, the use of agroind trial residues has been reported to be economically viable also adding sustainability the process.

As aforementioned, another approach to achieve yield improvement and consequer increase the viability of industrial production of fungal metabolites is the use of mut strains and genetic engineering [99]. The current knowledge of the metabolic pathw and secondary metabolism precursors allow to manipulate fungi as "real industrial factories" [103] and take advantage of the entire pigment gene cassette to improve pigm yield [104]. In this way, Liu et al. [99] managed to knock-out a cAMP phosphodiester gene in *M. purpureus* HJ11, which led to the accumulation of intracellular cAMP ca ing a stimulating effect in secondary metabolism that resulted in 2.3-fold increase pigment production.

4.4. Extraction Approach

Another phase important in yield improvement consists of the extraction step, wh helps in concentration and pre-purification of fungal pigments. Prior to the extracti it is necessary to take into consideration where the pigments produced are deposit Classically, the extraction procedure is usually accomplished by liquid-liquid extracti of the broth with medium polarity solvents such as ethyl acetate. This extraction wo well to obtain extrolytes, i.e., extracellular metabolites present in the broth or linked the external surface of fungal biomass. On some occasions, mycelial adhesion is verifi as reported for a water-soluble extracellular yellow pigment produced by a *Monas* in submerged fermentation. This effect was reversed furnishing sodium and potassi nitrate as nitrogen source to the fungus. Sodium nitrate is suggested to reduce the to amount of extracellular polysaccharides, increase extracellular proteins, and diminish viscosity of the fermentation broth, rising pigment recovery [16].

Although effective for extraction of metabolites produced in liquid cultures, et acetate is not a choice solvent in terms of toxicity. Non-toxic and easily available ethanol better choice for the extraction step, although can only be applied to solid state fermentati as it is water-miscible and cannot be utilized to extract aqueous liquid media. Ethanol w utilized for pigments extraction in the solid-state fermentation of *M. purpureus* M9 usi durian seed as substrate. Extractions were carried out at two temperatures (30 and 60 using a mixture of ethanol and water in different proportions (10:0; 9:1; 8:2; 7:3; 6:4 a 5:5). The most effective conditions for pigments recovery were achieved using the low ethanol:water ratios at 30 °C [105].

Occasionally the pigments remain inside the cells requiring disintegration and dis lution of the glucan-chitin complex of the wall cell to be recovered, therefore demandi alternative extraction procedures [106], while cellular lysis is necessary for recoveri intracellular metabolites. In this way, *T. amestolkiae* DPUA 1275 was subjected to an ternative extraction procedure to recover red pigments. The procedure was conduct with aqueous solutions of imidazolium salt instead of organic solvents, together w ultrasound-temperature-assisted mechanical cell disruption to enhance the recovery intracellular *T. amestolkiae* pigments [103].

Cell Pressurized Liquid Extraction technique was utilized to recover pigments p duced by mycelial biomass of *Talaromyces* sp. 30570 (CBS 206.89 B) isolated from the co reef of the Réunion island (France) and cultivated in PDB media containing complex ganic nitrogen sources like amino acids and proteins. Eco-friendly solvents were chosen the extraction (90 °C and 10 MPa) as water, methanol and/or ethanol. At the end, twe nitrogen-containing azaphilone red pigments were identified while known mycotoxi were not produced [13]. Two-phase aqueous extraction [107] was successfully tested for the extraction of pigments from *T. albobiverticillius*. These organic solvents free techniques guarantee good extraction yields without structural damage in the extracted compounds. Cell disruption methods for improved extraction of pigments from microorganisms were recently reviewed [108].

For pigments production, submerged fermentation is preferable, as it produces better yields, has lower risk of contamination and is easier to monitor when compared to cultivation in solid medium [109]. In addition, using submerged fermentation, it is possible to separate intracellular and extracellular pigments, soluble in the culture medium [102]. However, it is known that not all species of pigment-producing fungi have the ability to diffuse these pigments into the culture medium [73]. Among the techniques to increase the production of extracellular pigments, the design of mutant strains of *M. purpureus* [102], the addition of glycerol to the cultivation medium of *M. pilosus* MS-1 [110] and the establishment of a hyperosmotic environment to *M. ruber* CGMCC 10910 [82] were successfully utilized. The last two methods are related to the regulation of metabolism and gene expression caused by environmental stress.

5. Potential Applications of Azaphilones outside Food Sector

As in the food industry, azo dyes represent the most widely used chemical class of dyes in textiles production, an industrial sector that requires high amounts of stable colorants/pigments [111]. Textiles dying quality is also highly important for market competitiveness and consumer identification and public opinion have been driving an increase demand for natural pigments to replace synthetic dyes. In addition, change is necessary to avoid chronic effects in workers exposed to hazardous synthetic dyes during industrial processes. Once present in clothes, aromatic amines can be biotransformed by skin bacteria into aromatic amines, many of which are carcinogenic and can be absorbed by human skin [112]. Non-regulated aromatic amines were detected in a substantial number of colored textiles in a survey done in Switzerland raising questions on genotoxicity, dyes purity, consumer health risks, release of dyestuffs and dermo absorption [113]. Last, but not least, environmental pollution by effluents from textile industry cause multiple environmental harms.

Textile market can absorb some microbial dyes excluded from food applications by regulatory agencies [19]. Toxicity issues and growing preference for natural goods reached clothing sector and many brands are adapting themselves to meet the expectations for sustainable products. This demand increased, especially in millennials and Z-Generation group, as statistics proved to be alarming in global scale in terms of gas emission by textile industry, water contamination and pollution with industrial dyes [114,115]. Modern demands have been raising integrated practices, as well as international networks and partnerships to address sustainability issues and to look for solutions in the textile and clothing industry [115]. This behavior applies to the low-income clothing producing/exporting countries as well as the buyers' international market. The latter can impose restrictions to imported products containing artificial dyes that either are rejected or avoided by consumers due to the awareness of the unsustainable effluents generated in producing countries.

Cosmetics sector is another market that may incorporate azaphilone compounds in the future. Development of new strategies for on line sales, digital advice, and decentralization of distribution centers helped some cosmetic chains to grow even with the world economic problems associated to the COVID-19 pandemic [116]. The global cosmetics industry was valuated in over USD 380 billion in 2019 and is projected to reach USD 463 billion by 2027 [117]. Several facts contributed to the massive growth of this segment in the last period, such as increase in sale of personal care products, conquering an expressive number of male consumers, increasing number of make-up tutorials in social media and the search for well-being taking into consideration the connection of cosmetics and self-esteem increase [118]. It is also noteworthy that a new type of cosmetics is increasingly growing, named cosmeceuticals. Although regulation of cosmeceuticals was not fully addresse these products claim biological effects beyond cosmetic utility and many times are referr as cosmetic-pharmaceuticals hybrids.

This expansion in cosmetic market was accompanied by the aforementioned cons entious choice of safe, natural, and "not tested on animals" products [119]. Cosmeti and personal care products are usually directly applied to the skin in a daily basis, ma times associated to active ingredients to facilitate fastening or product penetration over t skin. Therefore, allergy and long-term toxicity have also been driving huge efforts for mc ernization in this area. In this way, long-lasting innovative natural color sources are al an important goal of cosmetics industry. Azaphilone metabolites comprise an importa part of the color pallet required by cosmetic industry and their reported biological effe make these compounds also good active components for cosmeceuticals formulatio Anti-inflammatory activity, related for some azaphilone [37,51] is a mechanism associat with anti-aging dermo-cosmetics [120], while antimicrobial activity [46] associated to co pigments can be helpful to extend shelf life of cosmetics.

6. Conclusions

The development of new pigments safe and effective to apply in foods, medicin textile and cosmetic industries is essential and welcome. Natural pigments are a gre alternative regarding not being related to toxic, allergic, and pollutant characteristics the most common synthetic dyes. Fungi azaphilone pigments are recognized as promisi candidates of colorants to substitute azo dyes in the food, cosmetics, and textile industr sectors, as long as safety and production issues are overcome.

Azaphilone research is proliferous and at least 101 new compounds of this class we reported between December 2019 and March 2021 from nine fungal genera (*Aspergill Chaetomium, Hypoxylon, Monascus, Muycopron, Penicillium, Phomopsis, Pleosporales,* and *laromyces*). Some of the new azaphilones exhibit complex chemical structures, and the biosynthesis have been studied to understand nutrients requirements for biomass produ tion and yield improvement. Also, several studies have been conducted to understar down-regulation of citrinin co-production.

Coloring properties and the natural origin are not the only features of azaphilon since antimicrobial, antioxidant, anti-inflammatory, and other properties related to the molecules have been widely reported. This potential can be explored in food or cosme processing to avoid microbial contamination or to furnish functional properties to food.

This review brought some strategies used to improve fermentation conditions, contr pigment production, and issues related to different fungal strains that produce azaphilo pigments, reported in the last two years. Future perspectives include more research th could allow azaphilone dyes to be regularized by the EU, US and other regulatory agenci so they can be plentiful incorporated in different technological innovative applications.

Funding: This research was funded by Fundação de Amparo à Pesquisa do Estado de Minas Gera (FAPEMIG PPM-00255-18), Conselho Nacional de Desenvolvimento Científico e Tecnológico (CN Grant 304922/2018-8), and National Institute of Science and Technology—INCT BioNat, (gran 465637/2014-0), Brazil.

Institutional Review Board Statement: Not applicable.

Informed Consent Statement: Not applicable.

Conflicts of Interest: The authors declare no conflict of interest.

References

1. Charlin, V.; Cifuentes, A. A general framework to study the price-color relationship in paintings with an application to Ma Rothko rectangular series. *Color Res. Appl.* **2021**, *46*, 168–182. [CrossRef]
2. Labrecque, L.I. Color research in marketing: Theoretical and technical considerations for conducting rigorous and impactful co research. *Psychol. Mark.* **2020**, *37*, 855–863. [CrossRef]

Sigurdson, G.T.; Tang, P.; Giusti, M.M. Natural Colorants: Food Colorants from Natural Sources. *Annu. Rev. Food Sci. Technol.* **2017**, *8*, 261–280. [CrossRef]
Meruvu, H.; dos Santos, J.C. Colors of life: A review on fungal pigments. *Crit. Rev. Biotechnol.* **2021**, 1–25. [CrossRef]
De Mejia, E.G.; Zhang, Q.; Penta, K.; Eroglu, A.; Lila, M.A. The Colors of Health: Chemistry, Bioactivity, and Market Demand for Colorful Foods and Natural Food Sources of Colorants. *Annu. Rev. Food Sci. Technol.* **2020**, *11*, 145–182. [CrossRef] [PubMed]
Ravi, N.; Keshavayya, J.; Mallikarjuna, M.; Kumar, V.; Zahara, F.N. Synthesis, spectral characterization, anticancer and cyclic voltammetric studies of azo colorants containing thiazole structure. *Chem. Data Collect.* **2021**, *33*, 100686. [CrossRef]
Benkhaya, S.; M'rabet, S.; El Harfi, A. Classifications, properties, recent synthesis and applications of azo dyes. *Heliyon* **2020**, *6*, e03271. [CrossRef]
Al Reza, M.S.; Hasan, M.M.; Kamruzzaman, M.; Hossain, M.I.; Zubair, M.A.; Bari, L.; Abedin, M.Z.; Reza, M.A.; Khalid-Bin-Ferdaus, K.M.; Haque, K.M.F.; et al. Study of a common azo food dye in mice model: Toxicity reports and its relation to carcinogenicity. *Food Sci. Nutr.* **2019**, *7*, 667–677. [CrossRef] [PubMed]
Bakthavachalu, P.; Kannan, S.M.; Qoronfleh, M.W. Food Color and Autism: A Meta-Analysis. In *Personalized Food Intervention and Therapy for Autism Spectrum Disorder Management*; Essa, M.M., Qoronfleh, M.W., Eds.; Springer: Berlin/Heidelberg, Germany, 2020; p. 700.
Su, C.H.; Tsai, C.H.; Chen, M.H.; Lv, W.Q. U.S. sustainable food market generation Z consumer segments. *Sustainability* **2019**, *11*, 3607. [CrossRef]
Morales-Oyervides, L.; Ruiz-Sánchez, J.P.; Oliveira, J.C.; Sousa-Gallagher, M.J.; Méndez-Zavala, A.; Giuffrida, D.; Dufossé, L.; Montañez, J. Biotechnological approaches for the production of natural colorants by *Talaromyces/Penicillium*: A review. *Biotechnol. Adv.* **2020**, *43*, 107601. [CrossRef] [PubMed]
Arikan, E.B.; Canli, O.; Caro, Y.; Dufossé, L.; Dizge, N. Production of bio-based pigments from food processing industry by-products (apple, pomegranate, black carrot, red beet pulps) using *Aspergillus carbonarius*. *J. Fungi* **2020**, *6*, 240. [CrossRef]
Lebeau, J.; Petit, T.; Fouillaud, M.; Dufossé, L.; Caro, Y. Alternative extraction and characterization of nitrogen-containing azaphilone red pigments and ergosterol derivatives from the marine-derived fungal *Talaromyces* sp. 30570 strain with industrial relevance. *Microorganisms* **2020**, *8*, 1920. [CrossRef]
Choe, D.; Song, S.M.; Shin, C.S.; Johnston, T.V.; Ahn, H.J.; Kim, D.; Ku, S. Production and characterization of anti-inflammatory *Monascus* pigment derivatives. *Foods* **2020**, *9*, 858. [CrossRef]
Dufossé, L. Red colourants from filamentous fungi: Are they ready for the food industry? *J. Food Compos. Anal.* **2018**, *69*, 156–161. [CrossRef]
Yang, S.-Z.; Huang, Z.-F.; Liu, H.Q.; Hu, X.; Wu, Z.Q. Improving mycelial morphology and adherent growth as well as metabolism of *Monascus* yellow pigments using nitrate resources. *Appl. Microbiol. Biotechnol.* **2020**, *104*, 9607–9617. [CrossRef] [PubMed]
Baek, S.; Mohanty, S.K.; Glambosky, M. COVID-19 and stock market volatility: An industry level analysis. *Financ. Res. Lett.* **2020**, *37*, 101748. [CrossRef]
Research and Market. The Global Market for Food Additives. 2021. Available online: https://www.researchandmarkets.com/ (accessed on 25 June 2021).
Venil, C.K.; Velmurugan, P.; Dufossé, L.; Devi, P.R.; Ravi, A.V. Fungal pigments: Potential coloring compounds for wide ranging applications in textile dyeing. *J. Fungi* **2020**, *6*, 68. [CrossRef] [PubMed]
BCC Research. BCC Research. Available online: https://www.bccresearch.com/ (accessed on 25 May 2021).
El-Borm, H.T.; Badawy, G.M.; El-Nabi, S.H.; Ahmed, W.A.; Atallah, M.N. Toxicity of sunset yellow FCF and tartrazine dyes on DNA and cell cycle of liver and kidneys of the chick embryo: The alleviative effects of curcumin. *Egypt. J. Zool.* **2020**, *74*, 43–55. [CrossRef]
Zellner, D.A.; Durlach, P. What is refreshing? An investigation of the color and other sensory attributes of refreshing foods and beverages. *Appetite* **2002**, *39*, 185–186. [CrossRef]
Pomirleanu, N.; Gustafson, B.M.; Bi, S. Ooh, that's sour: An investigation of the role of sour taste and color saturation in consumer temptation avoidance. *Psychol. Mark.* **2020**, *37*, 1068–1081. [CrossRef]
Salnikova, E.; Grunert, K.G. The role of consumption orientation in consumer food preferences in emerging markets. *J. Bus. Res.* **2020**, *112*, 147–159. [CrossRef]
Fortune Business Insights. Annatto Market Size, Share & COVID-19 Impact Analysis, by Type (Solvent Extraction & Emulsified Annatto and Aqueous Extraction Annatto), Application (Food Industry, Natural Fabric Industry, Cosmetic Industry, and Others), and Regional Forecasts, 2020—20. 2020. Available online: https://www.fortunebusinessinsights.com/ (accessed on 25 June 2021).
Market Data Forecast. Anthocyanins Market Analysis By Product Type (Cyanidin, Malvidin, Delphinidin, Peonidin, Others), Application Type (Food Beverage, Pharmaceutical Products, Personal Care, Others), and By Region (North America, Europe, Asia Pacific, Latin America, Middle E. 2020. Available online: https://www.marketdataforecast.com/ (accessed on 25 June 2021).
Market Data Forecast. Global Beetroot Powder Market by Application (Curries and Gravies, Food Colour, Soups and Coatings), by Type (Organic and Conventional), by Distribution Channel (Online Sales, Retailer Shops, Departmental Stores, Supermarket/Hypermarket), By Packaging Ca. 2020. Available online: https://www.marketdataforecast.com/ (accessed on 25 June 2021).
Allied Market Research. Carmine Market by Form (Powder, Liquid, and Crystal), Application (Dairy & Frozen Products, Food & Beverages, Cosmetics, Bakery & Confectionery, and Meat Products), and End User (Food Processing Companies, Beverage Industry, Catering Industry, and Cosmeti. 2019. Available online: https://www.alliedmarketresearch.com/ (accessed on 25 June 2021).

29. Research and Markets. Global Curcumin Market (2020 to 2027)—Size, Share & Trends Analysis Report. Available onl https://www.researchandmarkets.com/ (accessed on 25 June 2021).
30. Gao, J.-M.; Yang, S.-X.; Qin, J.-C. Azaphilones: Chemistry and Biology. *Chem. Rev.* **2013**, *113*, 4755–4811. [CrossRef]
31. Chen, C.; Tao, H.; Chen, W.; Yang, B.; Zhou, X.; Luo, X.; Liu, Y. Recent advances in the chemistry and biology of azaphilones. *R Adv.* **2020**, *10*, 10197–10220. [CrossRef]
32. Isbrandt, T.; Tolborg, G.; Ødum, A.; Workman, M.; Larsen, T.O. Atrorosins: A new subgroup of *Monascus* pigments fr *Talaromyces atroroseus*. *Appl. Microbiol. Biotechnol.* **2020**, *104*, 615–622. [CrossRef]
33. Isbrandt, T.; Frisvad, J.C.; Madsen, A.; Larsen, T.O. New azaphilones from *Aspergillus neoglaber*. *AMB Express* **2020**, *10*, [CrossRef] [PubMed]
34. Petersen, T.I.; Kroll-Møller, P.; Larsen, T.O.; Ødum, A.S.R. A Novel Class of Pigments in. Aspergillus. Patent No. WO2020094 8 November 2019.
35. El-Kashef, D.H.; Youssef, F.S.; Hartmann, R.; Knedel, T.-O.; Janiak, C.; Lin, W.; Reimche, I.; Teusch, N.; Liu, Z.; Proksch Azaphilones from the red sea fungus *Aspergillus falconensis*. *Mar. Drugs* **2020**, *18*, 204. [CrossRef] [PubMed]
36. Gu, L.; Sun, F.-J.; Li, C.-P.; Cui, L.-T.; Yang, M.-H.; Kong, L.-Y. Ardeemins and citrinin dimer derivatives from *Aspergillus terr* harbored in *Pinellia ternate*. *Phytochem. Lett.* **2021**, *42*, 77–81. [CrossRef]
37. Gao, W.; Chai, C.; Li, X.-N.; Sun, W.; Li, F.; Chen, C.; Wang, J.; Zhu, H.; Wang, Y.; Hu, Z.; et al. Two anti-inflammatory chlorina azaphilones from *Chaetomium globosum* TW1-1 cultured with 1-methyl-L-tryptophan and structure revision of chaephilon *Tetrahedron Lett.* **2020**, *61*, 151516. [CrossRef]
38. Wang, W.; Liao, Y.; Chen, R.; Hou, Y.; Ke, W.; Zhang, B.; Gao, M.; Shao, Z.; Chen, J.; Li, F. Chlorinated azaphilone pigments w antimicrobial and cytotoxic activities isolated from the deep sea derived fungus *Chaetomium* sp. NA-S01-R1. *Mar. Drugs* **2018**, 61. [CrossRef] [PubMed]
39. Song, C.; Ding, G.; Wu, G.; Yang, J.; Zhang, M.; Wang, H.; Wei, D.; Qin, J.; Guo, L. Identification of a Unique Azaphilone Produ by *Chaetomium globosum* Isolated from *Polygonatum sibiricum*. *Chem. Biodivers.* **2020**, *17*, e1900744. [CrossRef] [PubMed]
40. Sarmales-Murga, C.; Akaoka, F.; Sato, M.; Takanishi, J.; Mino, T.; Miyoshi, N.; Watanabe, K. A new class of dimeric prod isolated from the fungus *Chaetomium globosum*: Evaluation of chemical structure and biological activity. *J. Antibiot.* **2020**, 320–323. [CrossRef] [PubMed]
41. Wang, W.; Yang, J.; Liao, Y.-Y.; Cheng, G.; Chen, J.; Cheng, X.-D.; Qin, J.J.; Shao, Z. Cytotoxic Nitrogenated Azaphilones from Deep-Sea-Derived Fungus *Chaetomium globosum* MP4-S01-7. *J. Nat. Prod.* **2020**, *83*, 1157–1166. [CrossRef] [PubMed]
42. Wani, N.; Khanday, W.; Tirumale, S. Evaluation of anticancer activity of *Chaetomium cupreum* extracts against human bre adenocarcinoma cell lines. *Matrix Sci. Pharma* **2020**, *4*, 31. [CrossRef]
43. Yamada, T.; Muroga, Y.; Tanaka, R. New azaphilones, seco-chaetomugilins A and D, produced by a marine-fish-deriv *Chaetomium globosum*. *Mar. Drugs* **2009**, *7*, 249–257. [CrossRef]
44. Zu, W.-Y.; Tang, J.-W.; Hu, K.; Zhou, Y.-F.; Gou, L.-L.; Su, X.-Z.; Lei, X.; Sun, H.-D.; Puno, P.-T. Chaetolactam A, an Azaphil Derivative from the Endophytic Fungus *Chaetomium* sp. g1. *J. Org. Chem.* **2021**, *86*, 475–483. [CrossRef]
45. Zhang, X.-Y.; Tan, X.-M.; Yu, M.; Yang, J.; Sun, B.-D.; Qin, J.-C.; Guo, L.-P.; Ding, G. Bioactive metabolites from the desert pl associated endophytic fungus *Chaetomium globosum* (Chaetomiaceae). *Phytochemistry* **2021**, *185*, 112701. [CrossRef] [PubMed
46. Becker, K.; Pfütze, S.; Kuhnert, E.; Cox, R.J.; Stadler, M.; Surup, F. Hybridorubrins A–D: Azaphilone Heterodimers from Strom of *Hypoxylon fragiforme* and Insights into the Biosynthetic Machinery for Azaphilone Diversification. *Chem. A Eur. J.* **2021**, 1438–1450. [CrossRef] [PubMed]
47. Surup, F.; Narmani, A.; Wendt, L.; Pfütze, S.; Kretz, R.; Becker, K.; Menbrivès, C.; Giosa, A.; Elliott, M.; Petit, C.; et al. Identificat of fungal fossils and novel azaphilone pigments in ancient carbonised specimens of *Hypoxylon fragiforme* from forest soils Châtillon-sur-Seine (Burgundy). *Fungal Divers.* **2018**, *92*, 345–356. [CrossRef]
48. Lambert, C.; Pourmoghaddam, M.J.; Cedeño-Sanchez, M.; Surup, F.; Khodaparast, S.A.; Krisai-Greilhuber, I.; Voglmayr, Stradal, T.E.B.; Stadler, M. Resolution of the *Hypoxylon fuscum* complex (hypoxylaceae, xylariales) and discovery and biologi characterization of two of its prominent secondary metabolites. *J. Fungi* **2021**, *7*, 131. [CrossRef]
49. Wu, H.-C.; Chen, J.-J.; Wu, M.-D.; Cheng, M.-J.; Chang, H.-S. Identification of new pigments produced by the fermented rice the fungus *Monascus pilosus* and their anti-inflammatory activity. *Phytochem. Lett.* **2020**, *40*, 181–187. [CrossRef]
50. Nakashima, K.I.; Tomida, J.; Tsuboi, T.; Kawamura, Y.; Inoue, M. Muyocopronones A and B: Azaphilones from the endophy fungus *Muyocopron laterale*. *Beilstein J. Org. Chem.* **2020**, *16*, 2100–2107. [CrossRef]
51. Cain, J.W.; Miller, K.I.; Kalaitzis, J.A.; Chau, R.; Neilan, B.A. Genome mining of a fungal endophyte of *Taxus yunnanensis* (Chin yew) leads to the discovery of a novel azaphilone polyketide, lijiquinone. *Microb. Biotechnol.* **2020**, *13*, 1415–1427. [CrossRef]
52. Sabdaningsih, A.; Liu, Y.; Mettal, U.; Heep, J.; Wang, L.; Cristianawati, O.; Nuryadi, H.; Sibero, M.T.; Marner, M.; Radjasa, O et al. A new citrinin derivative from the Indonesian marine sponge-associated fungus *Penicillium citrinum*. *Mar. Drugs* **2020**, 227. [CrossRef] [PubMed]
53. Wei, Q.; Bai, J.; Yan, D.; Bao, X.; Li, W.; Liu, B.; Zhang, D.; Qi, X.; Yu, D.; Hu, Y. Genome mining combined metabolic shunti and OSMAC strategy of an endophytic fungus leads to the production of diverse natural products. *Acta Pharm. Sin. B* **2021**, 572–587. [CrossRef] [PubMed]
54. Li, H.-T.; Duan, R.-T.; Liu, T.; Yang, R.-N.; Wang, J.-P.; Liu, S.-X.; Yang, Y.B.; Zhou, H.; Ding, Z.-T. Penctrimertone, a bioact citrinin dimer from the endophytic fungus *Penicillium* sp. T2-11. *Fitoterapia* **2020**, *146*, 104711. [CrossRef] [PubMed]

Yang, Z.-J.; Zhang, Y.-F.; Wu, K.; Xu, Y.-X.; Meng, X.-G.; Jiang, Z.-T.; Ge, M.; Shao, L. New azaphilones, phomopsones A-C with biological activities from an endophytic fungus *Phomopsis* sp. CGMCC No.5416. *Fitoterapia* **2020**, *145*, 104573. [CrossRef]
Chen, S.; Liu, Z.; Chen, Y.; Tan, H.; Liu, H.; Zhang, W. Tersaphilones A-E, cytotoxic chlorinated azaphilones from the deep-sea-derived fungus *Phomopsis tersa* FS441. *Tetrahedron* **2021**, *78*, 131806. [CrossRef]
Cao, F.; Meng, Z.-H.; Wang, P.; Luo, D.-Q.; Zhu, H.-J. Dipleosporalones A and B, Dimeric Azaphilones from a Marine-Derived *Pleosporales* sp. Fungus. *J. Nat. Prod.* **2020**, *83*, 1283–1287. [CrossRef]
Bai, W.; Jing, L.-L.; Guan, Q.-Y.; Tan, R.-X. Two new azaphilone pigments from *Talaromyces albobiverticillius* and their anti-inflammatory activity. *J. Asian Nat. Prod. Res.* **2021**, *23*, 325–332. [CrossRef]
Choe, D.; Jang, H.; Jung, H.H.; Shin, C.S.; Johnston, T.V.; Kim, D.; Ku, S. In vivo anti-obesity effects of *Monascus* pigment threonine derivative with enhanced hydrophilicity. *J. Funct. Foods* **2020**, *67*, 103849. [CrossRef]
Skariyachan, S.; Pius, S.; Gopal, D.; Muddebihalkar, A.G. Natural lead molecules probably act as potential inhibitors against prospective targets of SARS-CoV-2: Therapeutic insight for COVID-19 from computational modelling, molecular docking and dynamic simulation studies. *Chemistry* **2020**. preprint. [CrossRef]
Youssef, F.S.; Alshammari, E.; Ashour, M.L. Bioactive alkaloids from genus *Aspergillus*: Mechanistic interpretation of their antimicrobial and potential SARS-CoV-2 inhibitory activity using molecular modelling. *Int. J. Mol. Sci.* **2021**, *22*, 1866. [CrossRef]
Zhang, S.-P.; Huang, R.; Li, F.-F.; Wei, H.-X.; Fang, X.-W.; Xie, X.-S.; Lin, D.-G.; Wu, S.-H.; He, J. Antiviral anthraquinones and azaphilones produced by an endophytic fungus *Nigrospora* sp. from *Aconitum carmichaeli*. *Fitoterapia* **2016**, *112*, 85–89. [CrossRef]
Roy, B.G. Potential of small-molecule fungal metabolites in antiviral chemotherapy. *Antivir. Chem. Chemother.* **2017**, *25*, 20–52. [CrossRef] [PubMed]
Kim, J.-Y.; Woo, E.-E.; Ha, L.S.; Ki, D.-W.; Lee, I.-K.; Yun, B.-S. Neuraminidase Inhibitors from the Fruiting Body of *Glaziella splendens*. *Mycobiology* **2019**, *47*, 256–260. [CrossRef]
Pavesi, C.; Flon, V.; Mann, S.; Leleu, S.; Prado, S.; Franck, X. Biosynthesis of azaphilones: A review. *Nat. Prod. Rep.* **2021**, *38*, 1058–1071. [CrossRef]
Chen, D.; Wang, Y.; Chen, M.; Fan, P.; Li, G.; Wang, C. Ammonium nitrate regulated the color characteristic changes of pigments in *Monascus purpureus* M9. *AMB Express* **2021**, *11*, 3. [CrossRef]
Gmoser, R.; Ferreira, J.A.; Lennartsson, P.R.; Taherzadeh, M.J. Filamentous ascomycetes fungi as a source of natural pigments. *Fungal Biol. Biotechnol.* **2017**, *4*, 4. [CrossRef] [PubMed]
Sánchez-Muñoz, S.; Mariano-Silva, G.; Leite, M.O.; Mura, F.B.; Verma, M.L.; Da Silva, S.S.; Chandel, A.K. Production of fungal and bacterial pigments and their applications. In *Biotechnological Production of Bioactive Compounds*; Elsevier: Amsterdam, The Netherlands, 2019; pp. 327–361.
Kalra, R.; Conlan, X.A.; Goel, M. Fungi as a Potential Source of Pigments: Harnessing Filamentous Fungi. *Front. Chem.* **2020**, *8*, 369. [CrossRef]
Lehto, S.; Buchweitz, M.; Klimm, A.; Straßburger, R.; Bechtold, C.; Ulberth, F. Comparison of food colour regulations in the EU and the US: A review of current provisions. *Food Addit. Contam. Part A* **2017**, *34*, 335–355. [CrossRef]
Hong, J.L.; Wu, L.; Lu, J.Q.; Zhou, W.B.; Cao, Y.J.; Lv, W.L.; Liu, B.; Rao, P.F.; Ni, L.; Lv, X.C. Comparative transcriptomic analysis reveals the regulatory effects of inorganic nitrogen on the biosynthesis of: *Monascus* pigments and citrinin. *RSC Adv.* **2020**, *10*, 5268–5282. [CrossRef]
Ning, Z.Q.; Cui, H.; Xu, Y.; Huang, Z.B.; Tu, Z.; Li, Y.P. Deleting the citrinin biosynthesis-related gene, ctnE, to greatly reduce citrinin production in *Monascus aurantiacus* Li AS3.4384. *Int. J. Food Microbiol.* **2017**, *241*, 325–330. [CrossRef]
Pandit, S.G.; Puttananjaiah, M.H.; Serva Peddha, M.; Dhale, M.A. Safety efficacy and chemical profiling of water-soluble *Talaromyces purpureogenus* CFRM02 pigment. *Food Chem.* **2020**, *310*, 125869. [CrossRef]
Lebeau, J.; Venkatachalam, M.; Fouillaud, M.; Petit, T.; Vinale, F.; Dufossé, L.; Caro, Y. Production and new extraction method of polyketide red pigments produced by ascomycetous fungi from terrestrial and marine habitats. *J. Fungi* **2017**, *3*, 34. [CrossRef]
Venkatachalam, M.; Zelena, M.; Cacciola, F.; Ceslova, L.; Girard-Valenciennes, E.; Clerc, P.; Dugo, P.; Mondello, L.; Fouillaud, M.; Rotondo, A.; et al. Partial characterization of the pigments produced by the marine-derived fungus *Talaromyces albobiverticillius* 30548. Towards a new fungal red colorant for the food industry. *J. Food Compos. Anal.* **2018**, *67*, 38–47. [CrossRef]
Lagashetti, A.C.; Dufossé, L.; Singh, S.K.; Singh, P.N. Fungal pigments and their prospects in different industries. *Microorganisms* **2019**, *7*, 604. [CrossRef]
Wang, J.; Huang, Y.; Shao, Y. From Traditional Application to Genetic Mechanism: Opinions on *Monascus* Research in the New Milestone. *Front. Microbiol.* **2021**, *12*, 10–13. [CrossRef]
Yuliana, A.; Singgih, M.; Julianti, E.; Blanc, P.J. Derivates of azaphilone *Monascus pigments*. *Biocatal. Agric. Biotechnol.* **2017**, *9*, 183–194. [CrossRef]
Daud, N.F.S.; Said, F.M.; Ramu, M.; Yasin, N.M.H. Evaluation of Bio-red Pigment Extraction from *Monascus purpureus* FTC5357. In *IOP Conference Series: Materials Science and Engineering*; Institute of Physics Publishing: Bristol, UK, 2020; Volume 736.
Qian, G.-F.; Huang, J.; Farhadi, A.; Zhang, B.-B. Ethanol addition elevates cell respiratory activity and causes overproduction of natural yellow pigments in submerged fermentation of *Monascus purpureus*. *LWT* **2021**, *139*, 110534. [CrossRef]
Zhou, B.; Tian, Y.; Zhong, H. Application of a two-stage agitation speed control strategy to enhance yellow pigments production by *Monascus anka* Mutant. *J. Microbiol. Biotechnol. Food Sci.* **2019**, *8*, 1260–1264. [CrossRef]

82. Chen, G.; Yang, S.; Wang, C.; Shi, K.; Zhao, X.; Wu, Z. Investigation of the mycelial morphology of *Monascus* and the expression pigment biosynthetic genes in high-salt-stress fermentation. *Appl. Microbiol. Biotechnol.* **2020**, *104*, 2469–2479. [CrossRef]
83. Parul; Thiyam, G.; Dufossé, L.; Sharma, A.K. Characterization of *Talaromyces purpureogenus* strain F extrolites and development production medium for extracellular pigments enriched with antioxidant properties. *Food Bioprod. Process.* **2020**, *124*, 143–1 [CrossRef]
84. Liu, J.; Wu, J.; Cai, X.; Zhang, S.; Liang, Y.; Lin, Q. Regulation of secondary metabolite biosynthesis in *Monascus purpureus* cofactor metabolic engineering strategies. *Food Microbiol.* **2021**, *95*, 103689. [CrossRef]
85. Li, L.; Chen, S.; Gao, M.; Ding, B.; Zhang, J.; Zhou, Y.; Yingbao, L.; Yang, H.; Wu, Q.; Chen, F. Acidic conditions induce t accumulation of orange *Monascus* pigments during liquid-state fermentation of *Monascus ruber* M7. *Appl. Microbiol. Biotechn* **2019**, *103*, 8393–8402. [CrossRef]
86. de Oliveira, F.; Pedrolli, D.B.; Teixeira, M.F.S.; de Carvalho Santos-Ebinuma, V. Water-soluble fluorescent red colorant producti by *Talaromyces amestolkiae*. *Appl. Microbiol. Biotechnol.* **2019**, *103*, 6529–6541. [CrossRef] [PubMed]
87. De Oliveira, F.; Lima, C.; Lopes, A.M.; Marques, D.; Druzian, J.I.; Júnior, A.P.; Santos-Ebinuma, V.C. Microbial colorants producti in stirred-tank bioreactor and their incorporation in an alternative food packaging biomaterial. *J. Fungi* **2020**, *6*, 264. [CrossR [PubMed]
88. Darwesh, O.M.; Matter, I.A.; Almoallim, H.S.; Alharbi, S.A.; Oh, Y.K. Isolation and optimization of *Monascus ruber* OMNRC45 red pigment production and evaluation of the pigment as a food colorant. *Appl. Sci.* **2020**, *10*, 8867. [CrossRef]
89. Keekan, K.K.; Hallur, S.; Modi, P.K.; Shastry, R.P. Antioxidant Activity and Role of Culture Condition in the Optimization Red Pigment Production by *Talaromyces purpureogenus* KKP Through Response Surface Methodology. *Curr. Microbiol.* **2020**, 1780–1789. [CrossRef]
90. da Costa, J.P.V.; de Oliveira, C.F.D.; Vendruscolo, F. Cheese whey as a potential substrate for *Monascus* pigments producti *AIMS Agric. Food* **2020**, *5*, 785–798. [CrossRef]
91. Virk, M.S.; Ramzan, R.; Virk, M.A.; Yuan, X.; Chen, F. Transfigured morphology and ameliorated production of six *Monasc* pigments by acetate species supplementation in *Monascus ruber* M7. *Microorganisms* **2020**, *8*, 81. [CrossRef]
92. de Oliveira, F.; Ferreira, L.C.; Neto, Á.B.; Teixeira, M.F.S.; Ebinuma, V.D.C.S. Biosynthesis of natural colorant by *Talaromy amestolkiae*: Mycelium accumulation and colorant formation in incubator shaker and in bioreactor. *Biochem. Eng. J.* **2020**, *1* 107694. [CrossRef]
93. Wang, C.; Lu, H.; Lan, J.; Zaman, K.H.A.U.; Cao, S. A Review: Halogenated Compounds from Marine Fungi. *Molecules* **2021**, 458. [CrossRef]
94. Lucas, E.M.F.; De Castro, M.C.M.; Takahashi, J.A. Antimicrobial properties of sclerotiorin, isochromophilone VI and pencoli metabolites from a Brazilian cerrado isolate of *Penicillium sclerotiorum* Van Beyma. *Braz. J. Microbiol.* **2007**, *38*, 785–789. [CrossR
95. Gomes, D.C.; Takahashi, J.A. Sequential fungal fermentation-biotransformation process to produce a red pigment from sclerotior *Food Chem.* **2016**, *210*, 355–361. [CrossRef]
96. Kuzikova, I.; Rybalchenko, O.; Kurashov, E.; Krylova, Y.; Safronova, V.; Medvedeva, N. Defense Responses of the Marine-Deriv Fungus *Aspergillus tubingensis* to Alkylphenols Stress. *Water. Air Soil Pollut.* **2020**, *231*, 271. [CrossRef]
97. Tolborg, G.; Ødum, A.S.R.; Isbrandt, T.; Larsen, T.O.; Workman, M. Unique processes yielding pure azaphilones in *Talaromy atroroseus*. *Appl. Microbiol. Biotechnol.* **2020**, *104*, 603–613. [CrossRef]
98. Morales-Oyervides, L.; Ruiz-Sánches, J.P.; Oliveira, J.C.; Sousa-Gallagher, M.J.; Morales-Martínez, T.K.; Albergamo, A.; Sal A.; Giuffrida, D.; Dufossé, L.; Montañez, J. Medium design from corncob hydrolyzate for pigment production by *Talaromy atroroseus* GH2: Kinetics modelling and pigments characterization. *Biochem. Eng. J.* **2020**, *161*, 107698. [CrossRef]
99. Liu, J.; Du, Y.; Ma, H.; Pei, X.; Li, M. Enhancement of *Monascus* yellow pigments production by activating the cAMP signalli pathway in *Monascus purpureus* HJ11. *Microb. Cell Factories* **2020**, *19*, 224. [CrossRef]
100. Oppong-Danquah, E.; Budnicka, P.; Blümel, M.; Tasdemir, D. Design of fungal co-cultivation based on comparative metabolom and bioactivity for discovery of marine fungal agrochemicals. *Mar. Drugs* **2020**, *18*, 73. [CrossRef]
101. Pang, G.; Sun, T.; Yu, Z.; Yuan, T.; Liu, W.; Zhu, H.; Gao, Q.; Yang, D.; Kubicek, C.P.; Zhang, J.; et al. Azaphilones biosynthe complements the defence mechanism of *Trichoderma guizhouense* against oxidative stress. *Environ. Microbiol.* **2020**, *22*, 4808–48 [CrossRef]
102. Liu, J.; Luo, Y.; Guo, T.; Tang, C.; Chai, X.; Zhao, W.; Bai, J.; Lin, Q. Cost-effective pigment production by *Monascus purpure* using rice straw hydrolysate as substrate in submerged fermentation. *J. Biosci. Bioeng.* **2020**, *129*, 229–236. [CrossRef]
103. Nielsen, J.C.; Nielsen, J. Development of fungal cell factories for the production of secondary metabolites: Linking genomics a metabolism. *Synth. Syst. Biotechnol.* **2017**, *2*, 5–12. [CrossRef] [PubMed]
104. Chatragadda, R.; Dufossé, L. Ecological and Biotechnological Aspects of Pigmented Microbes: A Way Forward in Developme of Food and Pharmaceutical Grade Pigments. *Microorganisms* **2021**, *9*, 637. [CrossRef] [PubMed]
105. Srianta, I.; Ristiarini, S.; Nugerahani, I. Pigments extraction from *Monascus*-fermented durian seed. In *IOP Conference Series: Ea and Environmental Science*; Institute of Physics Publishing: Bristol, UK, 2020; Volume 443.
106. de Oliveira, F.; Hirai, P.R.; Teixeira, M.F.S.; Pereira, J.F.B.; Santos-Ebinuma, V.C. *Talaromyces amestolkiae* cell disruption and colora extraction using imidazolium-based ionic liquids. *Sep. Purif. Technol.* **2021**, *257*, 117759. [CrossRef]

Lebeau, J.; Petit, T.; Fouillaud, M.; Dufossé, L.; Caro, Y. Aqueous two-phase system extraction of polyketide-based fungal pigments using ammonium-or imidazolium-based ionic liquids for detection purpose: A case study. *J. Fungi* **2020**, *6*, 375. [CrossRef] [PubMed]

Nemer, G.; Louka, N.; Vorobiev, E.; Salameh, D.; Nicaud, J.-M.; Maroun, R.G.; Koubaa, M. Mechanical Cell Disruption Technologies for the Extraction of Dyes and Pigments from Microorganisms: A Review. *Fermentation* **2021**, *7*, 36. [CrossRef]

Venkatachalam, M.; Shum-Chéong-Sing, A.; Dufossé, L.; Fouillaud, M. Statistical Optimization of the Physico-Chemical Parameters for Pigment Production in Submerged Fermentation of *Talaromyces albobiverticillius* 30548. *Microorganisms* **2020**, *8*, 711. [CrossRef] [PubMed]

Shi, J.; Zhao, W.; Lu, J.; Wang, W.; Yu, X.; Feng, Y. Insight into *Monascus* pigments production promoted by glycerol based on physiological and transcriptome analyses. *Process Biochem.* **2021**, *102*, 141–149. [CrossRef]

Benkhaya, S.; Rabet, S.M.; Harfi, A. A review on classifications, recent synthesis and applications of textile dyes. *Inorg. Chem. Commun.* **2020**, *115*, 107891. [CrossRef]

Chormey, D.S.; Zaman, B.T.; Maltepe, E.; Büyükpınar, Ç.; Bulgurcuoğlu, A.E.; Turak, F.; Erulaş, F.A.; Bakırdere, S. Simultaneous Determination of Harmful Aromatic Amine Products of Azo Dyes by Gas Chromatography–Mass Spectrometry. *J. Anal. Chem.* **2020**, *75*, 1330–1334. [CrossRef]

Crettaz, S.; Kämpfer, P.; Brüschweiler, B.J.; Nussbaumer, S.; Deflorin, O. Survey on hazardous non-regulated aromatic amines as cleavage products of azo dyes found in clothing textiles on the Swiss market. *J. Verbrauch. Leb.* **2020**, *15*, 49–61. [CrossRef]

De Angelis, M.; Amatulli, C.; Pinato, G. Sustainability in the Apparel Industry: The Role of Consumers' Fashion Consciousness. In *Sustainability in the Textile and Apparel Industries*; Muthu, S.S., Gardetti, M.A., Eds.; Springer: Berlin/Heidelberg, Germany, 2020.

Beyers, F.; Heinrichs, H. Global partnerships for a textile transformation? A systematic literature review on inter- and transnational collaborative governance of the textile and clothing industry. *J. Clean. Prod.* **2020**, *261*, 121131. [CrossRef]

Hoekstra, J.C.; Leeflang, P.S.H. Marketing in the era of COVID-19. *Ital. J. Mark.* **2020**, *2020*, 249–260. [CrossRef]

Allied Market Research. Cosmetics Market by Category (Skin and Sun Care Products, Hair Care Products, Deodorants & Fragrances, and Makeup & Color Cosmetics), Gender (Men, Women, and Unisex), and Distribution Channel (Hypermarkets/Supermarkets, Specialty Stores, Pharmacies, Onlin. 2021. Available online: https://www.alliedmarketresearch.com/ (accessed on 25 June 2021).

Draelos, Z.D. The Use of Cosmetic Products to Improve Self Esteem & Quality of Life. In *Essential Psychiatry for the Aesthetic Practitioner*; Evan, A.R., Richard, G.F., Eds.; John Wiley & Sons Ltd.: Hoboken, United States, 2021; pp. 34–41.

Monnot, A.D.; Towle, K.M.; Ahmed, S.S.; Dickinson, A.M.; Fung, E.S. An in vitro human assay for evaluating immunogenic and sensitization potential of a personal care and cosmetic product. *Toxicol. Mech. Methods* **2021**, *31*, 205–211. [CrossRef] [PubMed]

Resende, D.I.S.P.; Ferreira, M.; Magalhães, C.; Sousa Lobo, J.M.; Sousa, E.; Almeida, I.F. Trends in the use of marine ingredients in anti-aging cosmetics. *Algal Res.* **2021**, *55*, 102273. [CrossRef]

Journal of
Fungi

*icle

haracterization of a Biofilm Bioreactor Designed for the ingle-Step Production of Aerial Conidia and Oosporein by *eauveria bassiana* PQ2

ctor Raziel Lara-Juache [1], José Guadalupe Ávila-Hernández [2], Luis Víctor Rodríguez-Durán [3,*], riela Ramona Michel [1], Jorge Enrique Wong-Paz [1], Diana Beatriz Muñiz-Márquez [1], Fabiola Veana [1], yra Aguilar-Zárate [4], Juan Alberto Ascacio-Valdés [5] and Pedro Aguilar-Zárate [1,*]

1 Departamento de Ingenierías, Instituto Tecnológico de Ciudad Valles, Tecnológico Nacional de México, Carretera al Ingenio Plan de Ayala Km. 2, Colonia Vista Hermosa, Ciudad Valles, San Luis Potosí C.P. 79010, Mexico; 15690303@tecvalles.mx (H.R.L.-J.); mariela.michel@tecvalles.mx (M.R.M.); jorge.wong@tecvalles.mx (J.E.W.-P.); diana.marquez@tecvalles.mx (D.B.M.-M.); fabiola.veana@tecvalles.mx (F.V.)

2 Facultad de Estudios Profesionales Zona Huasteca, Universidad Autónoma de San Luis Potosí, Romualdo del Campo, No. 501, Rafael Curiel, Ciudad Valles, San Luis Potosí C.P. 79060, Mexico; jose94guada@hotmail.com

3 Unidad Académica Multidisciplinaria Mante, Universidad Autónoma de Tamaulipas, E. Cárdenas González No. 1201, Jardín, Ciudad Mante, Tamaulipas C.P. 89840, Mexico

4 Facultad de Ciencias Químicas, Universidad Autónoma de San Luis Potosí, Av. Dr. Manuel Nava 6, Zona Universitaria, San Luis Potosí, San Luis Potosí C.P. 78290, Mexico; mayra.aguilar@uaslp.mx

5 Facultad de Ciencias Químicas, Universidad Autónoma de Coahuila, Boulevard Venustiano Carranza s/n, República Oriente, Saltillo, Coahuila C.P. 25280, Mexico; alberto_ascaciovaldes@uadec.edu.mx

* Correspondence: luis.duran@docentes.uat.edu.mx (L.V.R.-D.); pedro.aguilar@tecvalles.mx (P.A.-Z.)

tion: Lara-Juache, H.R.; a-Hernández, J.G.; ríguez-Durán, L.V.; Michel, M.R.; ng-Paz, J.E.; Muñiz-Márquez, D.B.; na, F.; Aguilar-Zárate, M.; acio-Valdés, J.A.; Aguilar-Zárate, haracterization of a Biofilm eactor Designed for the gle-Step Production of Aerial idia and Oosporein by *Beauveria iana* PQ2. *J. Fungi* **2021**, *7*, 582. os://doi.org/10.3390/jof7080582

demic Editors: Laurent Dufossé Craig Faulds

eived: 1 June 2021
epted: 19 July 2021
lished: 21 July 2021

lisher's Note: MDPI stays neutral n regard to jurisdictional claims in lished maps and institutional affil- ons.

Abstract: *Beauveria bassiana* is an entomopathogenic fungus that is used for the biological control of different agricultural pest insects. *B. bassiana* is traditionally cultivated in submerged fermentation and solid-state fermentation systems to obtain secondary metabolites with antifungal activity and infective spores. This work presents the design and characterization of a new laboratory-scale biofilm bioreactor for the simultaneous production of oosporein and aerial conidia by *B. bassiana* PQ2. The reactor was built with materials available in a conventional laboratory. K_La was determined at different air flows (1.5–2.5 L/min) by two different methods in the liquid phase and in the exhaust gases. The obtained values showed that an air flow of 2.5 L/min is sufficient to ensure adequate aeration to produce aerial conidia and secondary metabolites by *B. bassiana*. Under the conditions studied, a concentration of 183 mg oosporein per liter and 1.24×10^9 spores per gram of support was obtained at 168 h of culture. These results indicate that the biofilm bioreactor represents a viable alternative for the production of products for biological control from *B. bassiana*.

Keywords: *Beauveria bassiana*; biological control; oosporein; spore production

1. Introduction

The production of spores and metabolites, such as antibiotics, enzymes, and pigments from filamentous fungi, has taken on global importance for the biological control of pests and biotechnological purposes [1–5]. These metabolites have been produced using fermentation systems, such as submerged fermentation (SmF) and solid-state fermentation (SSF), which differ in the nature of their operating conditions [6–9]. Traditionally, both processes (SmF and SSF) are vital for industrial spore production. SmF is used for the production of biomass and mycelium that will be inoculated into the solid substrate in SSF [10].

Beauveria bassiana is an entomopathogen and endophyte fungi used as a biocontrol agent against pest insects as a spore formulation [3,11]. It also produces enzymes and secondary metabolites, such as bassianin, tenellin, beauvericin, bassiacridin, and oosporein;

this last one has been characterized as a soluble red pigment with the formula $C_{14}H_{10}$(and there is great scientific interest in mass producing it for biotechnological appli tions [12,13].

The in vivo production of a reddish coloration has been reported in *Musca domestica* (at the end of infection by *B. bassiana* attributed to oosporein [14] as the principal molecu responsible for immune host suppression [15]. This molecule has attracted attention for antimicrobial activities against bacterial and fungal phytopathogens [13,16,17], but it al exhibits cytotoxic properties [12,13,18]. Conversely, it has been observed that the use fermented crude extracts (raw secondary metabolites) of *B. bassiana* benefits plant grow inhibits the development of diseases caused by plant pathogenic fungi, and contribu to the production of phenolic compounds [5]. Instead of the use of synthetic insecticid and fungicides, biological control using spores and even metabolites represent a sa and feasible alternative for the control of pests [19], but production systems should developed to obtain quality products. Currently, biofilm reactors represent a formidab strategy that combines SmF and SSF in one apparatus for the production of spores, bioma and secondary metabolites [20]. The attachment of aerial hyphae to the inert support a the release of secondary metabolites to the medium, similar to natural development, gi a better yield and quality of conidia [21], while the inert support can be reused, maki it simple, user friendly, and inexpensive [6]. However, in a bioreactor system, ma parameters should be determined, one of the most important being the volumetric ma transfer coefficient (K_La), to establish the aeration efficiency during the aerobic bioproce that depends on the shape, size, agitation speed, air flow rate, etc. of the reactor, but n on its volume [21,22]. The measurement of microbial growth is also a concern. Measuri microbial growth in solid fermentation is a difficult task due to the adhesion of the bioma to the solid support. Therefore, the measurement of CO_2 production is a feasible meth for its estimation [23].

The aims of the current study were as follows: to design and characterize a biofil bioreactor for the production and recovery of aerial conidia and oosporein by *Beauve bassiana* at lab scale and to evaluate the fungal growth by monitoring the CO_2 producti

2. Materials and Methods

2.1. Fungal Strain and Media

The *Beauveria bassiana* PQ2 strain was obtained from the Food Analysis Laborato Instituto Tecnológico de Ciudad Valles, Ciudad Valles, San Luis Potosí, México. The fung strain was cryopreserved using glycerol and skimmed milk at −20 °C and reactivated potato-dextrose agar medium for seven days at 27 °C.

2.2. Biofilm Reactor Setup

The whole bioprocess was carried out in a low-cost lab-made biofilm reactor (Figure The biofilm reactor consisted of a wide-mouth glass bottle with 1300 mL capacity (Sch Duran GLS 80, Mainz, Germany) and a 20 g stainless-steel pad (Scotch-Brite, 3 M, St. Pa MN, USA) as inert support added to the bottle headspace (Figure 1). Air was suppli by an air pump. The medium was recirculated by a peristaltic pump (CRODE, Cela Mexico).

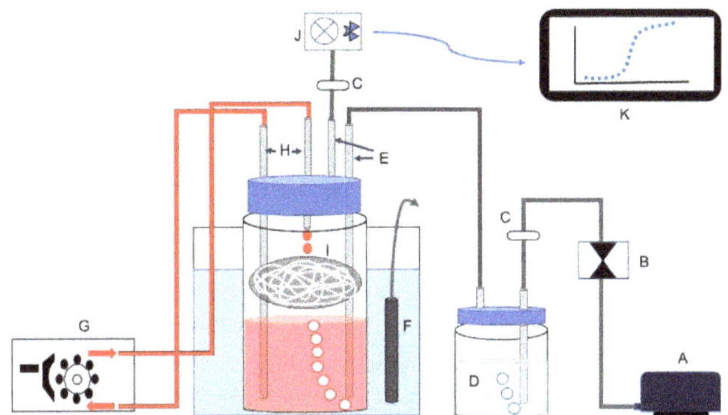

Figure 1. Scheme of biofilm bioreactor designed: (A) air pump; (B) air flow control; (C) air filter; (D) external bubbler; (E) air inlet and outlet; (F) thermostat; (G) peristaltic pump; (H) culture medium inlet and outlet; (I) inert support (metal structure packing); (J) wireless CO_2 sensor; (K) LabQuest interface for data recovery. The samples were taken at the culture media inlet port.

Volumetric oxygen transfer coefficients (K_La) were determined by three repetitions in both liquid and gaseous phases at three air flow rates (1.5, 2.0, and 2.5 L min^{-1}) at 30 ± 2 °C. For the liquid phase, the sodium sulfite oxidation method was used [24]. A solution composed of 600 mL of Na_2SO_3 (0.5 N) and $CuSO_4$ 0.001 M was used. The air inlet at different flow rates was started, and 2 mL samples from the solution were obtained in intervals of 1–8 h. To each sample, 3 mL of iodine (0.5 N) was added, and a titration with $Na_2S_2O_3$ (0.06 N) was performed applying a starch solution (10%) as indicator. K_La was obtained using Equation (1):

$$K_La = \frac{1}{C*} \times \frac{m \times N}{4 \times Vm}. \quad (1)$$

where C^* is the O_2 solubility, N is the concentration of Na_2SO_3, Vm is the volume of sample of Na_2SO_3, and m is the slope obtained by linear regression (Figure S1).

For the gaseous phase, air flows of 1.5, 2.0, and 2.5 L min^{-1} ($F^{in}{}_{O2}$) were tested using the gaseous O_2 sensor (Vernier, Beaverton, OR, USA) at the output flow ($F^{out}{}_{O2}$) in a solution composed of 600 mL of Na_2SO_3 (0.5 N) and $CuSO_4$ (0.001 M) [25]. Samples from O_2 were taken every 5 min for 480 min. After 480 min, the dissolved oxygen was determined by the Winkler method. The data obtained by the sensor and the Winkler method were used to obtain the gaseous phase balance with Equation (2):

$$K_La = \frac{F^{in}_{O2} - F^{out}_{O2}}{V \times (C^* - C_L)}. \quad (2)$$

where $F^{in}{}_{O2}$ is the molar flow rate of oxygen gas input, $F^{out}{}_{O2}$ is the molar outflow of oxygen gas, V is the reactor volume, C_L is the dissolved oxygen concentration in liquid phase, and C^* is the oxygen saturation concentration.

2.3. *Fungal Growth Evaluation, Aerial Conidia, and Red Pigment Production Dynamics*

Submerged fermentation was carried out in two lab-made biofilm reactors (1300 mL) with 600 mL of Czapek–Dox mineral medium (sucrose 22.5 g/L, yeast extract 6.0 g/L, KH_2PO_4 0.48 g/L, $MgSO_4$ 0.72 g/L, NH_4NO_3 0.06 g/L, and $CaCl_2$0.24 g/L) at pH of 6.0. The culture medium was autoclaved at 121 °C for 15 min and inoculated with *Beauveria bassiana* PQ2 at 1 × 10^6 spores/mL. Growth and pigment production were carried out in the following conditions: temperature of 30 ± 2 °C, air flow rate of 2.5 L/min, medium recirculation of 3.66 L/min (20 min, every 12 h), and duration of 168 h in the biofilm reactor for three repetitions. The oosporein and aerial conidia were monitored in the t bioreactors, while CO_2 evolution data were obtained from only one bioreactor.

Microbial growth was estimated indirectly by CO_2 production in the exit gases usin Go Direct® CO_2 analyzer (Vernier, Beaverton, OR, USA) controlled by LabQuest2 interf The data generated show the CO_2 production rate (Figure 2). The CO_2 production rate v integrated to obtain the CO_2 production. The CO_2 production was modeled as biom (mg CO_2/mL of liquid media) by the Verlhurts–Pearl logistic model following the meth proposed by Aguilar-Zárate et al. [26] (Equation (3)):

$$\frac{dCO2}{dt} = \mu CO2\left[1 - \frac{CO2}{CO2_{max}}\right]$$

where μ is the maximal specific CO_2 production rate and $CO2_{max}$ is the equilibrium va for CO_2 with $dCO2/dt = 0$. The solution to Equation (3) is shown as Equation (4):

$$CO_2(t) = \frac{CO2_{max}}{1 - \left(\frac{CO2_{max} - CO2_0}{CO2_0}\right)e^{-\mu t}}$$

$CO2_0$ is the value of CO_2 when $t = 0$. Square error values were minimized as a functi of $CO2_{max}$, $CO2_0$, and μ.

Samples were taken every 24 h to the end of fermentation, and oosporein and su concentrations were measured. Sugar consumption was measured by refractometry si the sucrose was the carbon source. Oosporein analyses were carried out by spectrop tometry as follows. Samples were filtered through a sterile Millipore membrane (0.45 µ (Minisart, Sartorius Stedim Biotech, Aubagne, France) and evaluated by spectrophotome at the wavelength of 430 nm. Data were compared against a standard calibration curve oosporein (0–31.25 ppm, purity ≥70%) provided by the Food Analysis Laboratory. T pigment production kinetics was modeled using the Luedeking–Piret model according Aguilar-Zarate et al. [27] with Equation (5):

$$\frac{d_{Oosp}}{dt} = Y_{Oosp/CO2}\frac{d_{CO2}}{d_t} + kCO2$$

where $Y_{Oosp/CO2}$ is the production coefficient and k (mg/h mg) is the secondary coeffici of oosporein production ($k > 0$) or destruction ($k < 0$). The solution to the previous equati is given below (Equation (6)):

$$Oosp = Oosp_0 + Oosp_{Oosp/CO2}\,(CO2 - CO2_0) + \frac{kCO2_{max}}{\mu}\,ln\left[\frac{CO2_{max} - CO2_0}{CO2 - CO2_0}\right]$$

with $Oosp_0$ being the value for oosporein when $CO_2 = CO2_0$.

The conidia production was evaluated at the end of the fermentation through recovery from the metal solid support of the biofilm reactor with a diluted sterile 0.0 (v/v) Tween 80 solution. The fungal conidia were counted with a Neubauer chamber usi a light microscope at 40×, and the spore yield was obtained by applying Equation (7):

$$\frac{COnidia}{g\ of\ support} = \left(\frac{Spores}{mL}\right) \times \frac{Volume\ in\ suspension\ of\ conidia\ recovered}{g\ of\ support}$$

2.4. Oosporein Characterization by HPLC Tandem Mass Spectrometry

The filtered total aqueous extract at the final point of fermentation was analyz by reversed-phase high-performance liquid chromatography (HPLC) equipped with autosampler (Varian ProStar 410, Walnut Creek, CA, USA), ternary pump (Varian ProS 230I, USA), and PDA detector (Varian ProStar 330, USA). A sample (5 µL) was inject into a Denali C-18 column (150 mm × 2.1 mm, 3.1 µm, Grace, Deerfield, IL, USA). T oven temperature was 30 °C. The elution gradient was formic acid (0.2 % v/v, solvent and acetonitrile (solvent B) with initial gradient course of 3% B, 5–15 min; 16% B linear, 15–45 min; and 50% B linear. The flow rate was 0.2 mL/min and the elution was monitored at 287 nm. Liquid chromatography–ion trap mass spectrometry (Varian 500-MS IT Mass Spectrometer, USA) equipped with an electrospray ion source was used. The MS analysis was performed in the negative mode $[M\text{-}H]^{-1}$ using nitrogen as the nebulizing gas and helium as the damping gas. The parameters of the ion source were as follows: spray voltage of 5.0 kV, capillary voltage of 90.0 V, and temperature of 350 °C. Full scan spectra were acquired in the m/z range 100–2000, and, subsequently, the MS/MS analyses were performed on a series of selected ions. The data were collected and processed using MS Workstation software (V 6.9).

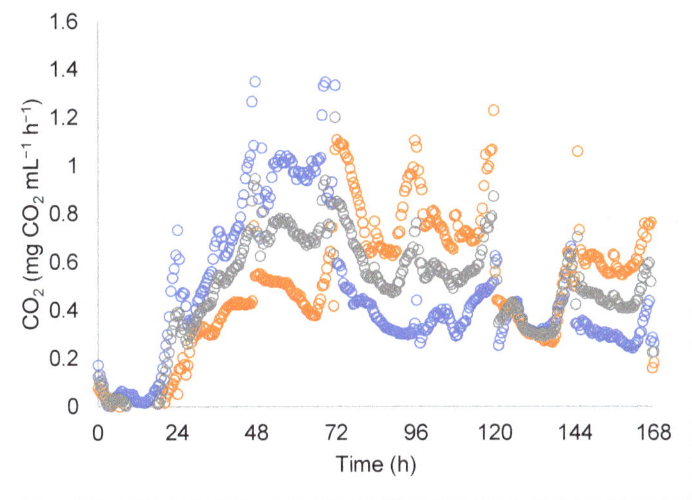

Figure 2. Evolution of CO_2 production during fermentation of *Beauveria bassiana* PQ2 for the production of aerial conidia and oosporein. The different colors of open circles are the repetitions.

2.5. Statistical Analysis

The biofilm reactor setup experiments were performed with three repetitions, and the results are presented as mean ± SD. A post hoc analysis was carried out by Tukey test (p = 0.05) for comparing the significant differences between airflow rates.

3. Results

The determination of K_La by the sodium sulfite oxidation method for both liquid and gaseous phases allowed the characterization of the biofilm reactor. Table 1 shows that 2.5 L/min represents the best flow rate for aeration conditions in both solid and gaseous phases.

Table 1. Bioreactor gas transfer in liquid and gaseous phases.

Flow Rate (L/min)	K_La Liquid Phase (min^{-1})	K_La Gaseous Phase (min^{-1})
1.5	0.99 ± 0.009 b	0.51 ± 0.261 b
2.0	1.16 ± 0.017 b	0.58 ± 0.102 b
2.5	2.10 ± 0.017 a	3.66 ± 0.394 a

Different letters indicate significant differences (Tukey test, p = 0.05).

In regard to the respiration activity of *B. bassiana* throughout CO_2 analysis, Figure 2 shows the evolution of CO_2 production along the fermentation process. A similar pattern in CO_2 evolution was seen during the 0–96 h period when submerged fermentation developed.

High variation was observed during the 96–168 h period due to the high respiratory activi mainly in the solid-state culture. At 48 and 69 h, a higher CO_2 production rate was observ (1.35 mg of CO_2 per mL of media per hour). The maximal production rate was 80.59 m CO_2 mL^{-1} (Table 2) at 168 h of culture (Figure 3). It could be considered that the maxim growth rate was reached at this stage [6]. The accumulated CO_2 is shown in Figure 3, a the data were obtained from the integration of the CO_2 evolution. The graph shows t microbial growth trend at a growth rate of μ = 0.04 h^{-1} as follows: At 24 h, a lag phase w found as the entire fermentation process (liquid and solid) began biomass formation on t inert support at 50 h. Then, an exponential phase was observed followed by constant sug consumption until the end of fermentation (168 h) (°$Brix_{initial}$ = 4.03, °$Brix_{final}$ = 0.55).

Table 2. Kinetic parameters of *Beauveria bassiana* PQ2 under biofilm bioreactor conditions.

Parameters	Units	Value
$CO_{2\ max}$	mg CO_2 mL^{-1}	80.59
μ	h^{-1}	0.04
Growth model	R^2	0.99
$Oosporein_{max}$	mgL^{-1}	183.0
Oosporein productivity	mg/L/h	1.09
Oosporein model	R^2	0.97
$Y_{Oosp/CO2}$	mg Oosporein/h × mg CO_2	0.02
Conidia recovery	Conidia/gram of support	1.24×10^9

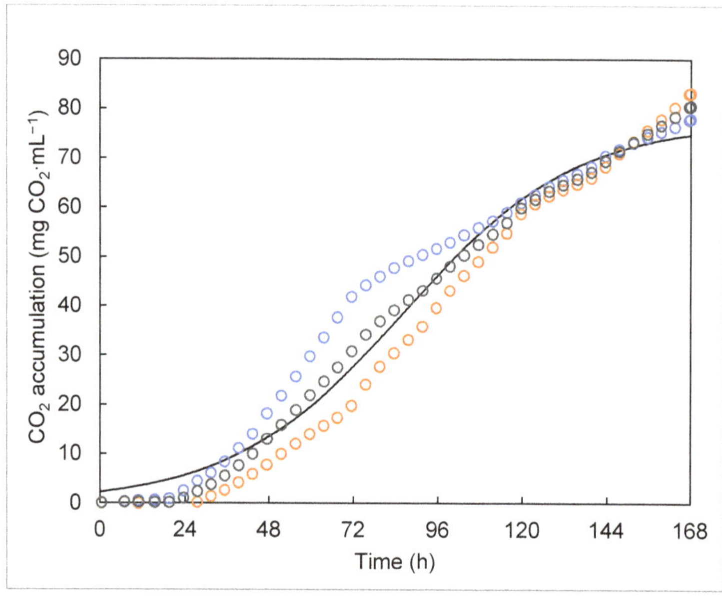

Figure 3. Accumulated CO_2 production during fermentation of *Beauveria bassiana* PQ2 in the biop cess for the production of aerial conidia and oosporein. Open circles are the experimental data, a the continuous line represents the data calculated by the model.

The production of aerial conidia did not present any problems using the metal str tured packing as an inert support (Figure S2), allowing us to obtain 1.24×10^9 conidia/gra of support at 168 h. Although part of the biomass was attached to the walls or intern parts of the biofilm bioreactor, its concentration was not considered in this study. T sterile conditions were confirmed, as there was no contamination of the culture mediu

The production of the water-soluble pigment oosporein was achieved after 72 h (Figure 4), reaching a maximum concentration of 183 mg/L^{-1} at the end of the fermentation. The whole process reached an oosporein productivity of 1.09 mg/L/h (Table 2). The oosporein concentration increased over time, as shown in Figure 4. The production of oosporein reached two peaks associated with the growth in SmF and SSF. The first peak was obtained after 96 h of culture (42.30 mg/mL) (Figure 4). As shown in Figure 2, during 0–96 h, the submerged culture was developed. The second peak was obtained at the end of the fermentation process with 183 mg/mL of oosporein. The oosporein yield was 0.02 mg of oosporein per hour per mg of CO_2.

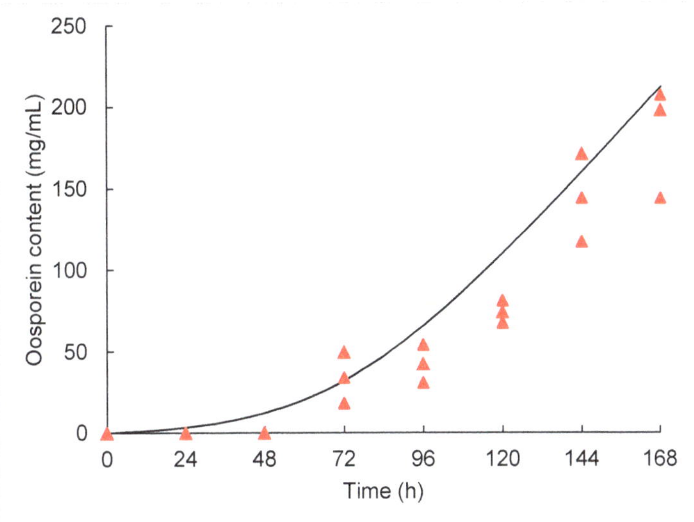

Figure 4. Kinetic production of oosporein by *Beauveria bassiana* PQ2 in biofilm bioreactor. Closed triangles are the experimental data, and the continuous line represents the data predicted by the model.

Five hundred milliliters of fermented extract were recovered after 168 h and then submitted to an HPLC-MS/MS analysis in order to characterize the secondary metabolites and corroborate the presence of oosporein produced by *B. bassiana* PQ2. The results show the presence of five ionized compounds (Figure 5). The fragmentation patterns of the four compounds did not allow the identification of the molecules. Oosporein was found as the major compound in the HPLC chromatogram at a retention time (R. T.) of 22.72 min. It was identified with m/z 305 (306 M. W.) (Table 3).

Table 3. Kinetic parameters of *Beauveria bassiana* PQ2 under biofilm bioreactor conditions.

Peak No.	R. T. (min)	M. W.	$[M-H]^-$ (m/z)	MS^2 Ion Fragment	Tentative Identity
1	3.40	283	282	150, 133	Unknown
2	5.92	244	243	200, 110	Unknown
3	12.84	291	290	254, 230, 214, 200, 128	Unknown
4	15.49	387	386	343, 299, 298, 286	Unknown
5	22.72	306	305	277, 262, 261, 249, 233, 217, 205, 189, 161	Oosporein

R. T., retention time; M. W., mass weight; m/z, mass-to-charge ratio; MS^2, tandem mass spectrometry.

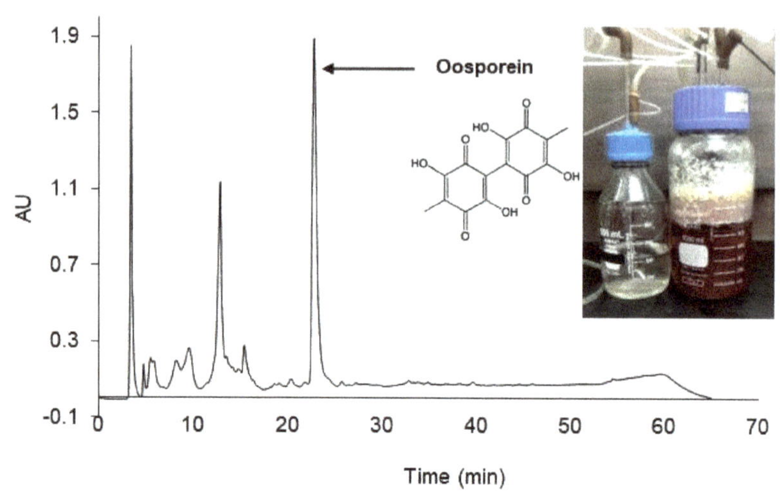

Figure 5. HPLC chromatogram of metabolites produced by *B. bassiana* PQ2 in the biofilm bioreac

4. Discussion

Biofilm reactors are a fermentation system for the production of value-added produ that combine SmF and SSF approaches, where the biomass adheres to an inert support [In this sense, the availability of oxygen in the solid fermentation (gaseous phase) w be exploited even when there is oxygen saturation in the liquid phase. In addition, biomass attachment to the support confers some advantages, such as the low viscosity the medium culture and easy downstream recovery of the products [29].

Oxygen availability is a critical factor for *B. bassiana* spore production [30–33]. T dissolved oxygen concentration is the result of the oxygen transfer rate (OTR) from gaseous phase to the liquid phase and the oxygen uptake rate (OUR). The volume mass transfer coefficient (K_La) is a parameter that allows characterizing the capacity o bioreactor to supply oxygen and is very important for the design, operation, and scal of bioreactors [34]. Therefore, the K_La of the biofilm bioreactor was determined by t methods, one physical method and one chemical method. Under the operating conditic studied, a K_La value of 0.99 to 2.10 min^{-1} was obtained by the chemical method, wh the physical method obtained values of 0.51 to 3.66 min^{-1}. These values are close those obtained for a circulating-bed biofilm reactor (0.017 s^{-1}) [35] and higher than th obtained for an up-flow concurrent packed-bed biofilm reactor (0.0013–0.012 s^{-1}) [Furthermore, the K_La values obtained for the biofilm reactor built for this work are wit the values (9.5–208 h^{-1}) measured for different commercial stirred-tank bioreactors us for the production of *B bassiana* blastospores [37].

Hence, the bioprocess using *B. bassiana* PQ2 was carried out using the aeration fl rate of 2.5 L/min, which also contributed to the high availability of oxygen at the liquid– interface [38].

Respirometry analysis showed the metabolic activity of *B. bassiana* PQ2 when adapti to the fermentation conditions (Figure 2). From 48 to 69 h, CO_2 production increased, wh was related to mycelial development and the spore yield that improved after 100 h of t cultivation process [10,26]. However, the high variation of CO_2 reported in Figur might indicate the need to control the operating conditions, such as temperature, pH, a nutrients; thus, in the future, it will be necessary to optimize them [10]. Figure 3 shows th the microbial growth data were analyzed by non-linear regression. Based on the analy the growth rate (μ) was considered as the parameter to prove that the samples came fr the same population. We obtained μ = 0.040 with a 95% of confidence interval of 0.039–0.C and standard error of 0.001. The results shown in Figure 3 are similar to those reported by Cruz-Barrera et al. [9] for *Trichoderma asperellum* Th204 during SSF, who found that the CO_2 accumulation indicated the lag phase occurred from 0 to 20 h, the exponential phase from 20 to 80 h, and the stationary phase from 100 to 160 h. Similarly, in the present work, CO_2 production was observed even at the end of the process, as the microorganisms continued to grow on the inert support, and it is an indicator of metabolic activity of *B. bassiana* mycelium [39]. Some authors have reported the respiratory activity as an indirect indicator of fungal growth from *Aspergillus niger* GH1 [26], *Trichoderma harzianum* IRDT22C [26], *Metarhizium anisopliae* strain CP-OAX [10], and *Metarhizium anisopliae* IBCB 425 [40], but there is no information available for *Beauveria bassiana*.

In the biofilm bioreactor, the presence of conidiophores and spherical conidia attached to the biofilm agglomeration of mycelia on the metal solid support was found. Unlike aerial conidia, blastospores produced in submerged culture had an oval shape and formed fine pellets due to pneumatic agitation. At the end of the fermentation process, the mycelia produced under submerged fermentation were disintegrated, probably by the production of proteases. Once the fermentation process was finished (168 h), the aerial conidia were harvested from the metal solid support (Figure S2). The yield obtained was 1.24×10^9 conidia/gram, which coincided with the reduction of CO_2 production. This condition is similar to that reported by Méndez-González et al. [10] for spore production by *M. anisopliae* in SSF. In a solid-state culture, results close to those obtained in this study using *B. bassiana* were reported by Kang et al. [30] using a packed-bed bioreactor. They achieved 1.1–1.2 $\times 10^{10}$ g^{-1} and 1×10^9 conidia/g on grain substrates [41] and 5.0×10^8 spores g^{-1} dry matter using rice husk [42]. These results serve as a comparison for the yield achieved from *B. bassiana* in SSF, indicating the biofilm bioreactor is a feasible tool for the production of conidia.

The pigment production is consistent with that reported by Ávila Hernández et al. [43], who mentioned that oosporein is the major pigment produced by *Beauveria bassiana* under submerged fermentation, which shows antimicrobial and insecticidal activities. The yield obtained (183 mg/L^{-1}) is close to the 270 mg/L^{-1} reported by Strasser et al. [44] from *B. brongniartii* in submerged culture. Amin et al. [45] produced red pigment from *B. bassiana* in submerged fermentation, reaching a yield of up to 480 mg/L. In addition, a combination of spores and pigment increases the insecticidal activity, which suggests that the use of the biofilm bioreactor developed in the present research would help to obtain infective units and metabolites in a single step for their possible use in the biological control of pests in further in vitro experiments. The production dynamics of oosporein (Figure 4) is similar to that of carbon dioxide (Figure 3) and may be another alternative for estimating fungal growth, because the use of certain metabolic products may offer more sensitive results [46].

The results obtained in the characterization of the compounds show five ionized peaks (Figure 5) where oosporein was the only relevant metabolite detected. It was identified by information reported in the literature, with the formula $C_{14}H_{10}O_8$ [1,15,16,44]. The negative ionization of MS analysis allowed the identification of a compound with m/z 305. The result agrees with that reported by Feng et al. [15], who characterized the production of oosporein in fungi. They mentioned that oosporein is only ionized in negative mode. It was not possible to identify peaks 1-4, and other metabolites, such as tenellin, bassianin, or beauvericin produced by *Beauveria* species, were not found. Strasser et al. [44] considered that the production of oosporein is constitutive. Hence, *B. bassiana* PQ2 produced oosporein independently of the culture medium or the bioreactor.

The results of this research represent a viable and novel way to produce aerial conidia and oosporein from *Beauveria bassiana* PQ2 using a biofilm reactor. Some authors have reported goods yields in protein production (hydrophobin II) by *Trichoderma reesei* [20]. In addition, conidia and secondary metabolites produced by *Aspergillus clavatus* in a biofilm reactor have been shown to be effective in mosquito (*Culex quinquefasciatus*) control [28].

5. Conclusions

The K_La of the biofilm bioreactor showed high oxygen transfer in both liquid a gaseous phases. The designed bioreactor ensured the development of SmF and SSF at t same time and the production of both aerial conidia and oosporein by *Beauveria bassia* PQ2. The CO_2 production dynamic allowed determining the fungal growth in the whc process. To our knowledge, this is the first report on the biomass estimation of *B. bassia* through respirometry and the production of oosporein in a biofilm bioreactor. It represer a viable alternative for obtaining value-added products from filamentous fungi.

Supplementary Materials: The following are available online at https://www.mdpi.com/article/ .3390/jof7080582/s1, Figure S1: Evolution of oxygen solubility at different air flow-rates quantifi by titration of liquid media with thiosulfate. Values of slope were used in Equation (1) for calculati Volumetric oxygen transfer coefficients (K_La) values, Figure S2: Invasion of metal solid support w mycelia and aerial conidia.

Author Contributions: H.R.L.-J. and J.G.Á.-H., conceptualization, methodology, and writing; L.V. D., writing, data curation, and formal analysis; M.R.M., writing and editing; J.E.W.-P., D.B.M.- and F.V., methodology and formal analysis; M.A.-Z., writing and review; J.A.A.-V., investigation a formal analysis; P.A.-Z., visualization, supervision, and funding acquisition. All authors have re and agreed to the published version of the manuscript.

Funding: This research was funded by TECNOLÓGICO NACIONAL DE MÉXICO, granted to C IT17A895-2021.

Conflicts of Interest: The authors declare no conflict of interest.

References

1. Da Costa Souza, P.N.; Bim Grigoletto, T.L.; Beraldo de Moraes, L.A.; Abreu, L.M.; Souza Guimarães, L.H.; Santos, C.; Ribe Galvão, L.; Gomes Cardoso, P. Production and chemical characterization of pigments in filamentous fungi. *Microbiology* **2016**, *1* 12–22. [CrossRef]
2. Neera, D.K.; Ramana, K.V.; Sharma, R.K. Optimization of *Monascus* pigment production and its antibacterial activity. *Int. J. Cu Res. Biosci. Plant Biol.* **2017**, *4*, 71–80. [CrossRef]
3. Barra-Bucarei, L.; González, M.G.; Iglesias, A.F.; Aguayo, G.S.; Peñalosa, M.G.; Vera, P.V. *Beauveria bassiana* multifunction as endophyte: Growth promotion and biologic control of *Trialeurodes vaporariorum*, (Westwood) (Hemiptera: Aleyrodidae) in toma *Insects* **2020**, *11*, 591. [CrossRef]
4. Stracquadanio, C.; Quiles, J.M.; Meca, G.; Cacciola, S.O. Antifungal activity of bioactive metabolites produced by *Trichoder asperellum* and *Trichoderma atroviride* in liquid medium. *J. Fungi* **2020**, *6*, 263. [CrossRef]
5. Soesanto, L.; Sari, L.Y.; Mugiastuti, E.; Manan, A. Cross application of entomopathogenic fungi raw secondary metabolites controlling fusarium wilt of chili seedlings. *Jurnal Hama Penyakit Tumbuhan Tropika* **2021**, *21*, 82–90. [CrossRef]
6. Manan, M.; Webb, C. Design aspects of solid state fermentation as applied to microbial bioprocessing. *J. Appl. Biotechnol. Bioe* **2017**, *4*, 91. [CrossRef]
7. Hussain, A.; Tian, M.-Y.; Ahmed, S.; Shahid, M. Current status of entomopathogenic fungi as mycoinecticides and th inexpensive development in liquid cultures. In *Zoologia*; García, M.D., Ed.; InTech: Rijeka, Croatia, 2012; pp. 103–122.
8. Pradeep, F.S.; Begam, M.S.; Palaniswamy, M.; Pradeep, B. Influence of culture media on growth and pigment production *Fusarium moniliforme* KUMBF1201 isolated from paddy field soil. *World Appl. Sci. J.* **2013**, *22*, 70–77. [CrossRef]
9. Cruz Barrera, M.; Gómez, M.I.; Serrato Bermúdez, J.C. Towards the production of fungal biocontrol candidates using in supports: A case of study of *Trichoderma asperellum* in a pilot fixed bed fermenter. *Biocontrol Sci. Technol.* **2019**, *29*, 162–1 [CrossRef]
10. Méndez-González, F.; Loera-Corral, O.; Saucedo-Castañeda, G.; Favela-Torres, E. Bioreactors for the production of biologic control agents produced by solid-state fermentation. In *Current Developments in Biotechnology and Bioengineering*; Pandey, Larroche, C., Soccol, C.R., Eds.; Elsevier: Amsterdam, The Netherlands, 2018; pp. 109–121.
11. Lohse, R.; Jakobs-Schönwandt, D.; Vidal, S.; Patel, A.V. Evaluation of new fermentation and formulation strategies for a hi endophytic establishment of Beauveria bassiana in oilseed rape plants. *Biological. Control* **2015**, *88*, 26–36. [CrossRef]
12. Valencia, J.W.A.; Gaitán Bustamante, A.L.; Jiménez, A.V.; Grossi-de-Sá, M.F. Cytotoxic activity of fungal metabolites from t pathogenic fungus *Beauveria bassiana*: An intraspecific evaluation of beauvericin production. *Curr. Microbiol.* **2011**, *63*, 3 [CrossRef] [PubMed]
13. Mao, B.-Z.; Huang, C.; Yang, G.-M.; Chen, Y.-Z.; Chen, S.-Y. Separation and determination of the bioactivity of oosporein fro *Chaetomium cupreum*. *Afr. J. Biotechnol.* **2010**, *9*. [CrossRef]

Mwamburi, L.A.; Laing, M.D.; Miller, R.M. Laboratory screening of insecticidal activities of *Beauveria bassiana* and *Paecilomyces lilacinus* against larval and adult house fly (*Musca domestica* L.): Research article. *Afr. Entomol.* **2010**, *18*, 38–46. [CrossRef]
Feng, P.; Shang, Y.; Cen, K.; Wang, C. Fungal biosynthesis of the bibenzoquinone oosporein to evade insect immunity. *Proc. Natl. Acad. Sci. USA* **2015**, *112*, 11365. [CrossRef] [PubMed]
Alurappa, R.; Bojegowda, M.R.M.; Kumar, V.; Mallesh, N.K.; Chowdappa, S. Characterisation and bioactivity of oosporein produced by endophytic fungus *Cochliobolus kusanoi* isolated from *Nerium oleander* L. *Nat. Prod. Res.* **2014**, *28*, 2217–2220. [CrossRef]
Nagaoka, T.; Nakata, K.; Kouno, K. Antifungal activity of oosporein from an antagonistic fungus against *Phytophthora infestans*. *Z. Nat. C* **2004**, *59*, 302–304. [CrossRef]
Ramesha, A.; Venkataramana, M.; Nirmaladevi, D.; Gupta, V.K.; Chandranayaka, S.; Srinivas, C. Cytotoxic effects of oosporein isolated from endophytic fungus *Cochliobolus kusanoi*. *Front. Microbiol.* **2015**, *6*, 870. [CrossRef]
Koch, E.; Ole Becker, J.; Berg, G.; Hauschild, R.; Jehle, J.; Köhl, J.; Smalla, K. Biocontrol of plant diseases is not an unsafe technology! *J. Plant Dis. Prot.* **2018**, *125*, 121–125. [CrossRef]
Khalesi, M.; Zune, Q.; Telek, S.; Riveros-Galan, D.; Verachtert, H.; Toye, D.; Gebruers, K.; Derdelinckx, G.; Delvigne, F. Fungal biofilm reactor improves the productivity of hydrophobin HFBII. *Biochem. Eng. J.* **2014**, *88*, 171–178. [CrossRef]
Musoni, M.; Destain, J.; Thonart, P.; Bahama, J.-B.; Delvigne, F. Bioreactor design and implementation strategies for the cultivation of filamentous fungi and the production of fungal metabolites: From traditional methods to engineered systems. *Biotechnol. Agron. Soc. Environ.* **2015**, *19*, 430–442.
Moutafchieva, D.; Popova, D.; Dimitrova, M.; Tchaoushev, S. Experimental determination of the volumetric mass transfer coefficient. *J. Chem. Technol. Metall.* **2013**, *48*, 351–356.
Aguilar-Zárate, P.; Wong-Paz, J.E.; Rodríguez-Duran, L.V.; Buenrostro-Figueroa, J.; Michel, M.; Saucedo-Castañeda, G.; Favela-Torres, E.; Ascacio-Valdés, J.A.; Contreras-Esquivel, J.C.; Aguilar, C.N. On-line monitoring of *Aspergillus niger* GH1 growth in a bioprocess for the production of ellagic acid and ellagitannase by solid-state fermentation. *Bioresour. Technol.* **2018**, *247*, 412–418. [CrossRef] [PubMed]
Cooper, C.; Fernstrom, G.; Miller, S. Performance of agitated gas-liquid contactors. *Ind. Eng. Chem.* **1944**, *36*, 504–509. [CrossRef]
Garcia-Ochoa, F.; Gomez, E. Bioreactor scale-up and oxygen transfer rate in microbial processes: An overview. *Biotechnol. Adv.* **2009**, *27*, 153–176. [CrossRef]
De la Cruz-Quiroz, R.; Roussos, S.; Aguilar, C.N. Production of a biological control agent: Effect of a drying process of solid-state fermentation on viability of *Trichoderma* spores. *Int. J. Green Tech.* **2018**, *4*, 1–6.
Aguilar-Zarate, P.; Cruz-Hernandez, M.A.; Montañez, J.C.; Belmares-Cerda, R.E.; Aguilar, C.N. Enhancement of tannase production by *Lactobacillus plantarum* CIR1: Validation in gas-lift bioreactor. *Bioprocess Biosyst. Eng.* **2014**, *37*, 2305–2316. [CrossRef] [PubMed]
Seye, F.; Bawin, T.; Boukraa, S.; Zimmer, J.-Y.; Ndiaye, M.; Delvigne, F.; Francis, F. Pathogenicity of *Aspergillus clavatus* produced in a fungal biofilm bioreactor toward *Culex quinquefasciatus* (Diptera: Culicidae). *J. Pestic. Sci.* **2014**, *39*, 127–132. [CrossRef]
Zune, Q.; Delepierre, A.; Gofflot, S.; Bauwens, J.; Twizere, J.C.; Punt, P.J.; Francis, F.; Toye, D.; Bawin, T.; Delvigne, F. A fungal biofilm reactor based on metal structured packing improves the quality of a Gla::GFP fusion protein produced by *Aspergillus oryzae*. *Appl. Microbiol. Biotechnol.* **2015**, *99*, 6241–6254. [CrossRef]
Kang, S.W.; Lee, S.H.; Yoon, C.S.; Kim, S.W. Conidia production by *Beauveria bassiana* (for the biocontrol of a diamondback moth) during solid-state fermentation in a packed-bed bioreactor. *Biotechnol. Lett.* **2005**, *27*, 135. [CrossRef] [PubMed]
Pham, T.A.; Kim, J.J.; Kim, K. Optimization of solid-state fermentation for improved conidia production of *Beauveria bassiana* as a mycoinsecticide. *Mycobiology* **2010**, *38*, 137–143. [CrossRef]
Mascarin, G.M.; Jackson, M.A.; Kobori, N.N.; Behle, R.W.; Dunlap, C.A.; Delalibera Júnior, Í. Glucose concentration alters dissolved oxygen levels in liquid cultures of *Beauveria bassiana* and affects formation and bioefficacy of blastospores. *Appl. Microbiol. Biotechnol.* **2015**, *99*, 6653–6665. [CrossRef] [PubMed]
Santa, H.S.D.; Santa, O.R.D.; Brand, D.; Vandenberghe, L.P.d.S.; Soccol, C.R. Spore production of Beauveria bassiana from agro-industrial residues. *Braz. Arch. Biol. Technol.* **2005**, *48*, 51–60. [CrossRef]
Damiani, A.L.; Kim, M.H.; Wang, J. An improved dynamic method to measure kLa in bioreactors. *Biotechnol. Bioeng.* **2014**, *111*, 2120–2125. [CrossRef] [PubMed]
Nogueira, R.; Lazarova, V.; Manem, J.; Melo, L.F. Influence of dissolved oxygen on the nitrification kinetics in a circulating bed biofilm reactor. *Bioprocess Eng.* **1998**, *19*, 441–449. [CrossRef]
Pérez, J.; Montesinos, J.L.; Gòdia, F. Gas–liquid mass transfer in an up-flow cocurrent packed-bed biofilm reactor. *Biochem. Eng. J.* **2006**, *31*, 188–196. [CrossRef]
García-Gutiérrez, C.; González-Maldonado, M.B.; Medrano-Roldán, H.; Solís-Soto, A. Study of the mixing conditions in bioreactor for blastospores production of *Beauveria bassiana*. *Rev. Colombiana Biotecnol.* **2013**, *15*, 47–54.
Kurt, T.; Marbà-Ardébol, A.-M.; Turan, Z.; Neubauer, P.; Junne, S.; Meyer, V. Rocking *Aspergillus*: Morphology-controlled cultivation of *Aspergillus niger* in a wave-mixed bioreactor for the production of secondary metabolites. *Microbial. Cell Factories* **2018**, *17*, 128. [CrossRef]
Pavlík, M.; Fleischer, P.; Fleischer, P.; Pavlík, M.; Šuleková, M. Evaluation of the carbon dioxide production by fungi under different growing conditions. *Curr. Microbiol.* **2020**, *77*, 2374–2384. [CrossRef]

40. Da Cunha, L.P.; Casciatori, F.P.; Vicente, I.V.; Garcia, R.L.; Thoméo, J.C. *Metarhizium anisopliae* conidia production in packed-b bioreactor using rice as substrate in successive cultivations. *Process Biochem.* **2020**, *97*, 104–111. [CrossRef]
41. Song, M.H.; Yu, J.S.; Kim, S.; Lee, S.J.; Kim, J.C.; Nai, Y.-S.; Shin, T.Y.; Kim, J.S. Downstream processing of *Beauveria bassiana* a *Metarhizium anisopliae*-based fungal biopesticides against *Riptortus pedestris*: Solid culture and delivery of conidia. *Biocontrol Technol.* **2019**, *29*, 514–532. [CrossRef]
42. Sala, A.; Artola, A.; Sánchez, A.; Barrena, R. Rice husk as a source for fungal biopesticide production by solid-state fermentat using B. bassiana and T. harzianum. *Bioresour. Technol.* **2020**, *296*, 122322. [CrossRef]
43. Ávila-Hernández, J.; Carrillo-Inungaray, M.; De-La-Cruz-Quiroz, R.; Wong-Paz, J.; Muñiz-Márquez, D.; Parra, R.; Agui C.; Aguilar-Zárate, P. *Beauveria bassiana* secondary metabolites: A review inside their production systems, biosynthesis, a bioactivities. *Mex. J. Biotechnol.* **2020**, *5*, 1–33. [CrossRef]
44. Strasser, H.; Abendstein, D.; Stuppner, H.; Butt, T.M. Monitoring the distribution of secondary metabolites produced by entomogenous fungus *Beauveria brongniartii* with particular reference to oosporein. *Mycol. Res.* **2000**, *104*, 1227–1233. [CrossR
45. Amin, G.A.; Youssef, N.A.; Bazaid, S.; Saleh, W.D. Assessment of insecticidal activity of red pigment produced by the fung *Beauveria bassiana*. *World J. Microbiol. Biotechnol.* **2010**, *26*, 2263–2268. [CrossRef]
46. Ng, H.E.; Raj, S.S.A.; Wong, S.H.; Tey, D.; Tan, H.M. Estimation of fungal growth using the ergosterol assay: A rapid tool assessing the microbiological status of grains and feeds. *Lett. Appl. Microbiol.* **2008**, *46*, 113–118. [CrossRef]

Journal of
Fungi

view

Iolecular Characterization of Fungal Pigments

riam S. Valenzuela-Gloria [1], Nagamani Balagurusamy [1], Mónica L. Chávez-González [2], Oscar Aguilar [3], erim Hernández-Almanza [1,*] and Cristóbal N. Aguilar [2,*]

1 School of Biological Sciences, Universidad Autónoma de Coahuila, Torreón 27000, Coahuila, Mexico; miriamvalenzuela@uadec.edu.mx (M.S.V.-G.); bnagamani@uadec.edu.mx (N.B.)
2 Bioprocesses and Bioproducts Research Group, BBG-DIA, Food Research Department, School of Chemistry, Universidad Autónoma de Coahuila, Saltillo 25280, Coahuila, Mexico; monicachavez@uadec.edu.mx
3 Tecnológico de Monterrey, Escuela de Ingeniería y Ciencias, Av. Eugenio Garza Sada 2501 Sur, Monterrey 64849, Nuevo León, Mexico; alex.aguilar@tec.mx
* Correspondence: ayerim_hernandez@uadec.edu.mx (A.H.-A.); cristobal.aguilar@uadec.edu.mx (C.N.A.)

Abstract: The industrial application of pigments of biological origin has been gaining strength over time, which is mainly explained by the increased interest of the consumer for products with few synthetic additives. So, the search for biomolecules from natural origin has challenged food scientists and technologists to identify, develop efficient and less consuming strategies for extraction and characterization of biopigments. In this task, elucidation of molecular structure has become a fundamental requirement, since it is necessary to comply with compound regulatory submissions of industrial sectors such as food, pharmaceutical agrichemicals, and other new chemical entity registrations. Molecular elucidation consists of establishing the chemical structure of a molecule, which allows us to understand the interaction between the natural additive (colorant, flavor, antioxidant, etc) and its use (interaction with the rest of the mixture of compounds). Elucidation of molecular characteristics can be achieved through several techniques, the most common being infrared spectroscopy (IR), spectroscopy or ultraviolet-visible spectrophotometry (UV-VIS), nuclear-resonance spectroscopy (MAGNETIC MRI), and mass spectrometry. This review provides the details that aid for the molecular elucidation of pigments of fungal origin, for a viable and innocuous application of these biopigments by various industries.

Keywords: fungi pigments; types; structure; molecular elucidation

tion: Valenzuela-Gloria, M.S.; gurusamy, N.; Chávez-González, .; Aguilar, O.; nández-Almanza, A.; Aguilar, . Molecular Characterization of gal Pigments. *J. Fungi* **2021**, *7*, 326. s://doi.org/10.3390/jof7050326

demic Editor: Laurent Dufossé

eived: 26 March 2021
epted: 18 April 2021
lished: 23 April 2021

lisher's Note: MDPI stays neutral ı regard to jurisdictional claims in lished maps and institutional affilions.

1. Introduction

At present, the colorants are employed by various industrial sectors such as textiles, pharmaceuticals, nutraceuticals, and most importantly as an additive in the food industry. The use of colorants by food products contributes widely to the visual impact to gain the attention and preference of the consumer, apart from adding value to the product. [1] The use of these agents has a number of disadvantages, including a lack of raw materials, differences in pigment extraction, and, most notably, the environmental effect of chemical syntheses used in the production of these additives. Recently, research has been focusing on the quest for new natural sources, and which has thrown up a myriad of potential sources such as plants, animals, bacteria, microalgae and even fungi.

Pigments derived from microbes have many benefits over those derived from plants or animals, including low environmental effects, viability, profitability, and ease of handling prior to, during, and after processing [2]. Fungi stand out among microorganisms of interest because of their ability to produce a broad spectrum of soluble pigments under a variety of conditions and substrates [3]. Because of their ease of cultivation in the laboratory and comparatively lean downstream operations that are readily scalable at pilot or plant scales, pigments extracted from filamentous fungi have piqued industrial interest [4].

Fungal pigments have the capacity to be a significant source of biopigments due to their high yield potential and ease of extraction. For example, the biomass production of

Chlamydomonas reinhardtii microalgae oscillates in the range of 2.0 g L^{-1} in dry biomass [whereas the biomass production of a filamentous fungus such as *Mucor circinelloides* cillates in a radius of 4.0 g L^{-1} in dry biomass [6]. About the fact that all previous valu are in crops without optimization assessment, it is possible to find a 1:2 relationship the production of dry biomass. According to Zhang et al. [7], *Monascus* will increase yields from 48.4 to 215.4 mg L^{-1} by optimizing the glutamic acid present in the cultu medium, which is equal to a 1:3.5 ratio [7]. As a result, understanding and characterizi the molecular structure of these pigments for safe and sustainable use becomes critica mass application is intended [2].

In general, the qualitative nature of the pigments are studied using comparative char or colorimeter or by use of spectrophotometry. However, qualitative color assessment or helps one to make assumptions; therefore, understanding and investigation of pigments the molecular level are needed for later use.

The present review provides an overview of fungal pigment production as well an analysis of the current analytical methodologies commonly used for the chemical a structural characterization of these new pigments and other additives in order to ga permission from the regulatory agencies in charge of regulating their use as additives the agro-food-pharmaceutical industries.

2. Fungi as Biopigment Producers

Fungi are one of the kingdoms of the Eukarya domain that can be found in almo any climate, especially in terrestrial ecosystems [8]. They play an important role in t nitrogen cycle because they are scavengers, decomposers, predators, pathogens, and ev parasites, and they can survive in symbiotic relationships with plants, algae, and anima among others. Some fungi genera have grown in importance over time due to their abili for industrial applications [8].

Fungi, especially filamentous fungi, have gained popularity due to their ability produce a diverse range of secondary metabolites that are important in the health, foo agricultural, and other sectors. Biopigments are one of them, and they are being stu ied because of their biodegradable nature, low production costs, wide range of colo and biological properties ranging from antioxidants to anticancer [6,8]. Most fungi pr duce water-soluble pigments that are suitable for industrial production since they a easy to scale-up in industrial fermenters and can be extracted without the use of orgar solvents [9].

Some of the most important fungal species or genera? for pigment production a found in the families, they are as follow: *Monascaceae, Trichocomaceae, Nectriaceae, Hypoc aceae, Pleosporaceae, Cordycipitaceae, Xylariaceae, Chaetomiaceae, Sordariaceae, Chlorociboriace Hyaloscyphaceae, Hymenochaetaceae, Polyporaceae, Ophiostomataceae, Tremellaceae, Neurosp raceae*, and *Tuberaceae* [7,9–11]. These metabolites are generated by *Monascus* spp. in gener through the polyketide pathway, which is directly linked to fatty acid biosynthesis. Thoug *Neurospora* spp. does so through the carotenoids' biosynthetic pathway. *Monascus* spp. commonly used as a model fungus for the assessment of its biosynthetic pathway at the lot level; but, owing to its difficulty, it has not yet been completely elucidated. On the oth hand, some species, such as *Fusarium* spp., can produce pigments through the polyketi and carotenoid biosynthetic pathways [4,7,9]. Table 1 shows several fungal species an their pigment production.

Table 1. More commonly produced pigments and their fungal species.

Pigment Produced		Fungal Species	Reference
Carotenoids	β-carotene	*Blakeslea trispora*	[10,12,13]
		Rhodosporidium spp.	
		Rhodotorula spp.	
		Penicillium spp.	
		Aspergillus giganteus	
		Sclerotium spp.	
		Sporodionolus pararoseus	
	Astaxanthin	*Phaffia rhodozyma*	[14,15]
		Haematococcus pluvalis	
		Agrobacterium aurantiacum	
	Lycopene	*Blakeslea* spp.	[9,13,15,16]
		Rhodotorula spp.	
		Mucorales spp.	
		Phycomyces spp.	
Polyketides	Anthraquinones	*Aspergillus* spp.	[9,17,18]
		Eurotium spp.	
		Emericella spp.	
		Fusarium spp.	
		Penicillium spp.	
		Mycospharella spp.	
		Microsporum spp.	
	Hydroxyanthraquinone	*Aspergillus* spp.	[9,17,19]
		Microsphaeropsis spp.	
		Geosmithia spp.	
		Trichoderma spp.	
		Verticicladiella spp.	
		Guignardia spp.	
	Naphthoquinones	*Chlorociboria* spp.	[9,20,21]
		Monascus spp.	
		Trichoderma spp.	
		Fusarium spp.	
	Azaphilones	*Monascus* spp.	[9,22]
		Penicillium spp.	
		Talaromyces spp.	
		Chaetomium spp.	

3. The Most Common Fungal Pigments and Their Properties

Obtaining pigments from natural sources is an activity that has been done for a long time, and in recent years has proven to be a great solution to avoiding the environmental impact caused by the production of synthetic colorants. Some of the most widely isolated and used natural pigments are carotenoids, anthocyanins, chlorophylls, phycobiliproteins, betalains, and also quinones. The natural origins of these pigments are diverse; howev microorganisms emerge due to their ease of cultivation and extraction, as well as th large genetic diversity [23].

Because of their high pigment production yields, fungi have managed to gain a nota position within the diverse range of microorganisms pigment producers. Which fu produce in the form of secondary metabolites under various stress conditions. Fun pigments can be categorized as carotenoids or polyketides depending on their chemi composition. The fungal polyketides are made up of tetraetides and octaetides, which fo eight C_2 units to form the polyketide chain, while carotenoids are made up of terpenoi which comprise forty carbons in their main chain [10,24].

Pigments derived from fungal metabolism not only have dyeing properties, but th also have a number of beneficial properties that enhance their effects, such as anti-oxid and antitumor activity to name a few examples [25]. *Aspergillus, Fusarium, Penicilli Trichoderma*, and *Monascus* are some of the most widespread pigment-producing fun genera. Some fungi can produce a variety of pigments depending on their growth con tions. To put it another way, the synthesis of pigments by fungi is critical to their grow since the production of these secondary metabolites is primarily a photoprotection mec nism by the microorganism. Different colorimetric ranges are obtained depending on genus and/or species, which end up being important descriptive characteristics of ea one. For example, the green color of *Penicillium*, the violet color of *Cortinarius*, the yell orange, and red color of *Monascus* are its distinguishing features. Many fungal pigments quinones or similar conjugated structures [7,13]. Quinones are produced by fungi throu the polyketide pathway. Fugigatin from *Aspergillus fumigatus* is another example, whi is a polyketide found in Nature. The latter can also synthesize two other pigments fr the same polyketide route; however, this occurs under different stress conditions; bo pigments are members of the hydroquinone family (auroglaucin and flavoglaucin) [2 Another clear example is the synthesis of pigments by *Monascus*, which has the ability synthesize six distinct polyketide pigments with colors ranging from yellow to red; wh monascin and ankaflavin have an amber color, while monascorubrin and rubropuncta have an orange color, and finally, monascorubramine and rubropuntamine is becom reddish color [9,15,16].

4. Problems of Mycotoxin Production in Pigment Producing Fungi

The production of pigments by filamentous fungi has piqued the attention of indus not only as a value-added commodity for biorefineries, but also as an alternative to sy thetic pigments due to the growing demand for natural pigments, especially by the fo industry [1].

One of the benefits of microbial pigments is that they are simple to produce. Or the strains are selected, optimal growth conditions for shorter generation period and hi yields can be achieved, as can the use of agro-industrial residues. The optimization the production and subsequent processes is strain specific [7,9]. Several experiments ha concentrated on the structure of the growth medium and difference in culture conditions increase the yield of these compounds [7,17,24], while others have focused on taxonom identification of fungi and chemical characterization of products before choosing the strains for industrial commercialization [27]. Figure 1 depicts a general short scheme the production of fungal pigments under established regulated conditions.

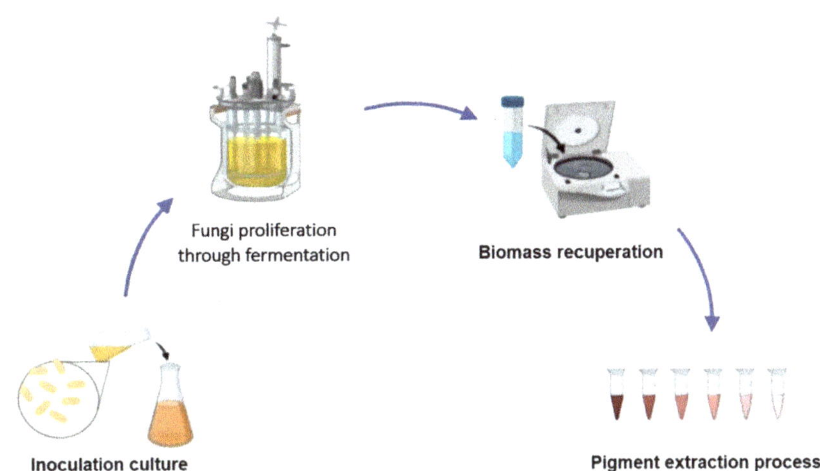

Figure 1. A general scheme for the production of fungal pigments under established controlled conditions.

The toxicity of the compounds, which occurs when the pigment is bound to a collection of interferents known as impurities, is a significant disadvantage that can impede processing or the establishment of a commercial process. Some strains not only contain useful pigments, but also a number of mycotoxins [12]. In general, safety regulatory authorities play a key role in overseeing the use of such strains based on previous cytotoxicity studies [11]. Several species of *Penicillium, Eurotium, Fusarium* are among the microorganisms that can produce toxic metabolites [28]. As a result, the possible production of mycotoxins is a significant issue that restricts the commercial use of these fungal strains. This problem, along with the increasing demand for natural colorant alternatives from both consumers and regulators, has prompted research and analysis of other genera of potential pigment-producing fungi. Thus, in order to expand the applications of pigments produced by fungi, researchers have focused on edible fungi that can naturally synthesize and secrete pigments with a lower propensity for toxic compound synthesis.

5. Downstream Processing of Fungal Biopigments

The next step is the isolation and recovery of produced pigments after the pigments are produced by various microorganisms. Through time, an infinite number of chemical solvents have been used to remove pigments and other forms of bioactive compounds. However, today's consumer expects a product free of chemical residues due to the health risks and environmental pollution that its toxicity can cause. Traditional extraction techniques necessitate the use of high temperatures and long exposure periods, which affects the stability of molecules [13,14]. Different methods for the recovery of metabolites of concern have been explored as a result of these new toxicity-free criteria and some international regulations. The use of a green solvent means shorter operation periods and less waste production. Carotenoids are among the pigments most widely produced by fungi. Table 2 shows some new and more environmentally sustainable carotenoid extraction processes.

Although the extraction process is not the focal point of the research in fungal pigments, it has earned the title of critical point and if it is not carried out properly and safely, it would be very difficult to proceed with the study at the same time. Molecular elucidation step, since the greatest possible precision in sample preparation is needed in this phase to obtain a stable and effective molecular elucidation.

Table 2. Some recent methods for the extraction of carotenoids.

Microorganism	Extraction Method	Carotenoid Produced	Reference
Rhodotorula toruloides NRRL Y-1091	Saponification with alcoholic KOH and hexane	Carotenoids	[29]
R. glutinis	Supercritical CO_2	β-carotene and torularhodin esters	[30]
R. glutinis ATCC 2527	Manosonication	Total carotenoids	[31]
R. glutinis ATCC 2527	Pulsed electric field	Total carotenoids	[32]
R. glutinis P4M422	Bead mill	Lycopene	[33]
Sporobolomyces ruberriums H110	French pressure cell	Torularhodin, torulene, β-carotene, γ-carotene	[34]
Blakeslea trispora	Steam-explosion-assisted	Total carotenoids	[35]
Phaffia rhodozyma NRRL Y-17268	Diatomaceous earth and enzymatic lysis	Total carotenoids	[36]

6. Molecular Elucidation of Fungal Pigments

Importance of Molecular Elucidation

Information of a compound's chemical and molecular structure is critical because describes the conformation of a chemical structure and allows for the establishment of t forms of relations that regulate the structures. Furthermore, it enables one to comprehen how certain atoms and bonds combine to form various functional groups. All of th knowledge enables one to understand and model the physical and chemical properties chemical structures. This knowledge, if obtained in a timely manner, would allow for th description of the pigments' stability behavior as well as the quest for better matrices contain them, thus extending their useful life.

One of the shortcomings of the chemical characterization of pigments of microbi origins is the lack of industry reference criteria that enable adequate characterizatio of the compounds formed by various microorganisms. This is a field of possibility f businesses and/or laboratories interested in producing microbial pigments, as well as critical challenge that must be completed before reaching a commercial level.

On the other hand, there are various analytical instruments available today to d termine the chemical structure of compounds. These methods allow us to determine th molecular weights of molecules, the types of bonds that engage in molecular conformatio the polarity of the structures, and their functional groups etc. The selection of the approa and/or set of analytical methodologies is an essential step in the proper description the structures. The most important analytical methodologies for elucidating the chemic structures of microbial pigments are described below.

7. Methodologies for Elucidation of Chemical Structure of Pigments

7.1. Spectrophotometric Methods

According to the Royal Spanish Academy, spectrophotometry is a scientific metho widely used to determine how much light a chemical compound absorbs and is base on the Beer-Lambert law. The Beer-lambert law is a combination of the other four law (Bouguer, Bunsen, Roscoe, and Beer) that enabled it to be enunciated. According Hardesty and Attili [37], "*the intensity of a monochromatic light beam incident perpendicular a sample decreases exponentially with sample concentration*". This law states that:

$$A = K * C \quad ($$

where, A = Sample absorbance; K= Constant of wavelength, which is fixed according the nature of the substance analyzed and the material of the cell used; and C = Samp concentration.

Then, if we look at this equation closely, we will see a similarity with the equation the line, which, since it lacks an interjection point with the coordinates (n), we can inf that it would pass through the origin of the coordinates (standard); where "K" is the slo of the line [38].

There are several modifications and adaptations to this process, and we will discuss some of the more widely used spectrophotometric methods for the analysis of fungal pigments below.

7.1.1. Ultraviolet Visible (UV-VIS)

The technique of ultraviolet-visible absorption spectrophotometry (UV-VIS) is based on the attenuation of electromagnetic radiation measurement by an absorbing substance [39]. This radiation has a spectral range of approximately 190-800 nm and varies from other similar regions in terms of energy levels and type of excitation [26,40].The wavelengths that can be detected using this method are classified as UVC, UVB, UVA, and visible; where the UVC wavelength is called short since it covers a range from 100 to 280 nm, UVB is known as the medium wavelength that covers 280 to 315 nm, UVA is known as the very long wavelength that covers 315 to 400 nm, and finally, the visible spectrum can be said to cover from 315 to 400 nm. Furthermore, of course should not forgetting to mention infrared light [38]. Reflection, absorption, and even interference both contribute to this attenuation. However, accurate calculations of this can be made using absorbance databases. In certain ways, the absorbance can be said to be equal to the concentration of the analyte to be measured and to the wavelength of the light as it passes through the sample during irradiation, as regulated by Beer's Law [25,40]. Since this is a linear relationship, it can be affected by a variety of factors, including the spectrophotometer's characteristics, photodegradation of molecules, the presence of dispersion or absorption interferences in the sample, fluorescent compounds in the sample, reactions between the analyte and the solvent, and the pH [25,26]. The short wavelength limit is caused by atmospheric gas absorption at ultraviolet wavelengths shorter than 180 nm. When a spectrometer is purged with nitrogen gas, this limit is raised to 175 nm. Working above 175 nm necessitates the use of a vacuum spectrometer and an ultraviolet light source [39].

UV-Vis is one of the most ubiquitous characterization and analytical techniques in science. Its use in materials research could be classified into two categories: (1) quantitative measurements of an analyte in the gas, liquid, or solid phase and (2) characterization of the optical and electronic properties of a material [39]. The first application for quantitative measurements derives from the linear relationship between absorbance and absorbent concentration. Which turns out to be easy and fast since the main requirement is only measuring absorbance or reflectance in a single wavelength. Since most molecular and solid materials have broad absorption characteristics, quantitative measurements are less sensitive to instrumental variables than other analytical methods [26,41]. The identification of an analyte, on the other hand, can be accomplished by matching the absorption spectrum of the unknown substance with graphs or tables of the spectra of known substances. Two or three analytes can be recognized in some cases [42].

Because of the above, this type of spectrophotometry can be used to characterize the absorption, transmission, and reflectance of a wide range of technologically significant materials, such as pigments. The substance in this case is a transparent host with intentional dopants or unintentional impurities that regulate optical absorption [26,40]. Molnàr et al. [43] described the chromophores present in the carotenoids isolated from *Sarcoscypha coccinea* and obtained a characteristic confirmation of such agents including 30, 40-didehydro-10,20-dihydro-β, ψ-carotene, 30,40-didehydro-10, 20-dihydro-β, ψ-carotene-20-one [43].

7.1.2. Infrared (IR)

Infrared spectrophotometry is the most widely used technique for pilot-level applications. This is done with the aid of an infrared spectrophotometer, which can distinguish molecular structures by producing wavelengths with a spectral range of 500 to 4000 nm. That is, the transmittance response for a given wavelength shows the difference in the links caused by photodegradation [38].

This method works by exposing a sample to ultraviolet radiation, which cau changes in the vibrational states of the sample's constituent molecules. Radiation absc tion by a sample indicates the type of bonds and functional groups present [25,44]. I useful to divide the infrared region into three regions called near infrared (NIR), m dle infrared (MIR), and far infrared (FIR) from the standpoint of instrumentation a applications [44,45]. The vast majority of traditional analytical applications of infrared sp troscopy are focused on the use of the middle infrared (4000-600 cm^{-1}) and near infrar allowing this method to be converted into a quantitative technique. The Fourier transfo technique, which converts a time domain spectrum to a frequency domain spectrum usi a mathematical operation, allows for the generation of spectra that are quick, accurate, a have elevated signal/noise (S/N) relationships [46].

There are several measurement techniques to obtain this type of spectra; howev some of the most common are described below:

Transmission: IR radiation is passed into the sample in this measurement proce recording the amount of energy absorbed by the sample. The IR spectrum is obtain using a reference experiment by comparing the radiation recorded after going throu the sample. With the proper accessories, this method analyzes gaseous, liquid, and so samples [47].

Reflection: Infrared radiation is reflected on the sample. The sample's molecu information is extracted by analyzing the reflected radiation and comparing it to incident radiation. To use this measuring tool, the sample must be reflective or moun on a reflective surface [44,45].

ATR mode: It is a sampling mode in which an infrared beam is projected onto a crys with a high refractive index. The beam reflects off the inside of the glass, generating evanescent wave that enters the sample. This must be in near proximity to the crys A portion of the evanescent wave's energy is absorbed, and the reflected radiation (conta ing chemical information about the sample) is directed to the detector. It is a very versa method for measuring liquid and solid samples without manually processing them [47

In order to gain as much detail as possible on the pigment in question, the methc previously described are usually used as a supplement. As in the case of Quijano-Orte et al. [48], both FTIR and ATR were used jointly to determine carotenoids present *Cucurbita* spp., where structural conformations typical of carotenoids could be determin for example the double bonding between carbons, deformation of CH_3 groups and ev the stable existence of CH_2 chains [48].

7.1.3. Combined Diffusion (Raman)

Raman spectroscopy is a high-resolution photonic technique that offers chemical a structural knowledge on virtually every organic or inorganic material or compound i couple of seconds, enabling it to be identified. As a result, the analysis of this method dependent on the evaluation of the light scattered by a material when a monochroma beam of light falls on it [49]. This is because a particular portion of light is inelastica dispersed, experiencing minor frequency shifts that are typical of the substance studi and regardless of the frequency of the incident light. It is an inspection procedure th is conducted directly on the sample to be examined without the need for any additior planning and without having any impact on the analyte surface [30,50].

Due to the simplicity of execution of this technique, it is used in many fields application, if not all, in response to its foundation on molecular vibrations, which ta place in any body. Among its multiple fields of application, pigments are positioned in important place, as organic compounds to determine the conformational macrocompone of these agents [51]. Nokkaew et al. [52] used the Raman technique to structurally ident the carotenoids present in crude palm oil through the identification of carbon-carb double bonds (C = C) and their positions in the terpenoid chain [52].

7.1.4. Mass Spectrometry (MS)

Mass spectrometry is a microanalytical method used to identify unknown substances, quantify known compounds, and elucidate molecule structure and chemical properties. It uses small volumes of sample to collect details such as the weight and, in some cases, the structure of the analyte [53]. As a result, the sample is ionized (and thereby destroyed) using different protocols. The electronic impact method is one of the most widely employed, as it involves bombarding the sample (which has traditionally been vaporized using a high vacuum and a heat source) with a high-speed current of electrons, and this is how the substance loses several electrons and fragments, producing various ions, radicals, and neutral molecules [31,32]. The ions (charged molecules or fragments) are then operated by an ion accelerator to a curved analyzer tube with a high magnetic field and then to a collector/analyzer where the impacts of those ions are collected as a function of their m/z ratio. Each compound is unique, because each compound can ionize and fragment in a different manner, and mass spectrometry uses this idea to classify each analyte [54,55].

We can use mass spectrometry to determine the chemical composition of samples, the composition of inorganic, organic, and biological compounds, the qualitative and quantitative composition of complex mixtures, the structure and composition of solid surfaces, and even about the isotopic ratios of atoms in the samples [32,34]. Chen et al. [56] employed LC-MS method to confirm the presence of flavonoids on three different types of pigmented cotton fibers.

Among the analytical techniques often used in mass spectrometry, chromatographic methods such as gas chromatography and liquid chromatography coupled to mass spectrometers and isotopic ratio mass spectrometry for the study of stable isotopes deserve special mention (C, N, H, O and S) [31,32,34].

The above is used to determine the molecular mass of peptides and proteins, the biomolecular elucidation of structures derived from natural agents such as proteins and also pigments, and the recognition of pairs of proteins that interact using affinity purified mass spectrometry. Proteins are an essential component of biological systems and studying proteins in depth is critical to understanding life and its processes [53].

7.2. Magnetic Spectrometry Methods

7.2.1. Nuclear Magnetic Resonance (NMR)

Nuclear Magnetic Resonance (NMR) is one of the most often used magnetic techniques for studying pigments of fungal origin. Whose basis is founded on the assumption that all nuclei with an odd number of protons and neutrons have a magnetic moment and inherent angular momentum or have a spin greater than zero [57]. As a result, this approach is primarily used to extract physical, chemical, electronic, and structural details on molecules.

By applying NMR on fungal pigments, it is possible to know different conformational aspects, such as the following [35,37]:

- To detect the number of protons, present in the pigment molecule (signal number).
- To identify the type of bond that these protons form, whether they are alkanes, alkenes, hydroxyl, aromatic, among others (signal position).
- Allows setting the number of protons generated in each peak/signal (signal intensity)
- To perform signal splitting, either in doublets, triplets and/or multiplets, which will allow in knowing the number of protons attached to the carbon that is next to the transporting carbon of the previously broken signal proton.

The use of this technique has helped over the years to explain some of the many structural details that pigments provide, especially those produced by fungi and some plants. NMR is called a low invasive procedure, which means that the effect it has on the molecular structure of pigments is comparatively low, allowing it to rule out the possibility that they are affected by energetic influences [58,59]. It is widely used in the study of the stability and reactivity of conformational isomers, which would be almost impossible to conduct without its inclusion. The proton ($_1$H), $_{13}$C, and $_{15}$N are among the most common nuclei for its realization [35,38,60].

7.2.2. Electron Spin Resonance (ESR)

Electron spin resonance (ESR) is routinely characterized as the amount of radiati absorbed by a static magnetic field; it is often referred to as "paramagnetic resonance" "paramagnetic electron resonance" [61,62]. Radicals are a special class of paramagne molecule. ESR spectroscopy can be used on these. As a result of the ESR techniqu existence, it is possible to apply it to the analysis of free radicals found in pigmen whether or not they are of fungal origin. Free radicals are well-known molecular speci that carry an unpaired electron in the paramagnetic group of the atomic orbital region molecular species. As a result of the above, these are unstable, reactive, and can also cau redox reactions with other molecules [42,63,64].

Because of its ability to detect free radical absorption, it has been used over tir to assess oxidative stability, copper chelating capability, and even radiation damage biological bodies when exposed to new technologies such as high pressures, electric pulses, ultrasound, and so on [53,65,66]. According to Gonçalves et al., the applicati of ESR in the melanins formed by *Aspergillus nidulans* is stated, with G values (2,0(identifying that the stability and vibrations produced by said pigments are linked to t presence of C=C and C=O groups [67]. In Figure 2, we can see a brief diagram of how t methods discussed in this section are implemented.

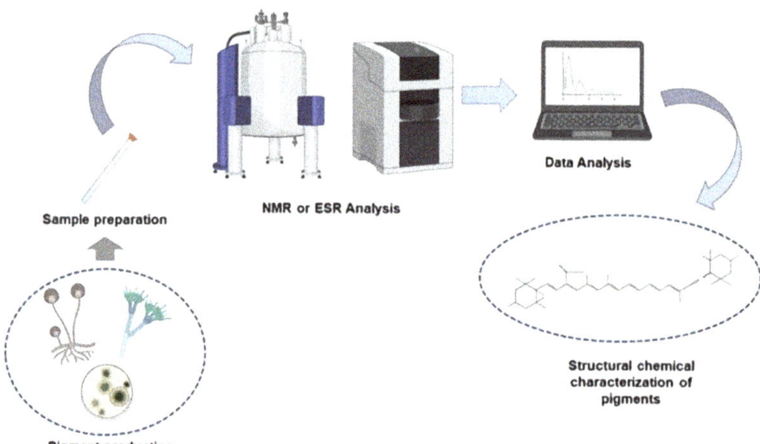

Figure 2. A general diagram of the fungal pigment analysis protocol using NMR and ESR.

7.3. Chromatographic Methods

7.3.1. High Performance Liquid (HPLC)

The chromatography principle essentially governs the separation of the componen of a mixture; this from a mobile phase fluid (gas, liquid, or supercritical fluid), which dra the sample at constant flow and pressure due to the use of a pump. This pressure flow lea to a column containing the stationary phase, which is either a solid or a liquid bound to solid [68]. As a result of the interaction between the components of the mixture and the tv phases, they allow separation; hence, at the end, it passes through a detector that produc a signal that allows both indicating the moment of appearance of the various componen that make up the sample and qualifying it quantitatively and qualitatively [69].

HPLC differs from other forms of chromatography in that it is not limited by t sample's volatility or thermal stability. The main explanation for this is that it is broad applicable for the study of macromolecules, such as fungal pigments in this case [43,7 Several authors have reported the use of this method in the molecular characterizatio of fungal pigments at various stages. Beginning with Lebeau et al. [21], who used U

VIS and HPLC-DAD (inverse) to identify and elucidate the conformation of two new pigments from *Fusarium oxysporum*, which they called wild-type purple naphthoquinone and bikaverin, respectively [21]. Venkatachalam et al. [22], reported the application of HPLC in some of its variants (PDA-ESI/MS) for the analysis of the pigments produced by *Talaromyces albobiverticillius* 30548 and reported twelve compounds of *Monascus* spp. Further, they characterized and elucidated a new compound called NGABA-PP-V (6-[(Z)-2-Carboxyvinyl] -N-GABA-PP-V) with a *cis* configuration at the C10-C11 carbon double bond [22]. Finally, Gonçalves et al. [71] contributed by identifying and quantifying the phenolic compounds gallic acid, catechin, chlorogenic acid, caffeic acid, and vanillin found in the ethyl extract of *Penicillium flavigenum* CML2965 using HPLC [71]. Since it is a non-limiting method, it is obvious that there are millions of works published in its use for the molecular elucidation of fungal pigments, which is why the examples above are just a few of the most recent.

7.3.2. Thin Layer (TLC)

Thin layer chromatography (TLC) is one of the oldest and most commonly used methods in the molecular analysis of biopigments, including those derived from fungi. This procedure involves chromatography in a stationary phase in the form of a thin layer, usually on aluminum, glass, and also plastic surfaces, where a standard solution is added and expressed as a narrow band on the thin layer of adsorbent as an indicator that it has been distributed evenly on the support/surface. This technique consists of chromatography carried out in a stationary phase in the form of a thin layer, commonly on aluminum, glass, and even plastic surfaces and with solvents as mobile phase. The separated molecules are detected after the evaporation of the solvent and by employing physical methods or chemical staining reagents [46,48,72,73].

Thin layer chromatography allows for the manipulation of extracts of different types, whether crude or pure; additionally, the equipment used is simple, inexpensive, and effective. And also, this method can be used quantitatively, where the components are isolated on the TLC tray, to be subsequently removed with an appropriate solvent, such as ethanol or acetone, and then lead to the application of a procedure using a chromogen, and thus be measured by spectrophotometry at a determined wavelength [74,75]. There are other ways for TLC quantification and detection, such as densitometry under the appropriate conditions, which is commonly used for the study of pigments, mostly anthocyanins and vitamin A precursors [76]. Another simple example is its application to the complete isolation of pigments, most notably those derived from organisms such as *Monascus*, which do not contain only one kind of pigment [77]. Finally, TLC has been used in conjunction with other approaches to increase the belief in carotenoid identifications. For example, Wang et al. [13] used both column chromatography and TLC to investigate the pigments of aeciospores of *Cronartium fusiforme* fungi [13].

8. Future Trends and Challenges

Certainly, the market for chemical-free pigments has increased significantly in industry and science. As a result, the extraction from biological origin has gained scientific interest, resulting in alternative processing and extraction from different strains of fungi. Since these have a number of significant advantages, such as low processing costs, high yields, and strong adaptability to shifting substrates, the exploitation of agro-industrial waste has emerged. This is how fungi are able to be considered sustainable industries for the production of pigments, where substantial progress has been made in the optimization of these processes by study and the implementation of different experiments. It is well known that different species of fungi collected from natural sources have a diverse chromatographic spectrum of pigments that are closely linked to a number of biological activities of importance, such as the antioxidant activity present in some pigments. Despite the fact that the area of optimizing processes that produce fungal pigments is progressing at a rapid pace. Another collection of investigations and projects is devoted solely to the structural, magnetic, and interactional information of them, i.e., their classification at the molecu level of fungal pigments, in which a variety of techniques are used, either individually collectively. To introduce these to a large scale, they must be not only commercially feasil but also safe, harmless, and controlled.

As a result, society is becoming more mindful of the environmental and health ha that chemically synthesized pigments can cause. The manufacture of fungal pigments taken a big step forward. However, it is already safe to state that said development is progress but, due to production conditions, it has not been possible to completely co existing consumer demand. Thus, achieving total availability by other means is one of industry's biggest obstacles. For example, the discovery and investigation of other gen and/or species of fungi that have been underutilized due to ignorance of the existence their secondary metabolites, which can be elucidated by either of the previously descrit techniques thus ensuring optimal process optimization.

Author Contributions: M.S.V.-G.: writing-review and editing. N.B. and M.L.C.-G.: resour writing-review. O.A.: resources, visualization. A.H.-A. and C.N.A.: conceptualization, supe sion, and project administration. All authors have read and agreed to the published versior the manuscript.

Funding: This research received no external funding.

Institutional Review Board Statement: Not applicable.

Informed Consent Statement: Informed consent was obtained from all subjects involved in the stu

Acknowledgments: Miriam S. Valenzuela-Gloria wants to thank the National Council of Scie and Technology (CONACYT-Mexico) for the financial support during her master's program Biochemical Engineering at the Autonomous University of Coahuila. The authors express th gratitude to Deepak Kumar Verma of the Agricultural and Food Engineering Department, Ind Institute of Technology Kharagpur, India, for his technical assistance, scientific correction, a language revision for the final versions of the manuscript.

Conflicts of Interest: The authors declare no conflict of interest.

References

1. Mapari, S.A.; Nielsen, K.F.; Larsen, O.T.; Frisvad, J.C.; Meyer, A.S.; Thrane, U. Exploring fungal biodiversity for the productior water-soluble pigments as potential natural food colorants. *Curr. Opin. Biotechnol.* **2005**, *16*, 231–238. [CrossRef] [PubMed]
2. Kumar, A.H.; Shankar Vishwakarma, J.; Singh, S.D.; Kumar, M. Microbial pigments: Production and their appli-cations in vari industries. *Int. J. Pharm. Chem. Biol. Sci.* **2015**, *5*, 203–212.
3. Abdulkadir, N.; Usman, N.A.H.M.; Gani, H.M.M.M. Bacterial Pigments and its Significance. *MOJ Bioequiv. Bioavailab.* **2017**, *4*, 1 [CrossRef]
4. Souza, P.N.D.C.; Grigoletto, T.L.B.; De Moraes, L.A.B.; Abreu, L.M.; Guimarães, L.H.S.; Santos, C.; Galvão, L.R.; Cardoso, P Production and chemical characterization of pigments in filamentous fungi. *Microbiology* **2016**, *162*, 12–22. [CrossRef]
5. Kong, Q.-X.; Li, L.; Martinez, B.; Chen, P.; Ruan, R. Culture of Microalgae Chlamydomonas reinhardtii in Wastewater for Biom Feedstock Production. *Appl. Biochem. Biotechnol.* **2009**, *160*, 9. [CrossRef] [PubMed]
6. Vicente, G.; Bautista, L.F.; Rodríguez, R.; Gutiérrez, F.J.; Sádaba, I.; Ruiz-Vázquez, R.M.; Torres-Martínez, S.; Garre, V. Biodie production from biomass of an oleaginous fungus. *Biochem. Eng. J.* **2009**, *48*, 22–27. [CrossRef]
7. Zhang, C.; Liang, J.; Yang, L.; Chai, S.; Zhang, C.; Sun, B.; Wang, C. Glutamic acid promotes monacolin K production a monacolin K biosynthetic gene cluster expression in Monascus. *AMB Express* **2017**, *7*, 22. [CrossRef]
8. Naranjo-Ortiz, M.A.; Gabaldón, T. Fungal evolution: Major ecological adaptations and evolutionary transitions. *Biol. Rev.* **20** *94*, 1443–1476. [CrossRef]
9. Venil, C.K.; Velmurugan, P.; Dufossé, L.; Devi, P.R.; Ravi, A.V. Fungal Pigments: Potential Coloring Compounds for Wide Rang Applications in Textile Dyeing. *J. Fungi* **2020**, *6*, 68. [CrossRef]
10. Arikan, E.B.; Canli, O.; Caro, Y.; Dufossé, L.; Dizge, N. Production of bio-based pigments from food processing indus by-products (apple, pomegranate, black carrot, red beet pulps) using *Aspergillus carbonarius*. *J. Fungi.* **2020**, *6*, 240. [CrossRef]
11. Fox, E.M.; Howlett, B.J. Secondary metabolism: Regulation and role in fungal biology. *Curr. Opin. Microbiol.* **2008**, *11*, 481–4 [CrossRef]
12. Gmoser, R.; Ferreira, J.A.; Lennartsson, P.R.; Taherzadeh, M.J. Filamentous ascomycetes fungi as a source of natural pigme *Fungal Biol. Biotechnol.* **2017**, *4*, 1–25. [CrossRef] [PubMed]

Wang, E.; Dong, C.; Park, R.F.; Roberts, T.H. Carotenoid pigments in rust fungi: Extraction, separation, quantification and characterisation. *Fungal Biol. Rev.* **2018**, *32*, 166–180. [CrossRef]
Gong, M.; Bassi, A. Carotenoids from microalgae: A review of recent developments. *Biotechnol. Adv.* **2016**, *34*, 1396–1412. [CrossRef]
Avalos, J.; Limón, M.C. Biological roles of fungal carotenoids. *Curr. Genet.* **2015**, *61*, 309–324. [CrossRef]
Feofilova, E.P.; Tereshina, V.M.; Memorskaya, A.S.; Dul'Kin, L.M.; Goncharov, N.G. Fungal lycopene: The biotechnology of its production and prospects for its application in medicine. *Microbiology* **2006**, *75*, 629–633. [CrossRef]
Lagashetti, A.C.; Dufossé, L.; Singh, S.K.; Singh, P.N. Fungal Pigments and Their Prospects in Different Industries. *Microorganisms* **2019**, *7*, 604. [CrossRef]
Velíšek, J.; Cejpek, K. Pigments of higher fungi—A review. *Czech J. Food Sci.* **2011**, *29*, 87–102. [CrossRef]
Adeboye, P.T.; Bettiga, M.; Olsson, L. The chemical nature of phenolic compounds determines their toxicity and induces distinct physiological responses in Saccharomyces cerevisiae in lignocellulose hydrolysates. *AMB Express* **2014**, *4*, 46. [CrossRef]
Mukherjee, G.; Mishra, T.; Deshmukh, S.K. Fungal Pigments: An Overview. In *Developments in Fungal Biology and Applied Mycology*; Springer Science and Business Media LLC: Berlin/Heidelberg, Germany, 2017; pp. 525–541.
Lebeau, J.; Petit, T.; Clerc, P.; Dufossé, L.; Caro, Y. Isolation of two novel purple naphthoquinone pigments concomitant with the bioactive red bikaverin and derivates thereof produced by Fusarium oxysporum. *Biotechnol. Prog.* **2019**, *35*, e2738. [CrossRef]
Venkatachalam, M.; Zelena, M.; Cacciola, F.; Ceslova, L.; Girard-Valenciennes, E.; Clerc, P.; Dugo, P.; Mondello, L.; Fouillaud, M.; Rotondo, A.; et al. Partial characterization of the pigments produced by the marine-derived fungus Talaromyces albobiverticillius 30548. Towards a new fungal red colorant for the food industry. *J. Food Compos. Anal.* **2018**, *67*, 38–47. [CrossRef]
Zhou, Z.-Y.; Liu, J.-K. Pigments of fungi (macromycetes). *Nat. Prod. Rep.* **2010**, *27*, 1531–1570. [CrossRef] [PubMed]
Gmoser, R.; Ferreira, J.A.; Taherzadeh, M.J.; Lennartsson, P.R. Post-treatment of fungal biomass to enhance pigment production. *Appl. Biochem. Biotechnol.* **2019**, *189*, 160–174. [CrossRef] [PubMed]
Goodwin, T.W. Fungal carotenoids. *Bot. Rev.* **1952**, *18*, 291–316. [CrossRef]
Latge, J.-P.; Chamilos, G. Aspergillus fumigatus and Aspergillosis in 2019. *Clin. Microbiol. Rev.* **2020**, *33*, 1–75. [CrossRef]
Mapari, S.A.; Meyer, A.S.; Thrane, U.; Frisvad, J.C. Identification of potentially safe promising fungal cell factories for the production of polyketide natural food colorants using chemotaxonomic rationale. *Microb. Cell Factories* **2009**, *8*, 24. [CrossRef]
Heo, Y.M.; Kim, K.; Kwon, S.L.; Na, J.; Lee, H.; Jang, S.; Kim, C.H.; Jung, J.; Kim, J.-J. Investigation of Filamentous Fungi Producing Safe, Functional Water-Soluble Pigments. *Mycobiology* **2018**, *46*, 269–277. [CrossRef]
Liu, Z.; Berg, C.V.D.; Weusthuis, R.A.; Dragone, G.; Mussatto, S.I. Strategies for an improved extraction and separation of lipids and carotenoids from oleaginous yeast. *Sep. Purif. Technol.* **2020**, *257*, 117946. [CrossRef]
Martínez, J.; Schottroff, F.; Haas, K.; Fauster, T.; Sajfrtová, M.; Álvarez, I.; Raso, J.; Jaeger, H. Evaluation of pulsed electric fields technology for the improvement of subsequent carotenoid extraction from dried Rhodotorula glutinis yeast. *Food Chem.* **2020**, *323*, 126824. [CrossRef]
Martínez, J.M.; Delso, C.; Aguilar, D.E.; Álvarez, I.; Raso, J. Organic-solvent-free extraction of carotenoids from yeast Rhodotorula glutinis by application of ultrasound under pressure. *Ultrason. Sonochem.* **2020**, *61*, 104833. [CrossRef]
Martínez, J.M.; Delso, C.; Angulo, J.; Álvarez, I.; Raso, J. Pulsed electric field-assisted extraction of carotenoids from fresh biomass of Rhodotorula glutinis. *Innov. Food Sci. Emerg. Technol.* **2018**, *47*, 421–427. [CrossRef]
Hernández-Almanza, A.; Navarro-Macías, V.; Aguilar, O.; Aguilar-González, M.; Aguilar, C.N. Carotenoids extraction from Rhodotorula glutinis cells using various techniques: A comparative study. *Indian J. Exp. Biol.* **2017**, *55*, 479–484.
Cardoso, L.; Jäckel, S.; Karp, S.; Framboisier, X.; Chevalot, I.; Marc, I. Improvement of Sporobolomyces ruberrimus carotenoids production by the use of raw glycerol. *Bioresour. Technol.* **2016**, *200*, 374–379. [CrossRef] [PubMed]
Wang, H.-B.; Zhang, L.-W.; Luo, J.; Yu, L.-J. Erratum to: Rapid and environmentally-friendly extraction of carotenoids from Blakeslea trispora. *Biotechnol. Lett.* **2015**, *37*, 2179–2180. [CrossRef] [PubMed]
Michelon, M.; Borba, T.D.M.D.; Rafael, R.D.S.; Burkert, C.A.V.; Burkert, J.F.D.M. Extraction of carotenoids from Phaffia rhodozyma: A comparison between different techniques of cell disruption. *Food Sci. Biotechnol.* **2012**, *21*, 1–8. [CrossRef]
Hardesty, J.H.; Attili, B. Spectrophotometry and the Beer-Lambert Law: An important analytical technique in chemistry. *Collin Coll. Dep. Chem.* **2010**, *9*, 76–99.
Filho, J.L.E.D.; Maia, P.C.D.A.; Xavier, G.D.C. Spectrophotometry as a tool for characterizing durability of woven geotextiles. *Geotext. Geomembr.* **2019**, *47*, 577–585. [CrossRef]
Lipson, R.H. Ultraviolet and Visible Absorption Spectrometers. *Encycl. Appl. Spectrosc.* **2009**, *1*, 353–380. [CrossRef]
Feng, Y.; Shao, Y.; Chen, F. Monascus pigments. *Appl. Microbiol. Biotechnol.* **2012**, *96*, 1421–1440. [CrossRef]
Patakova, P. Monascus secondary metabolites: Production and biological activity. *J. Ind. Microbiol. Biotechnol.* **2013**, *40*, 169–181. [CrossRef]
Laqua, K.; Melhuish, W.H.; Zander, M. Molecular Absorption Spectroscopy, Ultraviolet and Visible (Uv/Vis). *Pure Appl. Chem.* **1988**, *60*, 1449–1460. [CrossRef]
Molnár, P.; Ősz, E.; Turcsi, E.; Deli, J. Carotenoid composition of the mushroom Scarlet elf cup (Sarcoscypha coccinea). *Heliyon* **2019**, *5*, e01883. [CrossRef]

44. Moussa, S.A.-K.; Abdou, D.A.; Mohamed, G.A.; Abo-El-Seoud, M.A.; Eldin, A.-Z.K.; El-Mehalawy, A.A. Production of r pigment by Monascus purpureus NRRL1992 under submerged and solid-state fermentation. *Egypt. J. Microbiol.* **2018**, *53*, 83– [CrossRef]
45. Gunasekaran, S.M.; Poorniammal, R.M. Optimization of fermentation conditions for red pigment production from Penicillium under submerged cultivation. *Afr. J. Biotechnol.* **2008**, *7*, 1894–1898. [CrossRef]
46. Whetsel, K.B. Near-Infrared Spectrophotometry. *Appl. Spectrosc. Rev.* **1968**, *2*, 1–67. [CrossRef]
47. Jennings, G.; Bluck, L.; Wright, A.; Elia, M. The Use of Infrared Spectrophotometry for Measuring Body Water Spaces. *Clin. Ch* **1999**, *45*, 1077–1081. [CrossRef] [PubMed]
48. Quijano-Ortega, N.; Fuenmayor, C.A.; Zuluaga-Dominguez, C.; Diaz-Moreno, C.; Ortiz-Grisales, S.; García-Mahecha, M.; Gra S. FTIR-ATR Spectroscopy Combined with Multivariate Regression Modeling as a Preliminary Approach for Carotenoi Determination in Cucurbita spp. *Appl. Sci.* **2020**, *10*, 3722. [CrossRef]
49. Gerrard, D.L.; Bowley, H.J. Instrumentation for Raman Spectroscopy. In *Practical Raman Spectroscopy*; Springer Science a Business Media LLC: Berlin/Heidelberg, Germany, 1989; pp. 55–76.
50. Adadi, P.; Barakova, N.V.; Krivoshapkina, E.F. Selected Methods of Extracting Carotenoids, Characterization, and Hea Concerns: A Review. *J. Agric. Food Chem.* **2018**, *66*, 5925–5947. [CrossRef] [PubMed]
51. Movasaghi, Z.; Rehman, S.; Rehman, I.U. Raman Spectroscopy of Biological Tissues. *Appl. Spectrosc. Rev.* **2007**, *42*, 493–5 [CrossRef]
52. Nokkaew, R. Determination of carotenoids and dobi content in crude palm oil by spectroscopy techniques: Comparison of ram and ft-nir spectroscopy. *Int. J. Geomate* **2019**, *16*, 92–98. [CrossRef]
53. Rajawat, J.; Jhingan, G. Mass spectroscopy. In *Data Processing Handbook for Complex Biological Data Sources*; Elsevier BV: Amsterda The Netherlands, 2019; pp. 1–20.
54. Dole, M.; Cox, H.L.; Gieniec, J. Electrospray Mass Spectroscopy. In *Polymer Molecular Weight Methods*; Ezrin, M., Ed.; Advances Chemistry Series; American Chemical Society: Washington, DC, USA, 1973; pp. 73–84.
55. Fan-Chiang, H.-J.; Wrolstad, R.E. Anthocyanin Pigment Composition of Blackberries. *J. Food Sci.* **2006**, *70*, C198–C202. [CrossR
56. Chen, M.; Zhang, T.T.; He, L.; Wang, K.; Chen, Y. Qualitative analysis of cotton fiber pigment composition. *Text. Res. J.* **2021**, 456–463. [CrossRef]
57. Steyn, P.S.; Wessels, P.L.; Marasas, W.F. Pigments from fusarium moniliforme sheldon. *Tetrahedron* **1979**, *35*, 1551–1555. [CrossR
58. Rangasami, R.; Chidhara, S.; Chandrasekharan, A. Magnetic resonance imaging and magnetic resonance spectroscopy Salmonella meningoencephalitis. *J. Pediatr. Neurosci.* **2016**, *11*, 88–90. [CrossRef] [PubMed]
59. Shi, Q.; Wang, H.; Du, C.; Zhang, W.; Qian, H. Tentative Identification of Torulene *Cis/trans* Geometrical Isomers Isolat from *Sporidiobolus pararoseus* by High-Performance Liquid Chromatography—Diode Array Detection—Mass Spectrometry a Preparation by Column Chromatography. *Anal. Sci.* **2013**, *29*, 997–1002. [CrossRef]
60. Liu, C.; Cheng, Y.; Du, C.; Lv, T.; Guo, Y.; Han, M.; Pi, F.; Zhang, W.; Qian, H. Study on the wall-breaking method of carotenoi producing yeast *Sporidiobolus pararoseusand* the antioxidant effect of four carotenoids on SK-HEP-1 cells. *Prep. Biochem. Biotech* **2019**, *49*, 767–774. [CrossRef] [PubMed]
61. Kopáni, M.; Celec, P.; Danišovič, L.; Michalka, P.; Biró, C. Oxidative stress and electron spin resonance. *Clin. Chim. Acta* **2006**, 3 61–66. [CrossRef]
62. Barba, F.J.; Roohinejad, S.; Ishikawa, K.; Leong, S.Y.; Bekhit, A.A.E.-D.; Saraiva, J.A.; Lebovka, N. Electron spin resonance as a tc to monitor the influence of novel processing technologies on food properties. *Trends Food Sci. Technol.* **2020**, *100*, 77–87. [CrossR
63. Passos, M.L.; Saraiva, M.L.M. Detection in UV-visible spectrophotometry: Detectors, detection systems, and detection strategi *Measurement* **2019**, *135*, 896–904. [CrossRef]
64. Heikkilä, T.T.; Silaev, M.; Virtanen, P.; Bergeret, F.S. Thermal, electric and spin transport in superconductor/ferromagnet insulator structures. *Prog. Surf. Sci.* **2019**, *94*, 100540. [CrossRef]
65. Torgerson, D.; Skowronski, R.; Macfarlane, R. New approach to the mass spectroscopy of non-volatile compounds. *Bioche Biophys. Res. Commun.* **1974**, *60*, 616–621. [CrossRef]
66. Gessler, N.N.; Egorova, A.S.; Belozerskaya, T.A. Melanin pigments of fungi under extreme environmental conditions (Revie *Appl. Biochem. Microbiol.* **2014**, *50*, 105–113. [CrossRef]
67. Gonçalves, R.C.R.; Lisboa, H.C.F.; Pombeiro-Sponchiado, S.R. Characterization of melanin pigment produced by *Aspergill nidulans*. *World J. Microbiol. Biotechnol.* **2011**, *28*, 1467–1474. [CrossRef] [PubMed]
68. Grosser, K.; Van Dam, N.M. A Straightforward Method for Glucosinolate Extraction and Analysis with High-pressure Liqu Chromatography (HPLC). *J. Vis. Exp.* **2017**, *2017*, e55425. [CrossRef]
69. Sahu, P.K.; Ramisetti, N.R.; Cecchi, T.; Swain, S.; Patro, C.S.; Panda, J. An overview of experimental designs in HPLC meth development and validation. *J. Pharm. Biomed. Anal.* **2018**, *147*, 590–611. [CrossRef]
70. Dalal, R.C.; Henry, R.J. Simultaneous Determination of Moisture, Organic Carbon, and Total Nitrogen by Near Infrared Reflectan Spectrophotometry. *Soil Sci. Soc. Am. J.* **1986**, *50*, 120–123. [CrossRef]
71. Tavares, D.G.; Barbosa, B.V.L.; Ferreira, R.L.; Duarte, W.F.; Cardoso, P.G. Antioxidant activity and phenolic compounds of t extract from pigment-producing fungi isolated from Brazilian caves. *Biocatal. Agric. Biotechnol.* **2018**, *16*, 148–154. [CrossRef]
72. Esser, N.; Speiser, E. Introduction to Raman scattering at surfaces. In *Physics of Solid Surfaces*; Springer: Berlin/Heidelbe Germany, 2018; pp. 549–551. [CrossRef]

Poole, C.F. Thin-layer chromatography: Challenges and opportunities. *J. Chromatogr. A* **2003**, *1000*, 963–984. [CrossRef]
Stillwell, W. Membrane Reconstitution. In *An Introduction to Biological Membranes*; Elsevier BV: Amsterdam, The Netherlands, 2016; pp. 273–312.
Hostettmann, K.; Marston, A. Saponins. In *Saponins*; Cambridge University Press (CUP): Cambridge, UK, 1995; pp. 205–209.
Bates, C.J. Fat-Soluble. In *Vitamin D*, 2nd ed.; Elsevier Ltd.: Amsterdam, The Netherlands, 2005; pp. 159–172.
Blanc, P.; Loret, M.; Santerre, A.; Pareilleux, A.; Prome, D.; Prome, J.; Laussac, J.; Goma, G. Pigments of Monascus. *J. Food Sci.* **1994**, *59*, 862–865. [CrossRef]

Journal of
Fungi

Article

Seven New Cytotoxic and Antimicrobial Xanthoquinodins from *Jugulospora vestita*

Lulu Shao [1,2,3], Yasmina Marin-Felix [1,*], Frank Surup [1], Alberto M. Stchigel [4] and Marc Stadler [1,*]

1 Department Microbial Drugs, Helmholtz Centre for Infection Research, Inhoffenstrasse 7, 38124 Braunschweig, Germany; Lulu.Shao@helmholtz-hzi.de (L.S.); Frank.Surup@helmholtz-hzi.de (F.S.)
2 South China Botanical Garden, Chinese Academy of Sciences, Xingke Road 723, Tianhe District, Guangzhou 510650, China
3 School of Life Sciences, University of Chinese Academy of Sciences, Yuquanlu 19A, Beijing 100049, China
4 Mycology Unit, Medical School and IISPV, Universitat Rovira i Virgili, C/Sant Llorenç 21, 43201 Reus, Tarragona, Spain; albertomiguel.stchigel@urv.cat
* Correspondence: Yasmina.MarinFelix@helmholtz-hzi.de (Y.M.-F.); Marc.Stadler@helmholtz-hzi.de (M.S.)

Received: 18 August 2020; Accepted: 23 September 2020; Published: 25 September 2020

Abstract: During the course of a screening for novel biologically active secondary metabolites produced by the Sordariomycetes (Ascomycota, Fungi), the ex-type strain of *Jugulospora vestita* was found to produce seven novel xanthone-anthraquinone heterodimers, xanthoquinodin A11 (**1**) and xanthoquinodins B10–15 (**2–7**), together with the already known compound xanthoquinodin B4 (**8**). The structures of the xanthoquinodins were determined by analysis of the nuclear magnetic resonance (NMR) spectroscopic and mass spectrometric data. Moreover, the absolute configurations of these metabolites were established by analysis of the $^1H-^1H$ coupling constants, nuclear Overhauser effect spectroscopy (NOESY) correlations, and Electronic Circular Dichroism (ECD) spectroscopic data. Antifungal and antibacterial activities as well as cytotoxicity of all compounds were tested. Xanthoquinodin B11 showed fungicidal activities against *Mucor hiemalis* [minimum inhibitory concentration (MIC) 2.1 μg/mL], *Rhodotorula glutinis* (MIC 2.1 μg/mL), and *Pichia anomala* (MIC 8.3 μg/mL). All the compounds **1–8** displayed anti-Gram-positive bacteria activity (MIC 0.2–8.3 μg/mL). In addition, all these eight compounds showed cytotoxicity against KB 3.1, L929, A549, SK-OV-3, PC-3, A431, and MCF-7 mammalian cell lines. The six novel compounds (**1–3**, **5–7**), together with xanthoquinodin B4, were also found in the screening of other strains belonging to *Jugulospora rotula*, revealing the potential chemotaxonomic significance of the compound class for the genus.

Keywords: antimicrobial activity; cytotoxicity; secondary metabolites; Sordariales; xanthoquinodins

---

1. Introduction

Nowadays, the increasing drug resistance by bacterial and fungal pathogens and the decrease of new therapeutic agents and developmental candidates are a global hurdle [1]. This problem has led to strong demand to increase the chemical diversity of antibiotics and antifungals, with the fungi, whose secondary metabolites remain poorly studied, being the potential solution for this challenge [2]. Fungal secondary metabolites can also be employed as new and beneficial therapeutic agents, such as the cyathane diterpenoids found in the genus *Hericium* (Basidiomycota), which can be used to treat neurodegenerative diseases [3]. Another example is the cytotoxic compounds, which can hold a great potential for the cancer treatment since these can be combined with targeted therapy, achieving the delivery of the drug to cancer cell-specific genes or proteins or to tissue microenvironment of developing cancer [4]. Even though most anticancer drugs are retrieved from plants and bacteria [2], some natural substances from fungi are currently in the preclinical and clinical development stage, such as irofulven, which has been evaluated in phase I and II, showing promising results against brain and central nervous system, breast, colon, lung, ovarian, pancreas, and prostate cancers, as well as leukemia and sarcoma [5–7].

During the course of an ongoing project, rare and interesting members of the Sordariomycetes are being tested for the production of novel biologically active secondary metabolites, since this group of fungi has already been demonstrated to include prolific producer taxa [8–10]. According to Bills and Gloer [11], many important metabolites with practical potential have been discovered from the Sordariomycetes. Prominent examples are antibacterial antibiotics (cephalosporins and fusidic acid), the immunomodulatory drug cyclosporine, the ergot alkaloids, the anthelmintic cyclodepsipeptide PF1022A, and the serine palmitoyltransferase inhibitor myriocin that gave rise to the anti-inflammatory drug fingolimod. In addition, the potent antiparasitic nodulisporic acid, as well as the antimycotic sordarins, have also been found from species of Sordariomycetes [10,12].

Jugulospora vestita was initially described by Udagawa and Horie [13] as *Apiosordaria vestita* to accommodate a soil fungus isolated from Nepal. This species is characterized by ostiolate ascomata and two-celled ascospores, with pitted upper cells tending to appear reticulate [13]. In a recent phylogenetic study based on sequences of the internal transcribed spacer region (ITS), the nuclear rDNA large subunit (LSU), and fragments of ribosomal polymerase II subunit 2 (*rpb2*) and β-tubulin (*tub2*) genes, the type strain of this taxon was located far from the type species of the genus *Apiosordaria*, *A. verruculosa*, being placed in the monophyletic clade of the genus *Jugulospora* [14]. Therefore, the new combination *J. vestita* was proposed. The screening for novel biologically active secondary metabolites of the ex-type strain of this species led to the isolation of seven previously undescribed xanthoquinodins, together with the already known xanthoquinodin B4. Their structures were elucidated by one-dimensional and two-dimensional nuclear magnetic resonance (1D- and 2D-NMR) spectroscopy, and Electronic Circular Dichroism (ECD) spectra. Details of the isolation, structure elucidation, antimicrobial activity, and cytotoxicity of these new xanthoquinodins are presented herein.

2. Materials and Methods

2.1. General

Ultraviolet-Visible (UV/Vis) spectra were acquired using a UV-Vis spectrophotometer UV-2450 (Shimadzu, Kyoto, Japan). Optical rotations were recorded in methanol (MeOH) solution on a MCP 150 polarimeter at 20 °C (Anton-Paar Opto Tec GmbH, Seelze, Germany). ECD spectra were obtained on a J-815 spectropolarimeter (JASCO, Pfungstadt, Germany). High-resolution electrospray ionization mass spectra (HR-ESI-MS) were acquired with an Agilent 1200 Infinity Series HPLC-UV system (Agilent Technologies, Santa Clara, CA, USA) utilizing a C_{18} Acquity UPLC BEH column (2.1 × 50 mm, 1.7 μm: Waters, Milford, MA, USA), solvent A: H_2O + 0.1% formic acid; solvent B: acetonitrile (ACN) + 0.1% formic acid, gradient: 5% B for 0.5 min increasing to 100% B in 19.5 min, maintaining 100% B for 5 min, flow rate 0.6 mL min^{-1}, UV/Vis detection 190–600 nm) connected to an time-of-flight mass spectrometer (ESI-TOF-MS, Maxis, Bruker, Billerica, MA, USA) (scan range 100–2500 *m/z*, rate 2 Hz, capillary voltage 4500 V, dry temperature 200 °C). NMR spectra were recorded with an Avance III 500 spectrometer (Bruker, Billerica, MA, USA, ^{1}H-NMR: 500 MHz, and ^{13}C-NMR: 125 MHz).

2.2. Fermentation, Extraction, and Isolation

The ex-type strain of *Jugulospora vestita* CBS 135.91, which was isolated from soil from Nepal (https://wi.knaw.nl/page/fungal_table), was grown on potato dextrose agar (PDA; HiMedia, Mumbai, India) plates for 7 days at 23 °C; then, the fungal colonies on the culture medium were cut into pieces (1 × 1 cm) and transferred into two 250 mL Erlenmeyer flasks, each containing 100 mL of yeast-malt extract broth (YM broth; 4 g/L yeast extract, 10 g/L malt extract, 4 g/L D-glucose and pH 6.3 [3]). The seed fungus was incubated for 5 days at 23 °C under shake condition at 140 rpm. Fermentation was carried out in 20 × 1 L Erlenmeyer flasks, each containing 400 mL of YM broth and inoculated with 5.0 mL of the mycelial suspension and cultivated for 13 days at 26 °C on a rotary shaker at 108 rpm.

The mycelium and the supernatant were separated by filtration via gauze. The mycelium was macerated three times by acetone and put in an ultrasonic water bath for 30 min at 40 °C. The supernatant was mixed with 275 g of adsorbent resin (Amberlite XAD-16 N, Sigma-Aldrich, Deisenhofen, Germany) and stirred for 2 h. The Amberlite® resin was then filtered and eluted three times with acetone. The resulting acetone extracts were dried in vacuo at 40 °C and the remaining aqueous residue was diluted with the same amounts of ethyl acetate (EtOAc) and extracted three times. The mycelium and the supernatant extracts were combined according to their chromatographic homogeneity to afford 738 mg of an oily crude extract.

The total extract was dissolved in MeOH and subjected to preparative reverse phase HPLC (PLC 2020, Gilson, Middleton, WI, USA). As stationary phase, VP Nucleodur 100-5 C18 ec column (250 × 40 mm, 7 µm, Macherey-Nagel, Düren, Germany) was used, while the mobile phase consisted of: solvent A, deionized water; solvent B, ACN. Purification of the crude extract was performed by using a linear gradient elution of 35–80% aqueous ACN with 0.05% formic acid at a flow rate of 45 mL/min for 55 min, 80–100% solvent B in 5 min, and finally, isocratic elution at 100% solvent B for 5 min to afford compounds **2** (t_R: 13.4–13.5 min, 34 mg), **3** (t_R: 13.2–13.3 min, 13 mg), and other observed peaks (F1–F8). Compounds **8** (t_R: 10.6–10.8 min, 18.5 mg) and **7** (t_R: 11.4–11.5 min, 5.5 mg) were obtained from purification of fraction F4 with the elution gradient 65–75% solvent B for 23 min, followed by isocratic elution with 100% B for 10 min. With the same method as for fraction F4, compounds **4** (t_R: 11.6–11.7 min, 9.5 mg), **6** (t_R: 12.5–12.6 min, 2 mg), and **5** (t_R: 12.2–12.3 min, 3.3 mg) were obtained from fraction F5 and F6, as well as **1** (t_R: 13.5–13.6 min, 7.5 mg) from fraction F8.

For comparison of secondary metabolite production of strains of *Jugulospora*, *Jugulospora vestita* strain CBS 135.91 and *Jugulospora rotula* strains CBS 110112, CBS 110113, FMR 12691, and FMR 12781, as well as *Triangularia backusii* FMR 12439 (a species that was previously also placed in the genus *Jugulospora*), were grown on PDA at 23 °C and the well-grown cultures were cut into small pieces using a cork borer (7 mm). Subsequently, a 200 mL Erlenmeyer flask containing 100 mL of YM were inoculated using five of those pieces and incubated at 23 °C on a rotary shaker (140 rpm). The growth of the fungus was monitored by constantly checking the amount of free glucose using Medi-test Glucose (Macherey-Nagel, Düren, Germany), and the fermentation was terminated 3 days after glucose depletion. Then, the mycelium and the supernatant were separated by filtration via gauze. The mycelia were extracted one time with acetone in an ultrasonic bath at 40 °C for 30 min. The resulting acetone extracts were dried in vacuo at 40 °C. The remaining aqueous residues were diluted with the same amounts of ethyl acetate and extracted one time. The supernatants were extracted with the same amount of EtOAc. The solvents were dried in vacuo at 40 °C.

2.3. Spectral Data

2.3.1. Xanthoquinodin A11 (1)

Yellow, powder; $[\alpha]^{20}_D$ + 491° (*c* 0.005, MeOH); UV (MeOH) λ_{max} (log ε) 199 (4.6), 274 (3.8), 345 (4.4); CD (*c* 1.6 × 10^{-3} M, MeOH) λ_{max} ($\Delta\varepsilon$) 228 (+14.90), 271 (+3.15), 326 (+28.27), 378 (+4.99); ^{1}H-NMR and ^{13}C-NMR see Table 1; ESI-MS: *m/z* 643.20 $(M - H)^-$ and 645.26 $(M + H)^+$; high resolution electrospray ionisation mass spectrometry (HRESIMS) *m/z* 645.1965 $(M + H)^+$ (calculated for $C_{35}H_{33}O_{12}$, 645.1967).

2.3.2. Xanthoquinodin B10 (2)

Yellow, amorphous solid; $[\alpha]^{20}_D$ + 474° (*c* 0.005, MeOH); UV (MeOH) λ_{max} (log ε) 199 (4.5), 274 (3.8), 353 (4.3); CD (*c* 1.6 × 10^{-3} M, MeOH) λ_{max} ($\Delta\varepsilon$) 228 (+15.91), 264 (+2.69), 320 (+15.51), 359 (+14.38); ^{1}H-NMR and ^{13}C-NMR see Table 1; ESI-MS: *m/z* 643.19 $(M - H)^-$ and 645.24 $(M + H)^+$; HRESIMS *m/z* 645.1967 $(M + H)^+$ (calculated for $C_{35}H_{33}O_{12}$, 645.1967).

Table 1. One-dimensional nuclear magnetic resonance (1D-NMR) spectroscopic data for compounds **1–4** (in $CDCl_3$).

	1		**2**		**3**		**4**	
No	δ_C, Type	δ_H (*J* in Hz)	δ_C, Type	δ_H (*J* in Hz)	δ_C, Type	δ_H (*J* in Hz)	δ_C, Type	δ_H (*J* in Hz)
2	84.3, C	–	85.3, C	–	84.5, C	–	85.5, C	–
3	71.8, CH	4.25, dd (12.3, 5.3)	71.7, CH	4.46, dd (12.3, 5.0)	66.9, CH	4.49, dd (4.0, 2.0)	68.6, CH	4.74, dd (12.7, 5.1)
4	23.7, CH_2	2.15, m, H_a 2.08, m, H_b	23.9, CH_2	2.21, m, H_a 2.10, m, H_b	22.9, CH_2	2.18, m, H_a 2.00, m, H_b	32.7, CH_2	2.42, m, H_a 2.23, m, H_b
5	27.5, CH_2	2.65, m	27.7, CH_2	2.69, m	24.4, CH_2	2.84, m 2.44, m	65.9, CH	4.55, d (4.7)
6	178.1, C	–	178.5, C	–	180.2, C	–	174.0, C	–
7	101.1, C	–	101.6, C	–	100.2, C	–	101.8, C	–
8	186.5, C	–	186.7, C	–	186.9, C	–	188.0, C	–
9	104.9, C	–	105.4, C	–	105.3, C	–	105.8, C	–
10	156.8, C	–	160.1, C	–	160.2, C	–	160.1, C	–
11	117.3, C	–	114.2, CH	6.13, s	114.7, CH	6.15, s	114.3, CH	6.14, s
12	147.8, C	–	147.6, C	–	147.3, C	–	148.5, C	–
13	110.9, CH	6.07, s	115.3, C	–	115.2, C	–	115.5, C	–
14	158.4, C	–	154.8, C	–	153.6, C	–	155.0, C	–
15	169.6, C	–	169.8, C	–	171.0, C	–	169.2, C	–
1′	72.8, CH	5.96, s	72.7, CH	5.98, s	72.7, CH	5.98, s	72.7, CH	5.98, s
2′	136.6, C	–	136.5, C	–	136.6, C	–	136.5, C	–
3′	123.1, CH	6.89, s	123.2, CH	6.90, s	123.2, CH	6.90, s	123.3, CH	6.90, s
4′	147.6, C	–	147.7, C	–	147.8, C	–	147.8, C	–
5′	119.3, CH	6.81, s	119.3, CH	6.80, s	119.3, CH	6.80, s	119.4, CH	6.80, s
6′	161.7, C	–	161.6, C	–	161.7, C	–	161.5, C	–
7′	112.7, C	–	112.4, C	–	112.6, C	–	112.3, C	–
8′	185.8, C	–	185.3, C	–	185.2, C	–	184.8, C	–
9′	105.5, C	–	105.6, C	–	105.7, C	–	105.6, C	–
10′	186.0, C	–	186.1, C	–	186.4, C	–	186.4, C	–
11′	37.3, CH	4.76, dd (6.6, 1.0)	38.6, CH	4.81, dd (6.7, 0.7)	37.9, CH	4.76, dd (6.7, 1.0)	38.7, CH	4.83, dd (6.6, 0.9)
12′	131.9, CH	6.42, dd (8.5, 6.6)	131.8, CH	6.48, dd (8.4, 6.7)	131.6, CH	6.41, dd (8.4, 6.7)	131.7, CH	6.47, dd (8.4, 6.6)
13′	132.4, CH	6.05, dd (8.5, 1.0)	132.3, CH	6.08, dd (8.4, 0.7)	132.7, CH	6.07, dd (8.4, 1.0)	132.4, CH	6.08, d (8.4, 0.9)
14′	41.5, C	–	41.6, C		41.4, C	–	41.6, C	–
15′	35.1, CH_2	2.74, d (18.5), H_a 2.68, d (18.5), H_b	35.0, CH_2	2.79, d (18.0), H_a 2.67, d (18.0), H_b	35.0, CH_2	2.78, d (17.9), H_a 2.68, d (17.9), H_b	35.1, CH_2	2.80, d (18.0), H_a 2.69, d (18.0), H_b
16′	22.0, CH_3	2.38, s	22.1, CH_3	2.38, s	22.1, CH_3	2.38, s	22.1, CH_3	2.38, s
18′	173.1, C	–	173.1, C	–	173.0, C	–	173.1, C	–
19′	36.2, CH_2	2.22, m	36.2, CH_2	2.23, m	36.2, CH_2	2.22, m	36.2, CH_2	2.22, m
20′	18.4, CH_2	1.57, m	18.4, CH_2	1.58, m	18.4, CH_2	1.57, m	18.4, CH_2	1.57, m
21′	13.5, CH_3	0.86, t (7.4)	13.5, CH_3	0.86, t (7.4)	13.5, CH_3	0.86, t (7.4)	13.5, CH_3	0.86, t (7.4)
15-OCH_3	53.4, CH_3	3.67, s	53.3, CH_3	3.73, s	53.6, CH_3	3.74, s	53.4, CH_3	3.73, s
3-OH	–	14.14	–	14.26	–	–	–	–
6-OH	–	13.84	–	13.95	–	14.10	–	13.70
10-OH	–	11.76	–	11.05	–	11.24	–	10.91
6′-OH	–	11.58	–	11.40	–	11.55	–	11.30

2.3.3. Xanthoquinodin B11 (**3**)

Yellow, amorphous solid; $[\alpha]^{20}_D$ + 438° (*c* 0.005, MeOH); UV (MeOH) λ_{max} (log ε) 198 (4.5), 274 (3.6), 354 (4.1); CD (*c* 1.6 × 10^{-3} M, MeOH) λ_{max} ($\Delta\varepsilon$) 234 (−22.84), 263 (+2.78), 325 (+33.59), 353 (+19.33); ^{1}H-NMR and ^{13}C-NMR see Table 1; ESI-MS: *m/z* 643.19 $(M - H)^-$ and 645.24 $(M + H)^+$; HRESIMS *m/z* 645.1966 $(M + H)^+$ (calculated for $C_{35}H_{33}O_{12}$, 645.1967).

2.3.4. Xanthoquinodin B12 (4)

Yellow, crystalline solid; $[\alpha]^{20}_{D}$ + 442° (c 0.004, MeOH); UV (MeOH) λ_{max} (log ε) 199 (4.5), 275 (3.8), 355 (4.2); CD (c 1.5 × 10^{-3} M, MeOH) λ_{max} ($\Delta\varepsilon$) 228 (+1.85), 264 (+2.74), 324 (+22.52), 359 (+12.43); ^{1}H-NMR and ^{13}C-NMR see Table 1; ESI-MS: m/z 659.20 $(M - H)^-$ and 661.24 $(M + H)^+$; HRESIMS m/z 661.1915 $(M + H)^+$ (calculated for $C_{35}H_{34}O_{13}$, 661.1916).

2.3.5. Xanthoquinodin B13 (5)

Yellow, amorphous solid; $[\alpha]^{20}_{D}$ + 489° (c 0.001, MeOH); UV (MeOH) λ_{max} (log ε) 199 (4.6), 274(3.9), 354 (4.3); CD (c 1.6 × 10^{-3} M, MeOH) λ_{max} ($\Delta\varepsilon$) 228 (+12.74), 264 (+2.90), 323 (+15.04), 360 (+13.37); ^{1}H-NMR and ^{13}C-NMR see Table 2; ESI-MS: m/z 615.18 $(M - H)^-$ and 617.19 $(M + H)^+$; HRESIMS m/z 617.1650 $(M + H)^+$ (calculated for $C_{33}H_{29}O_{12}$, 617.1654).

Table 2. 1D-NMR spectroscopic data for compounds **5–7** (in $CDCl_3$).

	5		6		7	
No	δ_C, Type	δ_H (J in Hz)	δ_C, Type	δ_H (J in Hz)	δ_C, Type	δ_H (J in Hz)
2	85.3, C	–	84.9, C	–	87.3, C	–
3	71.8, CH	4.46, dd (12.5, 5.0)	80.8, CH	5.01, dd (7.6, 6.8)	73.8, CH	4.23, dd (10.8, 1.8)
4	23.9, CH_2	2.23, m 2.13, m	22.2, CH_2	2.42, m	25.6, CH_2	1.97, m 1.78, m
5	27.7, CH_2	2.70, m	27.7, CH_2	2.67, m	30.0, CH_2	2.69, m
6	178.5, C	–	175.2, C	–	177.2, C	–
7	101.6, C	–	38.6, CH_2	3.21, d (17.0) 3.04, d (17.0)	38.2, CH_2	3.20, s
8	186.8, C	–	194.1, C	–	195.7, C	–
9	105.4, C	–	105.8, C	–	105.9, C	–
10	160.1, C	–	160.1, C	–	160.1, C	–
11	114.3, CH	6.14, s	114.9, CH	6.16, s	114.1, CH	6.14, s
12	147.6, C	–	148.9, C	–	147.8, C	–
13	115.3, C	–	114.9, C	–	114.8, C	–
14	154.7, C	–	154.8, C	–	155.1, C	–
15	169.8, C	–	168.7, C	–	170.0, C	–
1′	73.0, CH	5.97, s	72.6, CH	5.98, s	72.6, CH	5.98, s
2′	136.4, C	–	136.5, C	–	136.5, C	–
3′	123.3, CH	6.90, s	123.2, CH	6.89, s	123.2, CH	6.90, s
4′	147.8, C	–	147.8, C	–	147.8, C	–
5′	119.4, CH	6.81, s	119.3, CH	6.80, s	119.3, CH	6.80, s
6′	161.7, C	–	161.8, C	–	161.7, C	–
7′	112.5, C	–	112.7, C	–	112.6, C	–
8′	185.6, C	–	184.8, C	–	185.4, C	–
9′	105.6, C	–	105.4, C	–	105.5, C	–
10′	185.9, C	–	186.6, C	–	186.2, C	–
11′	38.6, CH	4.79, d (6.7)	37.9, CH	4.65, dd (6.6, 0.9)	38.0, CH	4.73, dd (6.6, 0.6)
12′	131.8, CH	6.49, dd (8.5, 6.7)	131.4, CH	6.41, dd (8.5, 6.6)	131.4, CH	6.42, dd (8.5, 6.6)
13′	132.2, CH	6.09, d (8.5)	132.9, CH	6.09, d (8.5, 0.9)	132.9, CH	6.09, dd (8.5, 0.6)
14′	41.5, C	–	41.5, C	–	41.4, C	–
15′	35.0, CH_2	2.79, d (17.5) 2.68, d (17.5)	35.1, CH_2	2.79, d (18.2) 2.68, d (18.2)	35.1, CH_2	2.79, d (17.5) 2.67, d (17.5)
16′	22.1, CH_3	2.38, s	22.1, CH_3	2.38, s	22.1, CH_3	2.38, s
18′	170.4, C	–	173.1, C	–	173.1, C	–
19′	21.1, CH_3	2.02, s	36.2, CH_2	2.22, m	36.2, CH_2	2.21, m
20′	–	–	18.4, CH_2	1.57, m	18.4, CH_2	1.58, m
21′	–	–	13.5, CH_3	0.86, t (7.4)	13. 5, CH_3	0.86, t (7.4)
15-OCH_3	53.3, CH_3	3.74, s	53.8, CH_3	3.75, s	53.4, CH_3	3.73, s
3-OH	–	–	–	–	–	–
6-OH	–	13.95	–	–	–	–
10-OH	–	11.05	–	11.36	–	11.5
6′-OH	–	11.45	–	11.60	–	–

2.3.6. Xanthoquinodin B14 (**6**)

Yellow, amorphous solid; $[\alpha]^{20}_{D}$ + 508° (*c* 0.002, MeOH); UV (MeOH) λ_{max} (log ε) 199 (4.5), 275 (4.0), 361 (4.1); CD (*c* 1.6 × 10^{-3} M, MeOH) λ_{max} ($\Delta\varepsilon$) 233 (−12.96), 264 (+8.84), 323 (+26.05); ^{1}H-NMR and ^{13}C-NMR see Table 2; ESI-MS: *m/z* 643.20 $(M-H)^-$ and 645.24 $(M+H)^+$; HRESIMS *m/z* 645.1967 $(M+H)^+$ (calculated for $C_{35}H_{33}O_{12}$, 645.1967).

2.3.7. Xanthoquinodin B15 (**7**)

Yellow, amorphous solid; $[\alpha]^{20}_{D}$ + 425° (*c* 0.004, MeOH); UV (MeOH) λ_{max} (log ε) 199 (4.4), 207 (4.3), 276 (4.0), 361 (4.1); CD (*c* 1.5 × 10^{-3} M, MeOH) λ_{max} ($\Delta\varepsilon$) 207 (−35.18), 231 (−17.20), 264 (+10.67), 323 (+37.35); ^{1}H-NMR and ^{13}C-NMR see Table 2; ESI-MS: *m/z* 661.19 $(M-H)^-$ and 663.27 $(M+H)^+$; HRESIMS *m/z* 663.2072 $(M+H)^+$ (calculated for $C_{35}H_{35}O_{13}$, 663.2072).

2.3.8. Xanthoquinodin B4 (**8**)

Yellow, amorphous solid; $[\alpha]^{20}_{D}$ + 540.2° (*c* 0.005, MeOH); UV (MeOH) λ_{max} (log ε) 202 (4.4), 274 (3.9), 352 (4.3); CD (*c* 1.5 × 10^{-3} M, MeOH) λ_{max} ($\Delta\varepsilon$) 227 (+15.00), 263 (+1.74), 324 (+14.12), 357 (+14.60); ^{1}H-NMR (500 MHz, $CDCl_3$): δ_H 2.11 (1H, m, H-4a), 2.24 (1H, m, H-4b), 2.39 (3H, s, H-16′), 2.60 (1H, d, *J* = 18.0 Hz, H-15′b), 2.68 (2H, m, H-5), 2.73 (1H, d, *J* = 18.0 Hz, H-15′a), 3.73 (3H, s, 15-OCH_3), 4.4 (1H, dd, *J* = 12.5, 5.0 Hz, H-3), 4.55 (1H, s, H-1′), 4.83 (1H, d, *J* = 6.0 Hz, H-11′), 6.10 (1H, s, H-11), 6.47 (1H, d, *J* = 8.5 Hz, H-13′), 6.51 (1H, dd, *J* = 8.5, 6.5 Hz, H-12′), 6.78 (1H, s, H-5′), 6.80 (1H, s, H-3′), 11.05 (1H, s, 10-OH), 11:28 (1H, s, 6′-OH), 13.95 (1H, s, 6-OH);^{13}C NMR (125 MHz, $CDCl_3$): δ_C 22.0 (CH_3, 16′-CH_3), 23.8 (CH_2, CH_2-4), 27.7 (CH_2, CH_2-5), 35.3 (CH_2, CH_2-15′), 39.0 (CH, CH_2-11′), 42.8 (C, C-14′), 53.2 (CH_3, 15-OCH_3), 71.6 (CH, CH_2-3), 73.5 (CH, CH-1′), 85.3 (C, C-2), 101.6 (C, C-7), 105.0 (C, C-9), 105.3 (C, C-14′), 111.1 (C, C-7′), 114.1 (CH, CH-11), 115.4 (C, C-13), 119.0 (CH, CH-5′), 122.1 (CH, CH-3′), 131.9 (CH, CH-12′), 132.9 (CH, CH-13′), 140.6 (C, C-2′), 147.9 (C, C-4′), 148.0 (C, C-12), 160.0 (C, C-10), 161.7 (C, C-6′), 169.9 (C, C-15), 178.5 (C, C-6), 183.9 (C, C-8′), 186.7 (C, C-8), 188.2 (C, C-10′); ESI-MS: *m/z* 573.15 $(M-H)^-$ and 575.21 $(M+H)^+$; HRESIMS *m/z* 575.1547 $(M+H)^+$ (calculated for $C_{31}H_{27}O_{11}$, 575.1548).

2.4. Biological Assays

Compounds were tested for their antimicrobial activity against four fungi (*Candida albicans, Mucor hiemails, Pichia anomala, Rhodotorula glutinis* and *Schizosaccharomyces pombe*), four different Gram-positive bacteria (*Bacillus subtilis, Micrococcus luteus, Mycobacterium smegmatis* and *Staphylococcus aureus*), and three Gram-negative bacteria (*Chromobacterium violaceum, Escherichia coli* and *Pseudomonas aeruginosa*), using oxytetracycline as a positive control against Gram-positive and Gram-negative bacteria, while nystatin was used as an antifungal positive control. Besides, cytotoxicities of the compounds against seven mammalian cell lines (human endocervical adenocarcinoma KB 3.1, breast cancer MCF-7, lung cancer A549, ovary cancer SK-OV-3, prostate cancer PC-3, squamous cancer A431, and mouse fibroblasts L929) were determined by the microculture tetrazolium test (MTT) method, using epothilon B as the positive control. Both bioactivity assays were performed following our standard protocols [15].

2.5. Phylogenetic Study

The phylogenetic analysis was carried out based on the combination of the ITS, LSU, *rpb2*, and *tub2* sequences of the strain of *Jugulospora vestita* studied here and selected members belonging to the Sordariales, with *Camarops amorpha* SMH 1450 as outgroup. Each locus was aligned separately using MAFFT v. 7 [16], manually adjusted in MEGA v. 6.06 [17], and the individual gene phylogenies were checked for conflicts before the four gene datasets were concatenated [18,19]. The Maximum-Likelihood (ML) and Bayesian Inference (BI) methods were used in a phylogenetic analysis including the four loci concatenated as described by Hernández-Restrepo et al. [20]. Bootstrap support (bs) ≥ 70 and posterior probability values (pp) $\geq$ 0.95 were considered significant [21]. The alignment used in the phylogenetic analysis was deposited in TreeBASE (S26889).

3. Results and Discussion

3.1. Structure Elucidation of Compounds 1–7

In total, seven novel compounds (**1**–**7**) and xanthoquinodin B4 (**8**) [22], were isolated from the ex-type strain of *Jugulospora vestita* (Figure 1). Their structures were elucidated by 1D- and 2D-NMR spectroscopy (Supplementary Figures S1–S6, S8–S13, S15–S20, S22–S27, S29–S34, S36–S41, and S43–S48), HR-MS (Supplementary Figures S7, S14, S21, S28, S35, S42, and S49), and ECD spectra.

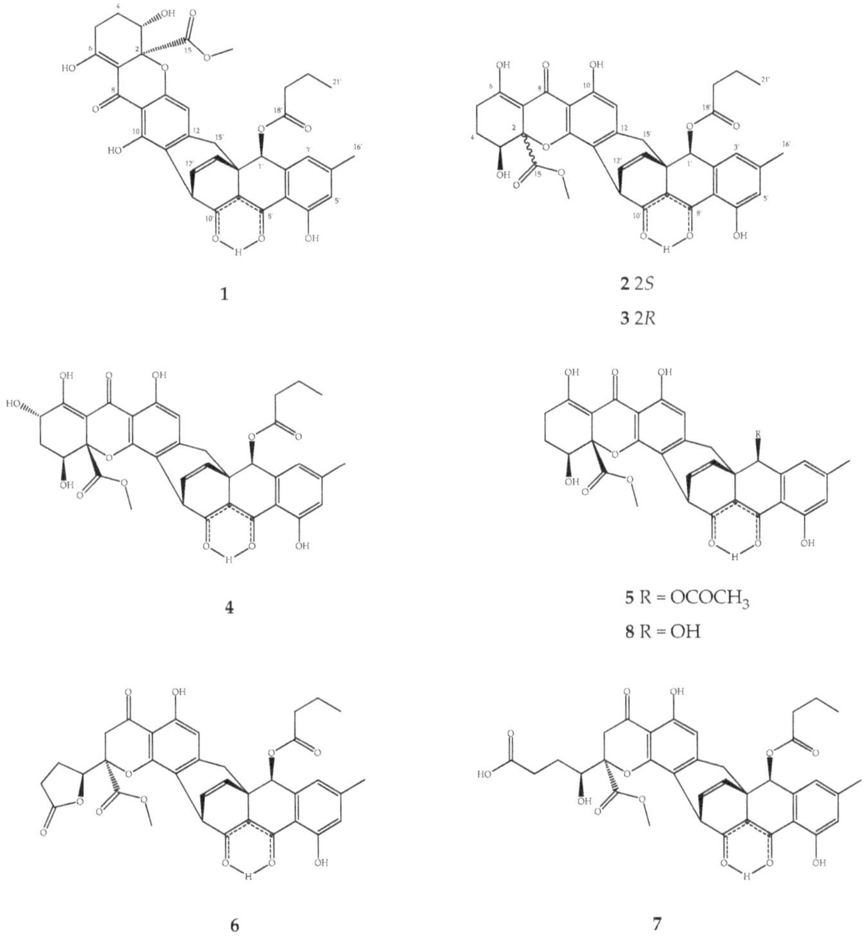

1

2 2*S*

3 2*R*

4

5 R = $OCOCH_3$

8 R = OH

6

7

Figure 1. Chemical Structures of Compounds **1**–**8**.

Compound **1** was obtained as a yellow powder and its molecular formula was established as $C_{35}H_{32}O_{12}$ (20 degrees of unsaturation) according to the mass ion peak at *m/z* 645.1967 $[M + H]^+$ in the HRESIMS spectrum. The ^{1}H- and ^{13}C-NMR spectra (Table 1), accompanied with heteronuclear single quantum coherence (HSQC) correlations, revealed signals of two methyl (δ_C 22.0, 13.5), one methoxy (δ_C 53.3), five sp^3 methylenes (δ_C 36.2, 35.1, 27.5, 23.7, and 18.4), three sp^3 methines

(δ_C 72.8, 71.8, and 37.3), five aromatic methines (δ_C 132.4, 131.9, 123.1, 119.3, and 110.9), seventeen sp^2 quaternary carbons, and two sp^3 quaternary carbons (δ_C 41.5 and 84.3 (oxygenated)). In the ^1H–^1H correlation spectroscopy (^1H–^1H COSY) spectrum, there were three isolated spin systems (H-3–H-4–H-5, H-11′–H-12′–H-13′, and H-19′–H-20′–H-21′). The heteronuclear multiple bond correlation (HMBC) spectrum showed correlations from H-21′ (δ_H 0.86) to C-20 and C-19′, from H-16′ (δ_H 2.38) to C-3′, C-4′, and C-5′, from H-3′ (δ_H 6.89) to C-1′, C-5′, C-7′, and C-16′, from H-5′ (δ_H 6.81) to C-3′, C-6′, C-7′, and C-16′, from H-1′ (δ_H 5.96) to C-2′, C-3′, C-7′, C-9′, C-13′, C-14′, C-15′, C-18′, and C-8′ (weak correlation), and from H-13′ (δ_H 6.05) to C-1′, C-9′, C-11′, C-14′, and C-10′ (weak correlation), above analysis, indicating an anthraquinone moiety (ABC-ring) with 1′-butyrate group. Moreover, the HMBC correlations from H-3 (δ_H 4.25) to C-2, C-4, C-5, and C-15, from H-5 (δ_H 2.65) to C-3, C-4, C-6, and C-7, from H-13 (δ_H 6.07) to C-9, C-11, and C-15′, and from 15-OCH_3 (δ_H 3.67) to C-15 revealed the rest part as the xanthone moiety (DEF-ring). In addition, the key HMBC correlations from H-15′ (δ_H 2.74 and 2.68) to C-1′, C-9′, C-13′, C-14′, C-11, C-12, and C-13, and from OH-10 (δ_H 11.76) to C-9, C-14, C-11, and C-11′ (weak correlation) indicated that a methylene (C-15′) linked these two moieties at C-12 and C-14′, as well as C-11 connected to C-11′.

Based on the combined above NMR analysis data and the molecular formula, the planar structure of **1** was elucidated as xanthone–anthraquinone heterodimer similar to xanthoquinodin A6 [21] and xanthoquinodin A9 [23]. The difference is at C-1′, where the hydroxyl is replaced by the butyl side chain of **1**. Noticeably, the β-keto-enol tautomeric system showed at C-8′ (δ_C 185.8) and C-10′ (δ_C 186.0), which both displayed keto carbonyl property in carbon chemical shift data. The $\Delta^{12',13'}$ double-bond was revealed as *Z* for the small coupling constant ($J_{H12'H13'}$ = 8.5 Hz). There are five chiral centers (C-2, C-3, C-1′, C-11′, and C-14′) in compound **1**, whose relative configuration was assigned by analysis of NOESY correlations and ^1H,^1H coupling constants. Since bridging carbons C-12′ and C-13′ must be on the same side, the relative configurations at C-11′ and C-14′ were deduced as *S* and *R*, respectively. In the NOESY spectrum, the strong intensity correlations of H-1′ with H_α-15′ and H_β-15′ indicated the *S** configuration at C-1′. The methyl ester group was in axial bond orientation positioned between the F and G ring. A diaxial orientation was deduced for both H-3 and H-4a due to the large coupling constant (J_{H3H4a} = 12.3 Hz) between these protons. For the assignment of absolute configuration of **1**, the ECD spectrum (Figure 2a) was measured, and showed a similar pattern as the spectrum of xanthoquinodin A6 [22], proving that both compounds possessed the same stereochemistry. On the basis of the above data, the absolute configuration of compound **1** was assigned as 2*S*, 3*S*, 1′*S*, 11′*S*, and 14′*R*, and named xanthoquinodin A11.

Compound **2** was obtained as a yellow amorphous solid. The molecular ion cluster at *m*/*z* 645.1967 $[M + H]^+$ in the HRESIMS spectrum indicated that the molecular formula of **2** was $C_{35}H_{32}O_{12}$ (20 degrees of unsaturation). The ^1H- and ^{13}C-NMR spectra, accompanied with HSQC correlations, revealed signals of two methyl (δ_C 22.1, 13.5), one methoxy (δ_C 53.3), five sp^3 methylenes (δ_C 36.2, 35.0, 27.7, 23.9, and 18.4), three sp^3 methines (δ_C 72.7, 71.7, and 38.6), five aromatic methines (δ_C 132.3, 131.8, 123.2, 119.3, and 114.2), seventeen sp^2 quaternary carbons, and two sp^3 quaternary carbons (δ_C 41.6 and 85.3 (oxygenated)). The same molecular formulae and the resemblance of NMR spectroscopic data of **1** and **2** (Table 1) suggested that they were isomers. The main differences between ^{13}C NMR spectrum of **1** and **2** were the upfield shifts at C-13 ($\Delta\delta$ −2.0) and C-14 ($\Delta\delta$ −3.6) in **2**, as well as the clearly downfield shifts at C-10 ($\Delta\delta$ +3.3) and C-11 ($\Delta\delta$ +3.3). Moreover, the strong HMBC correlations from OH-10 (δ_H 11.05) to aromatic methine δ_C 114.2 indicated that apart from the C-12–C-15′–C-14′ bridge, the two moiety (ABC-ring and FEG-ring) linked by C-13 connected to C-11′, which is similar to xanthoquinodin B series of structures. Meanwhile, the configurations of C-11′, C-14′, C-2, C-3, and 1′-butyrate were assigned as the same as those of **1**. Furthermore, the experimental ECD curves (Figure 2b) of **2** displayed the same as those previously given for xanthoquinodin B4 [23].

Therefore, the absolute configuration of **2** was assigned as 2*S*, 3*S*, 1′*S*, 11′*S*, and 14′*R*. The trivial name of xanthoquinodin B10 was given for compound **2**.

Compound **3** was obtained as a yellow amorphous solid with a molecular formula of $C_{35}H_{32}O_{12}$ (20 degrees of unsaturation) based on the mass ion peak at *m/z* 645.1966 [M + H]$^+$ in its HRESIMS spectrum. The ^{1}H- and ^{13}C-NMR spectra, accompanied with heteronuclear single quantum coherence (HSQC) correlations, revealed signals of two methyl (δ_C 22.1, 13.5), one methoxy (δ_C 53.6), five sp^3 methylenes (δ_C 36.2, 35.0, 24.4, 22.9, and 18.4), three sp^3 methines (δ_C 72.7, 66.9, and 37.9), five aromatic methines (δ_C 132.7, 131.6, 123.2, 119.3, and 114.7), seventeen sp^2 quaternary carbons, and two sp^3 quaternary carbons (δ_C 41.4 and 84.5 (oxygenated)).

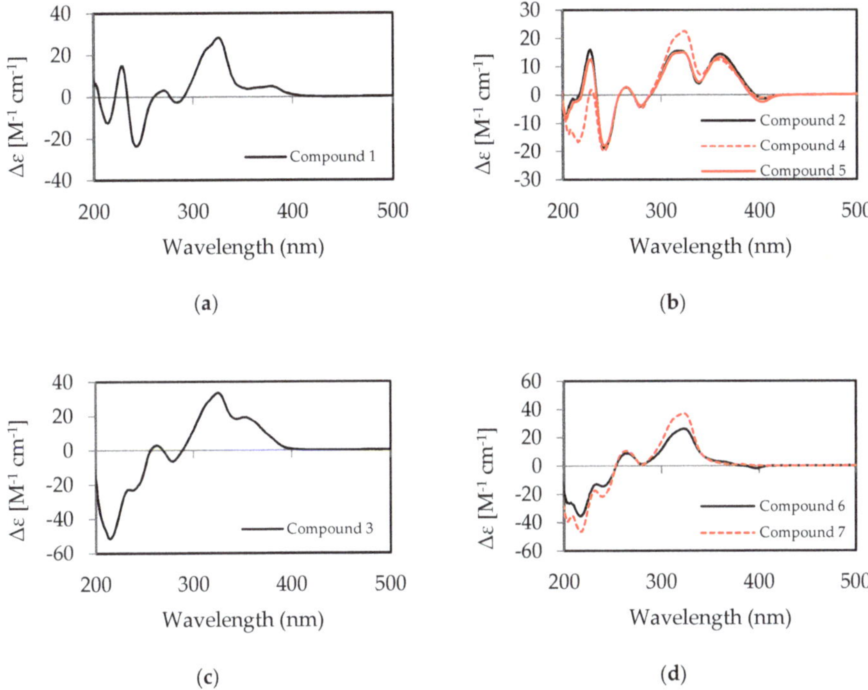

Figure 2. Electronic Circular Dichroism (ECD) spectra of compounds **1**–**7** measured in MeOH, (**a**) ECD spectrum of **1** (2*S*, 3*S*, 1′*S*, 11′*S*, and 14′*R*), (**b**) ECD spectra of **2** (2*S*, 3*S*, 1′*S*, 11′*S*, and 14′*R*), **4** (2*S*, 3*S*, 5*S*, 1′*S*, 11′*S*, and 14′*R*), and **5** (2*S*, 3*S*, 1′*S*, 11′*S*, and 14′*R*), (**c**) ECD spectrum of **3** (2*R*, 3*S*, 1′*S*, 11′*S*, and 14′*R*), (**d**) ECD spectra of **6** (2*R*, 3*S*, 1′*S*, 11′*S*, and 14′*R*) and **7** (2*R*, 3*S*, 1′*S*, 11′*S*, and 14′*R*).

The NMR spectroscopic data of **3** (Table 1) resembled those of **2**, with the main chemical shift difference (δ_C and δ_H) located at the positions 2–7 in the F-ring, which indicated epimerization at C-2 or C-3. The small coupling constant (4.0 and 2.0 Hz) of the C-3 methine proton demonstrated an equatorial bond of H-3. Therefore, configurations of chiral carbons C-2 and C-3 were opposite, indicating that compound **2** and **3** were epimers at C-2. In addition, the experimental ECD spectra of compound **2** (Figure 2b) and **3** (Figure 2c) were different at 200–250 nm, which displayed similar patterns to those differentiated for xanthoquinodin B4 and xanthoquinodin B5 [22]. Therefore, the absolute configuration of **3** was assigned as 2*R*, 3*S*, 1′*S*, 11′*S*, and 14′*R*. Xanthoquinodin B11 was the name chosen for compound **3**.

Compound **4** was obtained as a yellow crystalline solid. The mass ion peak at *m/z* 661.1915 [M + H]$^+$ in its HRESIMS spectrum indicated that the molecular formula of **4** was $C_{35}H_{33}O_{13}$ (20 degrees of unsaturation). The ^{1}H- and ^{13}C-NMR spectra, accompanied with HSQC correlations, revealed signals of two methyl (δ_C 22.1, 13.5), onrfde methoxy (δ_C 53.4), four sp^3 methylenes (δ_C 36.2, 35.1, 32.7, and 18.4), four sp^3 methines (δ_C 72.7, 68.6, 65.9, and 38.7), five aromatic methines (δ_C 132.4,

131.7, 123.3, 119.4, and 114.3), seventeen sp^2 quaternary carbons, and two sp^3 quaternary carbons (δ_C 41.6 and 85.5 (oxygenated)). The NMR spectroscopic data of **4** (Table 1) resembled those of **2**, except for the main chemical shift difference (δ_C and δ_H) located at the positions 3–6 in the F-ring, and C-5 (δ_C 65.9), which was oxygenated methine. Compared to **2**, there were one more oxygen and hydrogen atom in **4** according to the molecular formulae. On the basis of the above analysis, **4** was proposed as a new xanthoquinodin compound that possesses one additional hydroxyl substituent at the C-5. The configurations at C-2, C-3, C-1′, C-11′, and C-14′ were assigned as the same as those of **1** and **2** because of the similar corresponding NMR data (Table 1). The configuration at C-5 was confirmed as *S* based on the strong intensity ^{1}H–^{1}H COSY correlations of H-5 (δ_H 4.55, d (4.7)) with H_a-4 (δ_H 2.42, m). However, it barely showed correlations with H_b-4 (δ_H 2.23, m), indicating the dihedral right-angle of H-5–C-5–C-4–H_b-4 and the bond of H-5 at the axial orientation consistent with the axial bond of H_a-4 (δ_H 2.42, m), which showed large coupling constant (12.3 Hz) with H-3. Furthermore, compound **4** shared the similar experimental ECD curve with compound **2** (Figure 2b). Based on the above data, the absolute configuration of **4** was identified as 2*S*, 3*S*, 5*S*, 1′*S*, 11′*S*, and 14′*R*, and named xanthoquinodin B12.

Compound **5** was obtained as a yellow amorphous solid with a molecular formula of $C_{33}H_{28}O_{12}$ (20 degrees of unsaturation) according to the mass ion peak at *m/z* 617.1650 $[M + H]^+$ in its HRESIMS spectrum. The ^{1}H- and ^{13}C-NMR spectra, accompanied with HSQC correlations, revealed signals of two methyl (δ_C 35.0, 21.1), one methoxy (δ_C 53.3), three sp^3 methylenes (δ_C 35.0, 27.7, and 23.9), three sp^3 methines (δ_C 73.0, 71.8, and 38.6), five aromatic methines (δ_C 132.2, 131.8, 123.3, 119.4, and 114.3), seventeen sp^2 quaternary carbons, and two sp^3 quaternary carbons (δ_C 41.5 and 85.3 (oxygenated)). The NMR spectroscopic data of **5** (Table 2) were similar to those of **2** (Table 1), except for the absence of the butyrate group, replaced by acetate group at C-1′. In addition, the ECD spectrum of compound **5** showed the same pattern as those of compounds **2** and **4** (Figure 2b), corroborating that these compounds possessed the same stereochemistry. Besides, based on the same methods of analysis, the absolute configuration of **5** was proposed to be the same as that of **2**, being assigned as 2*S*, 3*S*, 1′*S*, 11′*S*, and 14′*R*. Compound **5** was named xanthoquinodin B13.

Compound **6** was obtained as a yellow amorphous solid. The mass ion peak at *m/z* 645.1967 $[M + H]^+$ in its HRESIMS spectrum indicated that the molecular formula of **6** was $C_{35}H_{32}O_{12}$ (20 degrees of unsaturation). The ^{1}H- and ^{13}C-NMR spectra, accompanied with HSQC correlations, revealed signals of two methyl (δ_C 22.1, 13.5), one methoxy (δ_C 53.8), six sp^3 methylenes (δ_C 38.6, 36.2, 35.1, 27.7, 22.2, and 18.4), three sp^3 methines (δ_C 80.8, 72.6, and 37.9), five aromatic methines (δ_C 132.9, 131.4, 123.2, 119.3, and 114.9), seventeen sp^2 quaternary carbons, and two sp^3 quaternary carbons (δ_C 41.5 and 84.9 (oxygenated)). The NMR spectroscopic data of **6** (Table 2) were similar to those of xanthoquinodin B6 [23], except for the substituent of C-1′, where the hydroxyl was replaced by the butyrate-like compound **1**. Furthermore, the NOESY correlations from H-1′ to H-15′ confirmed that 1′-hydroxyl was at the same orientation with the double bond of C-12′ and C-13′, consistent with all found xanthoquinodins. Likewise, the experimental ECD spectrum of compound **6** (Figure 2d) displayed a similar pattern at 200–250 nm to that of compound **3**, suggesting the same absolute configurations at C-2 and C-3, i.e., *R* and *S*, respectively. Thus, the absolute configuration of compound **6** was assigned as 2*R*, 3*S*, 1′*S*, 11′*S*, and 14′*R*, and the name given to it was xanthoquinodin B14.

Compound **7** was obtained as a yellow amorphous solid with a molecular formula of $C_{35}H_{34}O_{13}$ (19 degrees of unsaturation) based on the mass ion peak at *m/z* 663.2072 $[M + H]^+$ in its HRESIMS spectrum. The ^{1}H- and ^{13}C-NMR spectra, accompanied with HSQC correlations, revealed signals of two methyl (δ_C 22.1, 13.5), one methoxy (δ_C 53.4), six sp^3 methylenes (δ_C 38.2, 36.2, 35.1, 30.0, 25.6, and 18.4), three sp^3 methines (δ_C 73.8, 72.6, and 38.0), five aromatic methines (δ_C 132.9, 131.4, 123.2, 119.3, and 114.1), seventeen sp^2 quaternary carbons, and two sp^3 quaternary carbons (δ_C 41.4 and 87.3 (oxygenated)). The degrees of unsaturation and the observably different chemical shifts of C-2 to C-6 compared to compound **6** suggested the opening of the γ-lactone ring. The analysis of their NMR spectroscopic data (Table 2) and same experimental ECD curves (Figure 2d) revealed that **6** and **7** share the same absolute configuration as 2*R*, 3*S*, 1′*S*, 11′*S*, and 14′*R*, and **7** was named xanthoquinodin B15.

3.2. Antimicrobial and Cytotoxic Activities of Compounds 1–8

The eight compounds isolated showed antimicrobial activity against different fungi and/or different bacteria (Table 3).

Table 3. Minimum inhibitory concentration (MIC, μg/mL) of **1–8** against bacterial and fungal test organisms.

Test Organism	1	2	3	4	5	6	7	8	Positive Control
Schizosaccharomyces pombe	–	–	66.70	–	–	–	–	–	33.30 [1]
Pichia anomala	–	–	8.30	–	–	–	–	–	33.30 [1]
Mucor hiemalis	66.70	–	2.10	66.70	66.70	66.70	66.70	–	33.30 [1]
Candida albicans	–	–	16.70	–	–	–	–	–	33.30 [1]
Rhodotorula glutinis	–	–	2.10	66.70	–	–	–	–	16.70 [1]
Micrococcus luteus	4.20	2.10	2.10	2.10	2.10	8.30	4.20	8.30	0.80 [2]
Bacillus subtilis	0.40	0.40	0.20	0.80	0.20	4.20	4.20	2.10	8.30 [2]
Staphylococcus aureus	2.10	2.10	1.00	2.10	8.30	4.20	8.30	8.30	0.40 [2]
Mycobacterium smegmatis	–	–	–	–	–	–	–	–	1.70 [3]
Escherichia coli	–	–	–	–	–	–	–	–	3.30 [2]
Pseudomonas aeruginosa	–	–	–	–	–	–	–	–	0.40 [4]
Chromobacterium violaceum	–	–	–	–	–	–	–	66.70	0.80 [2]

[1] nystatin, [2] oxytetracycline, [3] kanamycin, [4] gentamicin, –: no inhibition observed under test conditions.

Only compound **3** exhibited antifungal activity against *C. albicans*, *Mu. hiemalis*, *P. anomala*, and *R. glutinis*, with MIC values in a range of 2.10–16.70 μg/mL. However, the other compounds **1**, **2**, and **4–8** were not active or showed only weak activity (MIC 66.70 μg/mL) against the four tested fungi. A previous study has reported significant antifungal effects of xanthoquinodins A6 and ketoxanthoquinodin A6 against *Co. truncatum* and *Cu. lunata*, and of xathoquinodins B4 and B5 against *Al. brassicicola*, *Co. gloeosporioides*, *Co. truncatum*, *Cu. lunata* and *Py. grisea* [23].

All the compounds **1–8** were active against the three tested Gram-positive bacteria, i.e., *Mi. luteus*, *B. subtilis*, and *S. aureus*, with MIC values in a range of 0.20–8.30 μg/mL, but inactive against *M. smegmatis*. As for the tested Gram-negative bacteria, only compound **8** showed weak activity (MIC 66.70 μg/mL) against *Ch. violaceum*. This agrees with previous studies that demonstrated antibacterial activity of different xanthoquinodins against different Gram-positive bacteria, while activity against any Gram-negative bacteria tested was not observed [23,24].

The cytotoxicity results demonstrated that compounds **1–8** were active against all seven mammalian cell lines (Table 4). Interestingly, all those xanthoquinodins exhibited significant selective cytotoxicity against A431 human squamous cancer cells and MCF-7 human breast cancer cells with half maximal inhibitory concentrations (IC_{50}) values in a range of 0.03–3.11 μM. Similarly, Anaya-Eugenio et al. reported that xanthoquinodin JBIR-99 exhibited high selective antiproliferative activity against PC-3 prostate cancer cells [25]. Compounds **1–3**, and **5**, which possess a F-ring (ρ-hydroxy hexatomic ring) and an ester group at C-1′, showed stronger cytotoxic activities against all tested cell lines, with IC_{50} values in a range of 0.03–1.46 μM, while compound **8** (possessing an OH group at C-1′ instead) displayed cytotoxic activities with IC_{50} values in a range of 0.10–4.70 μM. On the other hand, compounds **4**, **6**, and **7**, which possess a hydroxylated F-ring, γ-lactone ring, and a ring-opening respectively, showed cytotoxic activities against the tested cell lines, with IC_{50} values in a range of 1.03–18.6 μM. Cytotoxicity activities have been observed in different xanthoquinodins [22–24,26]. Sadorn et al. reported that xanthoquinodins A6, B4, and B5, which possessed an F-ring, showed stronger cytotoxicity against cell lines (NCI-H187 and Vero) than other xanthoquinodins with an open F-ring or a γ-lactone ring [23]. Furthermore, Chen et al. observed that xanthoquinodin A6 displayed significant cytotoxicity against all tested human cancer cell lines (HL-60, SMMC-7721, A-549, MCF-7, and SW480), with IC_{50} values in the range of 2.04–6.44 μM [22]. Therefore, the *p*-hydroxy hexatomic F-ring plays a key role in the structure–activity relationship of xanthoquinodins.

Moreover, some xanthoquinodins were reported to possess anticoccidial activity [27–29], as well as antimalarial activity [23]. Further studies need to be performed to confirm the bioactivity against coccidian protozoa and viruses of the new xanthoquinodins described here.

Table 4. Cytotoxicity of **1–8** against mammalian cell lines [half maximal inhibitory concentrations (IC_{50}): μM].

Compound	KB 3.1	L929	A549	SK-OV-3	PC-3	A431	MCF-7
1	0.20	1.13	0.28	0.14	0.71	0.05	0.06
2	0.19	1.46	0.40	0.13	1.06	0.06	0.04
3	0.15	0.98	0.26	0.15	0.65	0.05	0.04
4	5.76	13.48	9.39	4.09	8.18	2.27	1.97
5	0.19	1.12	0.29	0.14	0.62	0.06	0.03
6	3.42	10.56	18.63	3.88	5.43	3.11	1.10
7	3.47	9.97	11.48	3.93	4.98	2.87	1.03
8	1.06	4.70	1.22	0.47	3.14	0.16	0.10
epothilon B	0.00003	0.00051	0.00009	0.00009	0.00007	0.00005	0.00003

3.3. Comparison of Secondary Metabolite Production of Jugulospora spp.

During the course of the ongoing screening for novel biologically active secondary metabolites from the Sordariales, we observed that the ex-type strains of *Apiosordaria globosa* (CBS 110113) and *A. hispanica* (CBS 110112), now synonymyzed with *J. rotula* [14], produced similar chromatograms to the ex-type strain of *J. vestita*, with the six novel xanthoquinodins (**1–3**, **5–7**) and xanthoquinodin B4 also being present in both taxa. Moreover, we found the same novel compounds (except **4** and **7**) in two strains of *J. rotula* (FMR 12690 and FMR 12781), both isolated from soil samples, which is also the same substrate from which the ex-type strain of *J. vestita* and those of *A. globosa* and *A. hispanica* have been isolated. Even though all taxa belonging to the genus *Jugulospora* included in our study produced similar chromatograms, the production of the compounds was variable (see Figure 3). *Jugulospora rotula* CBS 110113 produced the xanthoquinodins in much larger amounts than the other strains, except for the compound **3**, which was produced by *J. vestita* CBS 135.91 as a major metabolite. On the other hand, *J. rotula* CBS 110112 produced much lower amounts of these compounds compared to the other four strains. *J. rotula* strains FMR 12690 and FMR 12781 produced higher quantity of compounds **5** and **8** and less of compounds **1–3** than the ex-type strain of *J. vestita*. Compound **7** was not detected in *J. rotula* FMR 12781, and compound **4** was not observed in any extracted strain of *J. rotula*. These data still rely on a limited number of experiments, and particularly the dependency of production on the culture medium and the time course of production remain to be studied further to get a better idea about their significance. However, they point towards the potential chemotaxonomic utility of xanthoquinodins in *Jugulospora* and allies.

The genus *Apiosordaria* is polyphyletic and scattered along the also polyphyletic family Lasiosphaeriaceae (order Sordariales) [30]. The main problem in the delimitation of lasiosphaeriaceous taxa is that the traditional circumscription based on the ascospore morphology is artificial, being that this an extremely homoplastic character not useful in predicting phylogenetic relationships [31,32]. Even though the structure of the ascomatal wall is clearly more useful for delimitation of some genera, it is not always useful [32]. Recently, *Apiosordaria* was synonymized with *Triangularia* and placed in the family Podosporaceae based on phylogenetic data [30]. However, several species of the genus, such as *A. microcarpa*, remain improperly taxonomically placed (see Figure 4). In that context, the new combination *Jugulospora vestita* has been recently proposed to accommodate *A. vestita*, in the family Schizotheciaceae, according to a phylogenetic study based on the ITS, LSU, *rpb2*, and *tub2* sequences [14].

In the same context, *A. globosa* and *A. hispanica* were synonymized with *J. rotula*. As we mentioned before, all these taxa now included in the genus *Jugulospora* produced similar chromatograms and compounds. However, the other species of *Apiosordaria* included in the screening study, i.e., *A. backusii* (now transferred to *Triangularia* [30]), produced completely different unrelated compounds (data not shown). Therefore, the production of secondary metabolites could be suitable as chemotaxonomic markers, as demonstrated before in other groups of fungi such as the Xylariales [8,33,34], helping to achieve a more natural classification of lasiosphaeriaceous taxa.

Finally, it is important to mention that some of the xanthoquinodins reported to date were found in members also of the Sordariales (which includes the genus *Jugulospora*), i.e., xanthoquinodins Al, A2, A3, Bl, B2, and B3 in *Humicola* sp. [28,29] and xanthoquinodins A4, A5, A6, B4, and B5 in *Chaetomium elatum* [22].

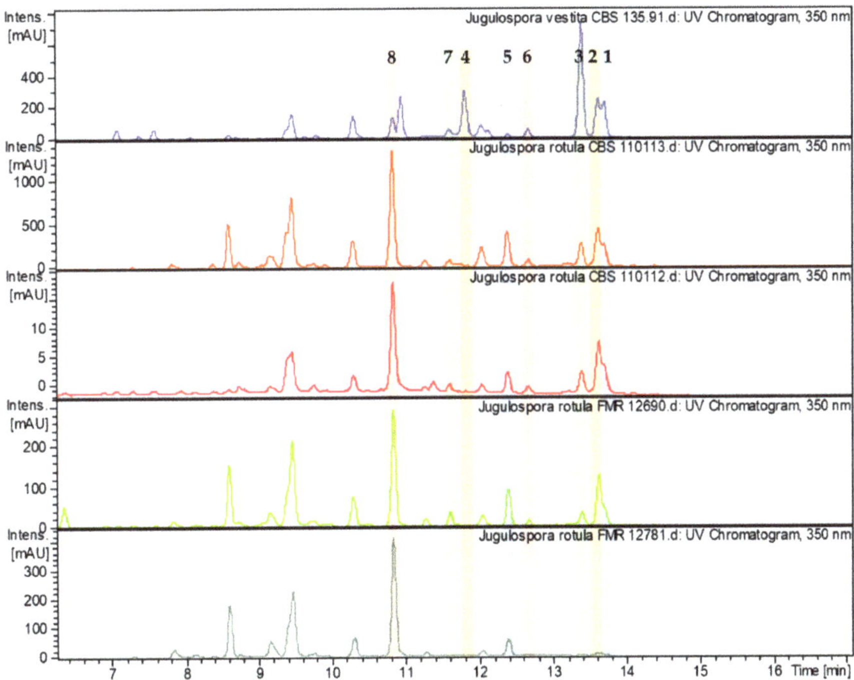

Figure 3. High performance liquid chromatography (HPLC) chromatograms (350 nm) of the ethyl acetate (EtOAc) extracts from the screened strains belonging to the genus *Jugulospora* with peaks of xanthoquinodins indicated by bold numbers referring to the molecules depicted in Figure 1.

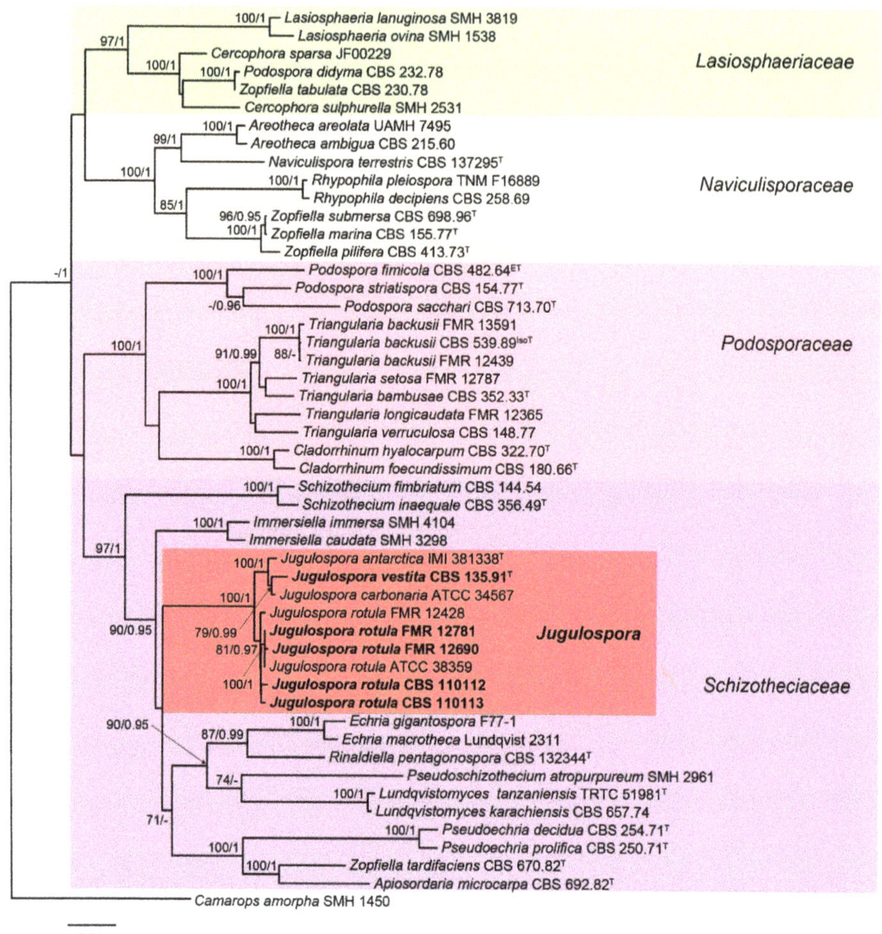

Figure 4. Randomized Axelerated Maximum Likelihood (RAxML) phylogram obtained from the combined sequences of the internal transcribed spacer region (ITS), the nuclear rDNA large subunit (LSU), and fragments of ribosomal polymerase II subunit 2 (*rpb2*) and β-tubulin (*tub2*) genes of selected strains belonging to the families Lasiosphaeriaceae, Naviculisporaceae, Podosporaceae, and Schizotheciaceae. *Camarops amorpha* SMH 1450 was used as an outgroup. Bootstrap support values ≥ 70/Bayesian posterior probability scores ≥0.95 are indicated along branches. Branch lengths are proportional to distance. Screened taxa in the present study are in **bold**. Ex-epitype, ex-isotype, and ex-type strains of the different species are indicated with ^{ET}, ^{IsoT}, and ^{T}, respectively. GenBank accession numbers were indicated in Marin-Felix et al. [14], as well as the methodology followed for performing the phylogenetic study. The major clades are marked in different colors for the sake of better readability.

4. Conclusions

The present study led to the isolation of eight cytotoxic and antimicrobial compounds, seven of which turned out to be new to science. Therefore, the potential of Sordariomycetes as prolific producers of bioactive secondary metabolites was again demonstrated here. Moreover, the production of these compounds by *Jugulospora vestita* and other strains belonging to the same genus, but not by other related genera of Sordariales, suggests that the production of secondary metabolites could be suitable as chemotaxonomic markers.

Supplementary Materials: The following are available online at http://www.mdpi.com/2309-608X/6/4/188/s1, Figure S1: ^{1}H NMR spectrum (500 MHz, chloroform-*d*) of xanthoquinodin A11 (**1**), Figure S2: ^{13}C NMR spectrum (126 MHz, chloroform-*d*) of xanthoquinodin A11 (**1**), Figure S3: HSQC NMR spectrum (500 MHz, chloroform-*d*) of xanthoquinodin A11 (**1**), Figure S4: COSY NMR spectrum (500 MHz, chloroform-*d*) of xanthoquinodin A11 (**1**), Figure S5: HMBC NMR spectrum (500 MHz, chloroform-*d*) of xanthoquinodin A11 (**1**), Figure S6: NOESY NMR spectrum (500 MHz, chloroform-*d*) of xanthoquinodin A11 (**1**), Figure S7: HRESIMS data of xanthoquinodin A11 (**1**), Figure S8: ^{1}H NMR spectrum (500 MHz, chloroform-*d*) of xanthoquinodin B10 (**2**), Figure S9: ^{13}C NMR spectrum (125 MHz, chloroform-*d*) of xanthoquinodin B10 (**2**), Figure S10: HSQC NMR spectrum (500 MHz, chloroform-*d*) of xanthoquinodin B10 (**2**), Figure S11: COSY NMR spectrum (500 MHz, chloroform-*d*) of xanthoquinodin B10 (**2**), Figure S12: HMBC NMR spectrum (500 MHz, chloroform-*d*) of xanthoquinodin B10 (**2**), Figure S13: NOESY NMR spectrum (500 MHz, chloroform-*d*) of xanthoquinodin B10 (**2**), Figure S14: HRESIMS data of xanthoquinodin B10 (**2**), Figure S15: ^{1}H NMR spectrum (500 MHz, chloroform-*d*) of xanthoquinodin B11 (**3**), Figure S16: ^{13}C NMR spectrum (125 MHz, chloroform-*d*) of xanthoquinodin B11 (**3**), Figure S17: HSQC NMR spectrum (500 MHz, chloroform-*d*) of xanthoquinodin B11 (**3**), Figure S18: COSY NMR spectrum (500 MHz, chloroform-*d*) of xanthoquinodin B11 (**3**), Figure S19: HMBC NMR spectrum (500 MHz, chloroform-*d*) of xanthoquinodin B11 (**3**), Figure S20: NOESY NMR spectrum (500 MHz, chloroform-*d*) of xanthoquinodin B11 (**3**), Figure S21: HRESIMS data of xanthoquinodin B11 (**3**), Figure S22: ^{1}H NMR spectrum (700 MHz, chloroform-*d*) of xanthoquinodin B12 (**4**), Figure S23: ^{13}C NMR spectrum (125 MHz, chloroform-*d*) of xanthoquinodin B12 (**4**), Figure S24: HSQC NMR spectrum (500 MHz, chloroform-*d*) of xanthoquinodin B12 (**4**), Figure S25: COSY NMR spectrum (500 MHz, chloroform-*d*) of xanthoquinodin B12 (**4**), Figure S26: HMBC NMR spectrum (700 MHz, chloroform-*d*) of xanthoquinodin B12 (**4**), Figure S27: NOESY NMR spectrum (500 MHz, chloroform-*d*) of xanthoquinodin B12 (**4**), Figure S28: HRESIMS data of xanthoquinodin B12 (**4**), Figure S29: ^{1}H NMR spectrum (700 MHz, chloroform-*d*) of xanthoquinodin B13 (**5**), Figure S30: ^{13}C NMR spectrum (125 MHz, chloroform-*d*) of xanthoquinodin B13 (**5**), Figure S31: HSQC NMR spectrum (700 MHz, chloroform-*d*) of xanthoquinodin B13 (**5**), Figure S32: COSY NMR spectrum (700 MHz, chloroform-*d*) of xanthoquinodin B13 (**5**), Figure S33: HMBC NMR spectrum (700 MHz, chloroform-*d*) of xanthoquinodin B13 (**5**), Figure S34: NOESY NMR spectrum (700 MHz, chloroform-*d*) of xanthoquinodin B13 (**5**), Figure S35: HRESIMS data of xanthoquinodin B13 (**5**), Figure S36: ^{1}H NMR spectrum (700 MHz, chloroform-*d*) of xanthoquinodin B14 (**6**), Figure S37: ^{13}C NMR spectrum (176 MHz, chloroform-*d*) of xanthoquinodin B14 (**6**), Figure S38: HSQC NMR spectrum (700 MHz, chloroform-*d*) of xanthoquinodin B14 (**6**), Figure S39: COSY NMR spectrum (700 MHz, chloroform-*d*) of xanthoquinodin B14 (**6**), Figure S40: HMBC NMR spectrum (700 MHz, chloroform-*d*) of xanthoquinodin B14 (**6**), Figure S41: NOESY NMR spectrum (700 MHz, chloroform-*d*) of xanthoquinodin B14 (**6**), Figure S42: HRESIMS data of xanthoquinodin B14 (**6**), Figure S43: ^{1}H NMR spectrum (500 MHz, chloroform-*d*) of xanthoquinodin B15 (**7**), Figure S44: ^{13}C NMR spectrum (125 MHz, chloroform-*d*) of xanthoquinodin B15 (**7**), Figure S45: HSQC NMR spectrum (500 MHz, chloroform-*d*) of xanthoquinodin B15 (**7**), Figure S46: COSY NMR spectrum (500 MHz, chloroform-*d*) of xanthoquinodin B15 (**7**), Figure S47: HMBC NMR spectrum (500 MHz, chloroform-*d*) of xanthoquinodin B15 (**7**), Figure S48: NOESY NMR spectrum (500 MHz, chloroform-*d*) of xanthoquinodin B15 (**7**), Figure S49: HRESIMS data of xanthoquinodin B15 (**7**).

Author Contributions: Conceptualization, Y.M.-F. and F.S.; methodology, F.S., L.S. and Y.M.-F.; software, M.S.; formal analysis, F.S., L.S. and Y.M.-F.; investigation, F.S., L.S. and Y.M.-F.; visualization, L.S. and Y.M.-F.; resources, M.S. and A.M.S.; data curation, L.S. and Y.M.-F.; writing—original draft preparation, L.S. and Y.M.-F.; writing—review and editing, A.M.S., F.S. and M.S. All authors have read and agreed to the published version of the manuscript.

Funding: A joint PhD training program fellowship from University of Chinese Academy of Sciences (UCAS) was awarded to L.S., and Y.M.-F. was supported by a postdoctoral fellowship from Alexander-von-Humboldt Foundation, Germany.

Acknowledgments: The authors wish to thank Christel Kakoschke for recording the NMR spectra and Wera Collisi for conducting the bioassays.

Conflicts of Interest: The authors declare no conflict of interest.

References

1. Cooper, M.A.; Shlaes, D. Fix the antibiotics pipeline. *Nature* **2011**, *472*, 32. [CrossRef] [PubMed]
2. Sandargo, B.; Chepkirui, C.; Cheng, T.; Chaverra-Munoz, L.; Thongbai, B.; Stadler, M.; Hüttel, S. Biological and chemical diversity go hand in hand: Basidiomycota as source of new pharmaceuticals and agrochemicals. *Biotechnol. Adv.* **2019**, *37*, 107344. [CrossRef]

3. Rupcic, Z.; Rascher, M.; Kanaki, S.; Köster, R.W.; Stadler, M.; Wittstein, K. Two new cyathane diterpenoids from mycelial cultures of the medicinal mushroom *Hericium erinaceus* and the rare species, *Hericium flagellum*. *Int. J. Mol. Sci.* **2018**, *19*, 740. [CrossRef]
4. Padma, V.V. An overview of targeted cancer therapy. *BioMedicine* **2015**, *5*, 19. [CrossRef]
5. Alexandre, J.; Raymond, E.; Kaci, M.O.; Brain, E.C.; Lokiec, F.; Kahatt, C.; Faivre, S.; Yovine, A.; Goldwasser, F.; Smith, S.L.; et al. Phase I and pharmacokinetic study of irofulven administered weekly or biweekly in advanced solid tumor patients. *Clin. Cancer Res.* **2004**, *10*, 3377–3385. [CrossRef]
6. Miyamoto, M.; Takano, M.; Kuwahara, M.; Soyama, H.; Kato, K.; Matuura, H.; Sakamoto, T.; Takasaki, K.; Aoyama, T.; Yoshikawa, T.; et al. Efficacy of combination chemotherapy using irinotecan and nedaplatin for patients with recurrent and refractory endometrial carcinomas: Preliminary analysis and literature review. *Cancer Chemother. Pharmacol.* **2018**, *81*, 111–117. [CrossRef]
7. Topka, S.; Khalil, S.; Stanchina, E.; Vijai, J.; Offit, K. Preclinical evaluation of enhanced irofulven antitumor activity in an ERCC3 mutant background by in vitro and in vivo tumor models. *AACR* **2018**, *78*, 3258.
8. Surup, F.; Kuhnert, E.; Lehmann, E.; Heitkämper, S.; Hyde, K.D.; Fournier, J.; Stadler, M. Sporothriolide derivatives as chemotaxonomic markers for *Hypoxylon monticulosum*. *Mycology* **2014**, *5*, 110–119. [CrossRef]
9. Helaly, S.E.; Thongbai, B.; Stadler, M. Diversity of biologically active secondary metabolites from endophytic and saprotrophic fungi of the ascomycete order Xylariales. *Nat. Prod. Rep.* **2018**, *35*, 992–1014. [CrossRef] [PubMed]
10. Wang, T.; Mohr, K.I.; Stadler, M.; Dickschat, J.S. Volatiles from the tropical ascomycete *Daldinia clavata* (Hypoxylaceae, Xylariales). *Beilstein J. Org. Chem.* **2018**, *14*, 135–147. [PubMed]
11. Bills, G.; Gloer, J. Biologically active secondary metabolites from the fungi. *Microbiol. Spectr.* **2016**, *4*. [CrossRef]
12. Vicente, F.; Basilio, A.; Platas, G.; Collado, J.; Bills, G.F.; González del Val, A.; Martín, J.; Tormo, J.R.; Harris, G.H.; Zink, D.L.; et al. Distribution of the antifungal agents sordarins across filamentous fungi. *Mycol. Res.* **2009**, *113*, 754–770. [CrossRef] [PubMed]
13. Udagawa, S.; Horie, Y. Two new species of terrestrial Ascomycetes from Eastern Nepal. In *Reports on the Cryptogamic Study in Nepal*; Otani, Y., Ed.; Miscellaneous Publication of the National Science Museum: Tokyo, Japan, 1982.
14. Marin-Felix, Y.; Miller, A.N.; Cano-Lira, J.F.; Guarro, J.; García, D.; Stadler, M.; Huhndorf, S.M.; Stchigel, A.M. Re-evaluation of the order Sordariales: Delimitation of Lasiosphaeriaceae s. str., and introduction of the new families Diplogelasinosporaceae, Naviculisporaceae and Schizotheciaceae. *Microorganisms* **2020**, *8*, 1430. [CrossRef] [PubMed]
15. Becker, K.; Wessel, A.C.; Luangsa-ard, J.J.; Stadler, M. Viridistratins A-C, antimicrobial and cytotoxic benzo[*j*]fluoranthenes from stromata of *Annulohypoxylon viridistratum* (Hypoxylaceae, Ascomycota). *Biomolecules* **2020**, *10*, 805. [CrossRef]
16. Katoh, K.; Standley, D.M. MAFFT multiple sequence alignment software v. 7: Improvements in performance and usability. *Mol. Biol. Evol.* **2013**, *30*, 772–780. [CrossRef]
17. Tamura, K.; Stecher, G.; Peterson, D.; Filipski, A.; Kumar, S. MEGA6: Molecular Evolutionary Genetics Analysis version 6.0. *Mol. Biol. Evol.* **2013**, *12*, 2725–2729. [CrossRef]
18. Mason-Gamer, R.; Kellogg, E. Testing for phylogenetic conflict among molecular data sets in the tribe Triticeae (Gramineae). *Syst. Biol.* **1996**, *45*, 524–545. [CrossRef]
19. Wiens, J.J. Testing phylogenetic methods with tree congruence: Phylogenetic analysis of polymorphic morphological characters in phrynosomatid lizards. *Syst. Biol.* **1998**, *47*, 427–444. [CrossRef]
20. Hernández-Restrepo, M.; Groenewald, J.Z.; Elliott, M.L.; Canning, G.; McMillan, V.E.; Crous, P.W. Take-all or nothing. *Stud. Mycol.* **2016**, *83*, 19–48. [CrossRef]
21. Alfaro, M.E.; Zoller, S.; Lutzoni, F. Bayes or bootstrap. A simulation study comparing the performance of Bayesian Markov chainMonte Carlo sampling and bootstrapping in assessing phylogenetic confidence. *Mol. Biol. Evol.* **2003**, *20*, 255–266. [CrossRef]
22. Chen, G.D.; Chen, Y.; Gao, H.; Shen, L.Q.; Wu, Y.; Li, X.X.; Li, Y.; Guo, L.D.; Cen, Y.Z.; Yao, X.S. Xanthoquinodins from the endolichenic fungal strain *Chaetomium elatum*. *J. Nat. Prod.* **2013**, *76*, 702–709. [CrossRef] [PubMed]
23. Sadorn, K.; Saepua, S.; Boonyuen, N.; Boonruangprapa, T.; Rachtawee, P.; Pittayakhajonwut, P. Antimicrobial activity and cytotoxicity of xanthoquinodin analogs from the fungus *Cytospora eugeniae* BCC42696. *Phytochemistry* **2018**, *151*, 99–109. [CrossRef] [PubMed]

24. Tantapakul, C.; Promgool, T.; Kanokmedhakul, K.; Soytong, K.; Song, J.; Hadsadee, S.; Jungsuttiwong, S.; Kanokmedhakul, S. Bioactive xanthoquinodins and epipolythiodioxopiperazines from *Chaetomium globosum* 7s-1, an endophytic fungus isolated from *Rhapis cochinchinensis* (Lour.) Mart. *Nat. Prod. Res.* **2018**, *34*, 494–502. [CrossRef] [PubMed]
25. Anaya-Eugenio, G.D.; Rebollar-Ramos, D.; Gonzalez, M.D.C.; Raja, H.; Mata, R.; Carcache de Blanco, E.J. Apoptotic activity of xanthoquinodin JBIR-99, from *Parengyodontium album* MEXU 30054, in PC-3 human prostate cancer cells. *Chem. Biol. Interact.* **2019**, *311*, 108798. [CrossRef] [PubMed]
26. Ueda, J.Y.; Takagi, M.; Shin-ya, K. New xanthoquinodin-like compounds, JBIR-97, -98 and -99, obtained from marine sponge-derived fungus *Tritirachium* sp. SpB081112MEf2. *J. Antibiot.* **2010**, *63*, 615–618. [CrossRef]
27. Tabata, N.; Tomoda, H.; Matsuzaki, K.; Omura, S. Structure and biosynthesis of xanthoquinodins, anticoccidial antibiotics. *J. Am. Chem. Soc.* **1993**, *115*, 8558–8564. [CrossRef]
28. Tabata, N.; Suzumura, Y.; Tomoda, H.; Masuma, R.; Haneda, K.; Kishi, M.; Iwai, Y.; Omura, S. Xanthoquinodins, new anticoccidial agents produced by *Humicola* sp. *J. Antibiot.* **1993**, *46*, 749–755. [CrossRef]
29. Tabata, N.; Tomoda, H.; Iwai, Y.; Omura, S. Xanthoquinodin B3, a new anticoccidal agent produced by *Humicola* sp. FO-888. *J. Antibiot.* **1995**, *49*, 267–271. [CrossRef]
30. Wang, X.W.; Bai, F.Y.; Bensch, K.; Meijer, M.; Sun, B.D.; Han, Y.F.; Crous, P.W.; Samson, R.A.; Yang, F.Y.; Houbraken, J. Phylogenetic re-evaluation of *Thielavia* with the introduction of a new family Podosporaceae. *Stud. Mycol.* **2019**, *93*, 155–252. [CrossRef]
31. Miller, A.N.; Huhndorf, S.M. A natural classification of *Lasiosphaeria* based on nuclear LSU rDNA sequences. *Mycol. Res.* **2004**, *108*, 26–34. [CrossRef]
32. Miller, A.N.; Huhndorf, S. Multi-gene phylogenies indicate ascomal wall morphology is a better predictor of phylogenetic relationships than ascospore morphology in the Sordariales (Ascomycota, Fungi). *Mol. Phylogenetics Evol.* **2005**, *35*, 60–75. [CrossRef] [PubMed]
33. Sir, E.B.; Kuhnert, E.; Lambert, C.; Hladki, A.I.; Romero, A.I.; Stadler, M. New species and reports of *Hypoxylon* from Argentina recognized by a polyphasic approach. *Mycol. Prog.* **2016**, *15*, 42. [CrossRef]
34. Kuhnert, E.; Sir, E.B.; Lambert, C.; Hyde, K.D.; Hladki, A.I.; Romero, A.I.; Rohde, M.; Stadler, M. Phylogenetic and chemotaxonomic resolution of the genus *Annulohypoxylon* (Xylariaceae) including four new species. *Fungal Divers.* **2017**, *85*, 1–43. [CrossRef]

Journal of
Fungi

iew

ıfety Evaluation of Fungal Pigments for Food Applications

endran Poorniammal [1,*], Somasundaram Prabhu [2], Laurent Dufossé [3,*] and Jegatheesh Kannan [1]

1 Department of Natural Resource Management, Horticultural College and Research Institute, Tamil Nadu Agricultural University (TNAU), Periyakulam 625 604, India; kannan.j@tnau.ac.in
2 Department of Plant Protection, Horticultural College and Research Institute, Tamil Nadu Agricultural University (TNAU), Periyakulam 625 604, India; prabhu.s@tnau.ac.in
3 Laboratoire de Chimie et Biotechnologie des Produits Naturels (CHEMBIOPRO), Université de La Réunion, ESIROI Agroalimentaire, 15 Avenue René Cassin, F-97400 Sainte-Clotilde, France
* Correspondence: r.poornii@tnau.ac.in (R.P.); laurent.dufosse@univ-reunion.fr (L.D.); Tel.: +33-668-73-19-06 (L.D.)

Abstract: Pigments play a major role in many industries. Natural colors are usually much safer when compared to synthetic colors and may even possess some medicinal benefits. Synthetic colors are economical and can easily be produced compared to natural colors. In addition, raw plant materials for natural colors are limited and season dependent. Microorganisms provide an alternative source for natural colors and, among them, fungi provide a wide range of natural colorants that could easily be produced cheaply and with high yield. Along with pigment, some microbial strains are also capable of producing a number of mycotoxins. The commercial use of microbial pigments relies on the safety of colorants. This review provides a toxicity evaluation of pigments from fungal origins for food application.

Keywords: fungal pigments; mycotoxins; safety evaluation; pigment toxicity

tion: Poorniammal, R.; Prabhu, ufossé, L.; Kannan, J. Safety uation of Fungal Pigments for 1 Applications. *J. Fungi* **2021**, *7*, https://doi.org/10.3390/)90692

demic Editor: David S. Perlin

ived: 16 June 2021
pted: 23 August 2021
ished: 26 August 2021

lisher's Note: MDPI stays neutral regard to jurisdictional claims in ished maps and institutional affilns.

1. Introduction

Due to the global increase in processed food production, the market for fresh food is predicted to grow more in coming years. Modern consumers have become more nutrition and health conscious and have a growing interest in *(i)* where their food comes from and *(ii)* in food additives [1]. Food labeling has increased interest in understanding the physiological needs of the body. Consumers tend to shy away from chemical compounds such as food additives, viz., antioxidants, preservatives and colors. They search for natural additives in natural foods that are safe and good for them. On the other hand, consumers have a growing awareness of natural products where toxicants may also be present [2].

Historically, throughout the world, natural pigments have been used for many purposes. The advent of synthetic dyes reduced the use of natural pigments to a greater extent. Recently, dyes derived from natural sources are gaining importance, since some synthetic dyes have been reported to have carcinogenic effects [3,4]. Natural pigments gained interest due to worldwide concern over the use of eco-friendly and biodegradable materials. Demand for natural colorants, especially yellow and red pigments, israpidly increasing worldwide. Production of safe and natural pigments from natural resources are mainly being focused on in food, textile and pharmaceutical industries, due to the serious environmental and safety problems caused by many artificial synthetic pigments [5].

Many natural pigments are from plant or microorganism-based origins. There are a number of major drawbacks to plant pigments, viz., non-availability throughout the year, pigment stability and solubility. Exploitation of plants on a large scale may lead to a loss of valuable species. For these reasons, the pigments from plant sources are not considered viable. Microorganisms, viz., fungi, bacteria, algae and actinomycetes, are a reliable and readily available alternative source of natural pigments [6,7]. Microorganisms are advantageous over plant for pigments production because of easy and rapid multiplication i low cost medium, easy processing and growth independent from weather conditions [

2. Synthetic Food Colors

Color is an important factor as far as food is concerned, as it plays a major role the taste and perception of food, along with flavor and texture. It is a known fact t consumers will probably reject food that does not look attractive. To make food m appealing to customers, manufacturers add color to retain the food's natural look, as far possible. Natural appearance is always preferable to anything that looks unusually colo Even though many foods can contain added artificial colors, most consumers believe t the color of the food is its natural color.

The first synthetic organic dye, discovered by William Henry Perkin in 1856, w a purplish lilac color named "mauve". Similar organic aniline dyes were synthesiz representing every color and tint of the rainbow, and these were used for food colori with little or few tests regarding their safety. Aniline and coaltar-based colors exhibi significant toxicity, which prompted regulators to examine the exact use of synthetic col in the food industry [9].

In recent years, an astonishing amount of the food we eat is processed. To increase shelf life and appearance of food, additives and colors are added, which make them uns for consumption. On average, processed food accounts for about 70% of the diet of U residents. This includes soft drinks, confectionery, packaged bread, buns, biscuits, cak preserved meat products, instant soups, noodles, packaged pizzas, pies and packag meals. The U.S. consumption of processed food is about forty times more than the diet Indian residents [10].

Health Hazards of Synthetic Food Colors

Risk analysis helps to evaluate the adverse effects of synthetic agents in food. I global context, food colors are of major concern with regard to the possible adverse effe of additives. In the mid-1980s, a possible link between tartrazine and hyperactivity children was suggested [11].

1. The azo-dye group of colorants consists of bright colors and is widely used in the fo industry. Increasing attention on these dyes revealed that they were potential carci gens, occurring in the intestines' microbiota, after their azo reduction to carcinoge metabolites [12]. Even at low levels of ingestion, permitted food colors, viz., ponce tartrazine and sunset yellow, provoked allergic reactions in many individuals. Comm allergic responses were urticaria, dermatitis, angioedema and the exacerbation of as matic symptoms [13]. A symptom of glossitis was reported due to the consumpti of a very high level of ponceau 4R in a particular brand of aniseed [14]. Hypertens children aged between 2 and 14 years were diagnosed with irritability, restlessness a sleep disturbance due to high levels of tartrazine. Ever rising demand for the use natural colorants has replaced the use of synthetic dyes in food [15].

3. Microbial Pigments

Natural pigments obtained from plants, animals and microorganisms are eco-frienc and have usually low or no toxicity [16,17]. The many disadvantages of using plants a animals prevent them from large-scale exploitation [18]. However, advantages of microb pigments help to utilize their immense potential in various fields [19,20]. Even thou the cost of microbial β-carotene production is several times more expensive, it can s compete with synthetic dyes in terms of it being natural and safe [21,22].

Microbial cells that produce color are referred to as microbial pigments producers [2 They produce a wide range of colors (Figure 1) and are mostly water-soluble [24,2 Natural pigments are mainly used as color additives or intensifiers; moreover, they a used as antioxidants and antibiotics (Figure 2). Due to indiscriminate use of synthetic col and contrary reports on the safety of synthetic dyes, there is an important need to ident safe colorants from natural pigments. Microbial pigments have several advantages, viz., yield, cost efficiency, stability and ease of downstream processing compared to pigments from plant or animal origins [26,27].

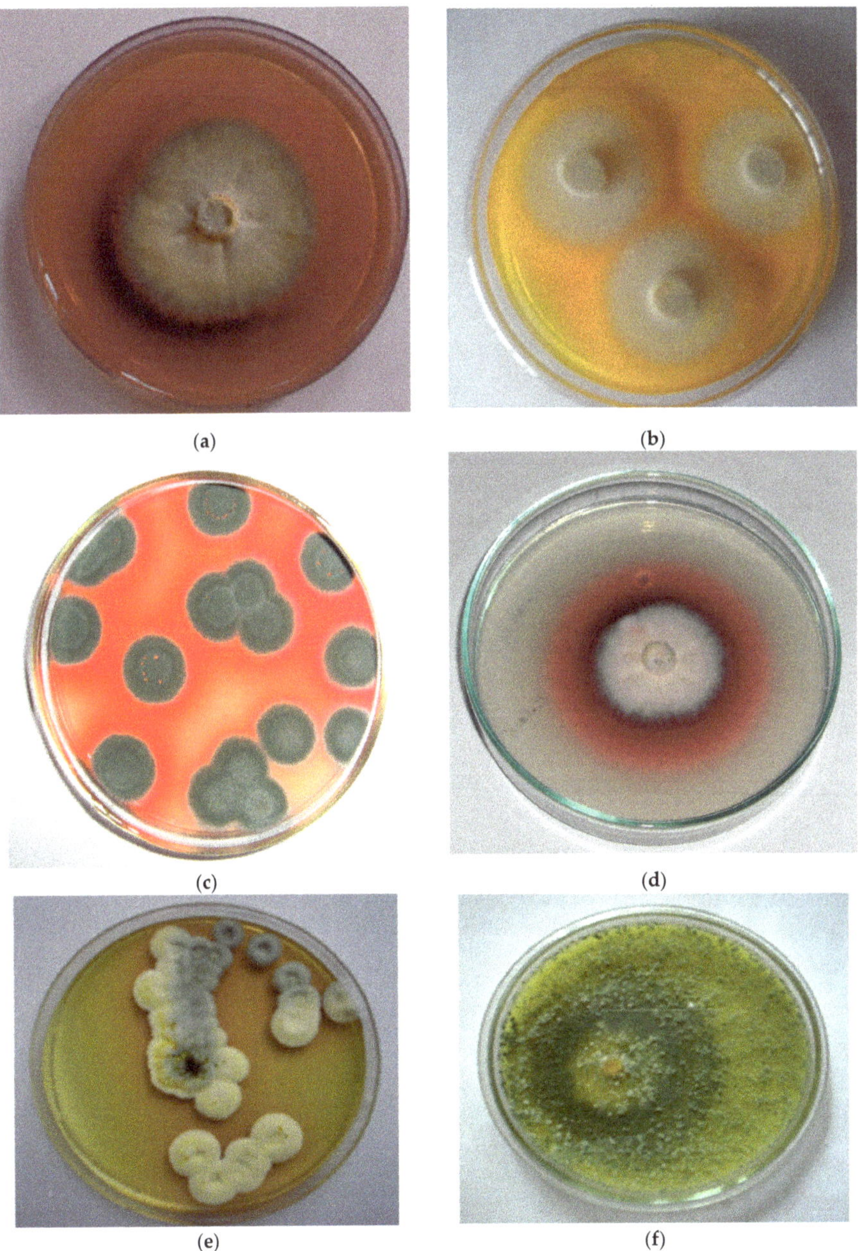

Figure 1. Pigments produced by different fungi: (**a**) *Chaetomium* sp. producing red pigment; (**b**) *Thermomyces* sp. producing yellow pigment; (**c**) *Penicillium purpurogenum* producing red pigment; (**d**) *Fusarium* sp. producing red pigment; (**e**) *Penicillium purpurescens* producing brown pigment (**f**) *Trichoderma* sp. producing yellow pigment.

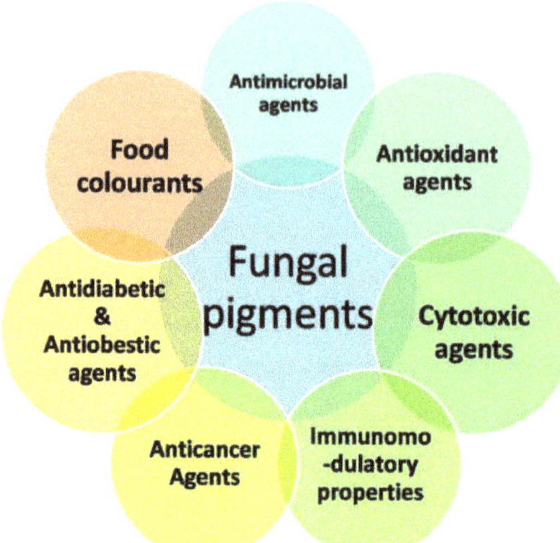

Figure 2. Applications of fungal pigments in the food industry.

Among pigment-producing microbes, fungi produce a wide range of water-soluble b pigments that have a variety of functions. Pigments extracted from fungi that are isolat from soil have various industrial applications. Filamentous fungi, viz., *Monascus, Aspergill Penicillium, Neurospora, Eurotium, Drechslera* and *Trichoderma* [28–30] are potential produce of bio-pigments. The pigments include carotenoids, melanins, flavins, phenazines, quinon monacins and indigo [31]. Hence, they are the subject of many studies.

Recently, fungal pigments have been used for textile dyes, food colorants, antimicr bial and anticancer applications. They are also natural without having undesirable effec on the environment. Many scientific researchers have proved that pigments from so fungi are a safer alternative to synthetic colorants, and there is good scope for industri application [32,33].

4. Fungal Pigments and Toxicity Evaluation

Most fungi produce pigments along with mycotoxins. The presence of mycotoxins pigments restricts the application of pigments as an additive in the food industry [16,3 The European Union and the United States prohibit the consumption of *Monascus* pi ments that are produced along with citrinin toxin, which poses a challenge over its sa use [35]. In short, natural pigments are a potential source of colorants that are eco-friend biodegradable, antimicrobial and have antioxidant properties. Apart from food additiv they are also used in cosmetics, pharmaceuticals and drug applications [36] (Table 1).

Table 1. Important fungal pigments and their safety evaluation.

Fungus	Pigment(s)	Color	Mycotoxin(s)	Safety Evaluation	Biological Activity	Reference(s)
Aspergillus carbonarius	Melanins	Yellow	Not described/not found up to now	Subacute toxicity study	Antioxidant	[37]
Blackeslea trispora	ß-carotene	Red-orange	Aflatoxin Mycotoxin	Genotoxicity and subacute toxicity study	Antioxidant, anticancer, suppression of cholesterol synthesis	[38,39]
Fusarium graminearum	Rubrofusarin	Red	Fumonisins, Zearalenone, Fusaric Acid, Fusarins and Beauvericins	Cytotoxic in colon cells	Antimicrobial, antiallergic, phytotoxicic	[40–42]
Fusarium fujikuroi	Fusarubin	Orange	Fumonisins, Zearalenone, Fusaric Acid, Fusarins and Beauvericins	Cytotoxic against leukemia cells	Anticancer, antimicrobial	[42,43]
Fusarium oxysporum	Bikaverin	Red	Fumonisins, Zearalenone, Fusaric Acid, Fusarins and Beauvericins	Cytotoxic against tumour cells, apoptosis suppressor	Antimicrobial, antitumour	[44]
Monascus purpureus	Monascorubramine rubropunctamine	Red	Citrinin	Acute oral toxicity	Antihypertensive metabolite	[45]
Monascus anka	Ankaflavin and Monascin	Yellow	No coproduction of toxin	Acute oral toxicity	Antibacterial, antitumor and immunosuppressive	[46]
Monascus ruber	Monascorubrin and rubropunctatin	Orange-red	No coproduction of toxin	Oral toxicity	Anti-inflammatory, anticancer andantihyperlipidemic activities	[47]
Penicillium purpurogenum	Azaphilone Purpurogenone Mitorubrino Mitorubrin	Brick red pigment Yellow-orange Orange-red Yellow	No coproduction of toxin	Brine shrimp *Artemia salina* study	Pharmaceutical and food industry	[3,48,49]
Penicillium europium	Benzoquinone	Pinkish red	Nontoxic	Subacute toxicity study	Antimicrobial	[50]
Penicillium resticulosum	Not described/not found up to now	Red	Nontoxic	Subacute toxicity study	Antimicrobial	[51]

Table 1. *Cont.*

Fungus	Pigment(s)	Color	Mycotoxin(s)	Safety Evaluation	Biological Activity	Reference(s)
Penicillium aculeatum	Ankaflavin	Yellow	Nontoxic, selective toxicity in cancer cells	Cytotoxicity study	Antimicrobial	[52]
Talaromycespurpureogenus	Purpuride, monascorubrin, purpurquinone-A, ankaflavin,	Yellow and red	Nontoxic	Subacute toxicity study	Antioxidant	[53]
Thermomyces sp.	Napthoquinone	Yellow	Nontoxic	Subacute toxicity study	Antioxidant, antimicrobial and food industry	[54]
Trichoderma viride	Emodin Viridol	Brown Yellow	Nonphytotoxic	Phytotoxicity assay	Antimicrobial	[28]
Scytalidium cuboideum	Xylindein	Red	Nontoxic	Zebrafish toxicity study	UV resistant	[55]
Rhodotorula glutinis	β-carotene, torulene and torularhodin	Red and orange	Nontoxic	Standard subchronic toxicity study	Antioxidant, Antimicrobial Food and feed additive	[56]
Rhodotorula gracilis	β-carotene, torulene and torularhodin	Red and orange	Nontoxic	Standard subchronic toxicity study	Antimicrobial	[57,58]
Yarrowia lipolytica	β-carotene	Brown	Nontoxic	Genotoxicity models and a standard subchronic toxicity study	Antimicrobial	[59]

4.1. *Aspergillus carbonarius*

Aspergillus carbonarius, an Ascomycota fungus of the family Aspergillaceae, is capable of producing a yellow-colored pigment in its biomass. It does not produce any antinutrients or mycotoxins [60]. It has been exploited for large-scale production of polygalacturonase and is capable of temperature tolerance by UV irradiation when grown in shake-flask cultures. During the growth phase, a yellow colored pigment is accumulated in its biomass and has the potential to be used as a food colorant [61].

Toxicity studies in both sexes of albino rats at acute and subacute doses of the pigment revealed that feeding of fungal biomass did not show any mortality in rats and there are no significant differences in food intake or organ and body weight. When comparing treated and untreated rats, hematological parameters, serum enzymes lactate dehydrogenase (LDH), alkaline phosphatase (ALP), alanine aminotransferase (ALT or ALAT) and cholesterol assay also remain normal [62].

4.2. *Blackslea trispora*

Blakesleatrispora is a Zygomycetes fungus of the order Mucorales, family Choanephoraceae. It is capable of undergoing both sexual and asexual reproduction through the production of zygospores and sporangiospores. The fungus does not produce any toxic compounds; hence, it is of industrial interest as a source of β-carotene for commercial exploitation [63,64]. β-carotene from *B. trispora* was the first authorized microbial food colorant in the European Union. It is efficient and can achieve the highest yield of all trans β-carotene at the expense of other structurally related carotenoids [33]. The process production was improved over a number of years, producing carotenoid contents of up to 20% dry weight [65,66].

The safety assessment of β-carotene, derived from *B. trispora,* has revealed no genotoxicity or subacute toxicity for 4weeks [38,67]. A subchronic toxicity study of 90 days was performed with oral administration of F344.Rats of both sexes showed no adverse effects on their biological systems [39]. β-carotene derived from the *B. trispora* at a 5.0% dietary level, equivalent to 3127 mg/kg/day and 3362 mg/kg/day for male and female rats, caused no adverse effects. The findings revealed that the daily intake of synthetic β-carotene from *B. trispora* by human beings is a negligible toxicological hazard [68].

4.3. *Fusarium* sp.

Fusarium are Ascomycota fungi that belong to the order Hypocreales, family Nectriaceae. They produce a wide range of fungal pigments that are structurally and functionally diverse. However, among the *Fusarium* sp., *Fusarium graminearum* (red naphthoquinone pigment, *rubrofusarin*) [40], *Fusarium fujikuroi* (orange carotenoids pigment, fusarubin) [43,69] and *Fusarium oxysporum* (red naphthoquinone pigment, bikaverin) [44] are the major pigment-producing fungi. Secondary metabolites from these fungi contain numerous toxic compounds, viz., fumonisins, zearalenone, fusaric acid, fusarins and beauvericins [70].

Toxicity analysis revealed that the red dimeric naphthoquinone pigment from *F. oxysporum*-contaminated products affects human health. Recently, red naphthoquinone pigment has often been reported as a mycotoxin. However, naphthoquinone pigment was not genotoxic according to a DNA synthesis assay. Biotechnological approaches and intelligent screening of the toxic metabolite pathway of the pigment from *Fusarium* sp. will be helpful in producing the pigment for food coloring [41,42].

4.4. *Monascus* sp.

Monascus sp. are fungi placed under order Eurotiales, family Monascaceae. There are many species in this genus, among which *M. purpureus* (monascorubramine and rubropunctamine) [45], *M. anka* (ankaflavin and monascin) [46] and *M. ruber* (monascorubrinandrubropunctatin) [47] are of greatest significance to the food industry. Traditionally, Monascus pigments were produced on rice using solid-state microbial fermentation. Synonyms for this food product include, Hon-Chi, Hong Qu, Dan Qu, Anka, Ankak rice,

Beni-Koji, red koji, red Chinese rice, red yeast rice and red mold rice (RMR). RMR w utilized as a food colorant in traditional Chinese medicine for more than 1000 years [71

Chinese, as well as other East Asian people, have confirmed the safety of red yeast r The European Food Safety Authority (EFSA) and the United States excluded red yeast r on the list of permissible food additives, due to complex secondary metabolites [72,73]. T toxigenic strain of *Monascus purpureus* is capable of producing nephrotoxic and hepatoto mycotoxin citrinin, which limits the wide application of the pigment [74].

For more than a thousand years, pigments produced by *Monascus* sp. were lega used as food colorants in South East Asia, even though they were demonstrated to ha physiological effects. There are numerous toxicological data available on this Monasc red pigment.

A genetically modified industrial strain, *M. purpureus* SM001 isolated in China, is capal of producing pigment without citrinin, which is the best *Monascus* pigment producer.Th results in the prolonged safety of *Monascus*-related products and their application [75].

4.5. *Penicillium* sp.

Penicillium are Ascomycota fungi belonging to the order Eurotiales, family Tricho maceae. They are capable of producing many pigments. *Penicillium* are ubiquitous sap phytic soil fungi, present wherever organic material is available. Several species are capal of producing highly toxic mycotoxins. Some species of the genus *Penicillium* are capal of producing antibiotics, while some other species are used in cheese making; howev pigment production by these fungi is less well known [48,76]. Patents contain informati about acute oral toxicity in mice. A 90-day subchronic toxicological study found acu dermal irritation, acute eye irritation, antitumor activity, micronucleus test in mice, AM test (*Salmonella typhimurium* reverse mutation assay) and an estimation of antibiotic activi including results of estimation of five mycotoxins [77].

Penicillium purpurogenum is capable of producing an azaphilone-like pigment. It s cretes a brick red pigment during growth, which generally diffuses into commonly us media. However, violet pigment (PP-V) and orange pigment (PP-O) were also report by altering culture conditions [78]. The production of pigment from *Penicillium* is mo efficient and profitable than any other microorganism. It secretes enzymes and pigmen out of the cell and the secreted pigment is water-soluble and relatively stable; thus, it easily purified [79].

Toxicity studies of *P. purpurogenum* DPUA 1275 on brine shrimp, *Artemia salina*, show antimicrobial effects and absence of toxicity to go along with pigment production. It al does not produce any known mycotoxins and is nonpathogenic to humans. It is a potenti strain for the production of food pigments [80]. Although many species of *Penicillium* a found to produce pigments, only a few toxicological studies have been conducted.

Penicillium europium, isolated from forest soil, is capable of producing a pinki pigment by using longifolene as a sole carbon source. A toxicity study on albino ra revealed that the pigment had no toxic effect on rats. Synthesized pigments from *P. europiu* could be used in food, feed and pharmaceutical industries. Apart from the food indust it could be used for various industrial applications, viz., dyes for textile and non-texti substrates such as paper, leather, paints and cosmetics. Moreover, as it is non cytotoxic, t pigment could be a potential replacement for hazardous synthetic dyes [50,81].

Penicillium resticulosum is capable of producing red pigments. An evaluation of t subacute toxicity of oral exposure on the synthesized pigment on adult male and fema mice for 28 days, using a pigment dose of up to 500 mg kg^{-1} body weight daily, h no effect on body weight, organ weight, or the activity of lactate dehydrogenase (LD alkaline phosphatase (ALP), alanine aminotransferase (ALT or ALAT) enzymes or bloc urea nitrogen (BUN) levels. However, mice taking the pigment over 500 mg·kg^{-1} bo weight daily showed fatty degeneration and mild necrosis of liver cells, indicating th doses under 500 mg·kg^{-1} body weight were safe for daily consumption [51,82].

Penicillium aculeatum produces a yellow (ankaflavin) pigment under submerged fermentation. Cytotoxicity studies of the pigment interacting with human colon carcinoma cell lines (HCT116) and human prostatic carcinoma cell lines (PC3) exhibited apoptosis and cell cycle inhibition at lower concentrations. An assay of human erythrocytes and human embryonic kidney (HEK-293) cell lines showed the least cytotoxicity atfor highest concentrations tested. Displaying selective cytotoxicity is an important property for an ideal anticancer drug [52].

4.6. *Talaromyces purpureogenus*

Talaromyces purpureogenus (basionym: *Penicillium purpureogenum*), is capable of producing yellow and red pigments under submerged fermentation. Pigments from *T. purpureogenus* CFRM02 are non toxic to *Artemia franciscana* (brine shrimp). In a single-dose acute toxicity study, (50, 300, 1000 and 2000 mg/kg body weight) conducted on female Wistar rats, there was no evidence of adverse effects on body weight and mortality after 14 days. Subacute studies (250–1000 mg/kg body weight) showed no significant changes in food intake, body weight gain and relative weight of vital organs after 28 days. Furthermore, a histopathological examination of the liver and kidney was normal. There were no significant changes in serum enzyme activities in the treated and control groups (acute and subacute). Safety efficacy of the pigment from *T. purpureogenus* CFRM02 is suggested for application in food and nutraceuticals [53]. This potential strain has resulted in in-depth studies of some strains of Talaromyces species, viz., *Talaromyces aculeatus*, *T. funiculosum*, *T. pinophilus* and *T. purpurogenum*. They are capable of producing *Monascus*-like polyketide azaphilone pigments, with or without coproducing citrinin or any other known mycotoxins [83].

4.7. *Thermomyces* sp.

Thermomyces are Ascomycota fungi ofthe order Eurotiales, family Trichocomaceae, and they are capable of producing a yellow pigment. They are thermophilic and hemicellulose degraders. One strain was isolated from soil in Kodaikanal, Dindugul District, Tamil Nadu, India. The yellow pigment can scavenge reactive oxygen species (ROS) induced by chemicals and UV rays and reduce the DNA damage by their antioxidant capabilities [84]. The bright pigmentation varies from yellow to red, depending on the growth, temperature, age and substrate [85].Food and beverages fortified by the yellow pigment recorded high antioxidant properties, antimicrobial properties and color stability [86].

Toxicological studies on the *Thermomyces* pigment using albino mice that orally ingested *Thermomyces* sp. pigment for 28 days showed no alterations in red blood cells, white blood cells, haemoglobin, organ weight orhistopathology of the liver or kidneys [54] (Figure 3). Apart from pigmentation, in vivo antioxidant activity was also observed [87,88]. Therefore, pigments from *Thermomyces* sp. could be safe for the food industry.

4.8. *Trichoderma viride*

Trichoderma viride is an Ascomycota fungus belonging to the order Hypocreales, family Hypocreaceae. It is capable of producing a brown colored pigment, identified as furfural (only one study and this should be confirmed) [89]. It is mostly used as a biological control agent against soil-borne plant pathogenic fungi. It is also capable of synthesizing emodin, a yellow pigment. *T. polysporum* is the only fungi capable of producing emodin in culture media. A phytotoxicity assay proved the nontoxic nature of *T. viride* on the germination of *Phaseolus aureus* Roxb. The pigment, i.e., furfural, produced by *Trichoderma viride*, has possible wide applications in various industries [89].

4.9. *Scytalidium cuboideum*

Scytalidium cuboideum, an Ascomycota fungus belonging to the order Helotiales, family Chaetomiaceae, is isolated from wood and is capable of coloring wood. It produces a red pigment, identified as xylindein.

Toxicity studies of the red pigment in zebrafish revealed a relatively high LD50 val making it unlikely to affect humans. Pure solidified pigments from *S. cuboideum* did demonstrate toxicity. However, significant mortality was associated with impure fun metabolites containing pigments. Hence, xylindein is considered as an environmenta safe pigmentation for future applications [55,90].

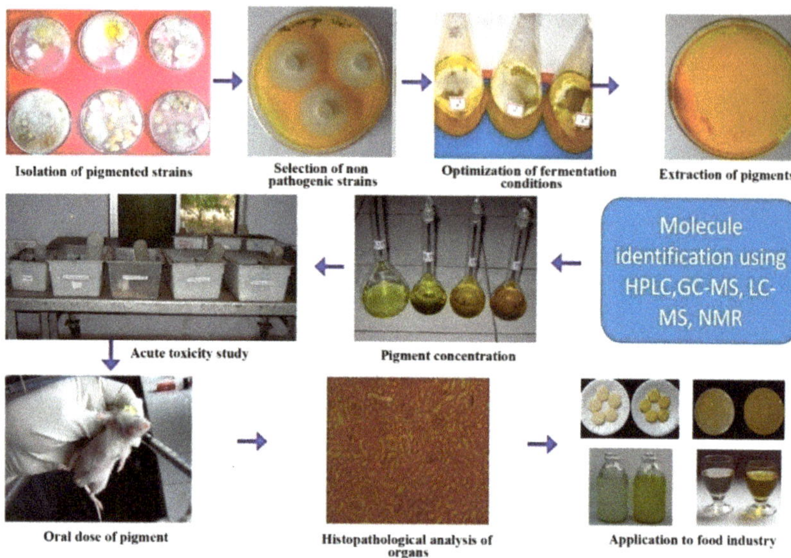

Figure 3. Toxicity evaluation of *Thermomyces* sp. pigment using an animal model.

4.10. *Neurospora crassa*

Neurospora crassa is a type of red bread mold of the phylum Ascomycota in order Sordariales, family Sordariaceae. It is a genetically and biochemically well-studi eukaryotic microorganism [91,92]. It is capable of producing polyketide and carotenc fungal pigments that are yellow to orange-red pigments widely used as food colora Research on *Neurospora* sp. has generally recognized it as safe after more than two centur with no record of mycotoxin production. For the production of ethanol, biomass a pigments, it can be grown rapidly on industrial residuals and lignocelluloses.

4.11. Other Fungal Pigments

The Cordycipitaceae family have promising pigment-producing genera such as *Torrubie Cordyceps, Beauveria, Hyperdermium* and *Lecanicillium. Beauveria bassiana* produces tenellin a *Beauveria brongniartii* produces bassianin. *Beauveria bassiana* produces pyridovericin, blood-r dibenzoquinone and pyridomacrolidin, along with mycotoxin. *Torubiella* produces torrub lones and *Lecanicillium aphanocladii* produces a pigment, along with oosporein [93]. T molecule has a wide range of bioactivities from antifungal, antimicrobial and phytotoxic effe to growth inhibition in plants. Additionally, kidney damage and even death were noticed poultry exposed to oosporein.

5. Yeast

Yeast synthesizes a variety of commercially important carotenoids, viz., carotene, to lene, torularhodin and astaxanthin. Genera *Rhodotorula, Sporobolomyces* and *Phaffia* are pot tial sources of pigments. Carotenoids from yeast are used in pharmaceutical, chemical, fo and feed industries. They are also an important precursor for vitamin A synthesis. Ap from coloring, it also has antioxidant and possible tumor-inhibiting properties [94,95].

5.1. *Rhodotorula* sp.

Rhodotorula a pigment-producing yeasts from division Basidiomycota are capable of synthesizing carotenoid pigments. Other species such as *R. gracilis*, *R. rubra* and *R. graminis* are also capable of producing pigments. The main compounds produced by *Rhodotorula glutinis* (red yeasts) are torulene and torularhodin, with a minute quantity of β-carotene [96,97].

A toxicological evaluation of *R. gracilis* CFR-1, using acute doses in freeze-dried form at 0.5–6.0 g/kg of body weight (*w*/*w*), did not show any toxic symptoms or mortality in adult rats. Dietary intake at low concentrations of 0.1–2.0% level (*w*/*w*) for 14 weeks did not induce any significant change in food intake and body weight gain in experimental rats compared to control animals [56,57].

Toxicity studies of β-carotene, torulene and torularhodin produced by *R. glutinis* conducted on rats demonstrated that they can be used as safe food additives. Dry powdered *R. glutinis* NCIM 3353 yeast biomass was added to the fodder of rats. The rats had protective effects against precancerous lesions of the liver induced by N-nitrosodimethylamine [98].

5.2. *Yarrowia lipolytica*

Yarrowia lipolytica is an Ascomycetous yeast and the only described species of the genus Yarrowia. It is widespread in nature, has many industrial uses and is important in the food industry and medical field. It is used as an alternative source of β-carotene. The major constituents of the carotenoids include all-trans-β-carotene with small amounts of 9-cis β-carotene, 15-cis β-carotene, 13-cis β- carotene and others [99].

Toxicity studies of β-carotene from *Y. lipolytica* in genotoxicity models and a standard subchronic rat study revealed no significant difference, compared with commercial products. An extracellular nontoxic pyomelanin pigment from *Y. lipolytica* was found to have antioxidant and noncytotoxic properties for two mammalian cell lines, viz., mouse fibroblast (NIH3T3) and human keratinocytes (HaCaT). Purified pyomelanin has a significant sun protection factor (SPF) value, highlighting its potential as a UVfilter in cosmetic preparations. Biomass of this yeast was defined as a safe novel food by the European Food Safety Authority [100].

6. Conclusions

In recent decades, natural pigments have been extensively used as colorants in food, pharmaceutical, cosmetic and textile industries. Several fungal strains are known for pigment production, while many fungi have not been systematically explored for their pigment-producing capability. Therefore, there is a need to explore novel and safe pigments using appropriate tools and techniques. Many pigments, including those with antibiotic-like properties, need to be studied for selective toxicity so that they can be produced commercially for human use. Fungal pigments open many new avenues in the production of textiles for medical use. This provides an extensive area of exploration to identify natural, eco-friendly pigments for diverse applications to satisfy public demand. In addition, biotechnological approaches help to produce pigments on a large scale with low cost, high yield and easy extraction without mycotoxins. The manufacture of fungal pigments has taken a big step to promote eco-friendly pigments. A literature search reveals the application of fungal pigments in the food and health care industries. These fungal pigments need to pass toxicity tests, quality tests and regulatory approval before their final entry into the market as food colorants or as drugs. Moreover, toxicology testing for most of the fungal pigments was not available. Another major impediment is that the funding required to carry out the necessary safety studies on such food additives is not available. For the above pigments to be feasible, testing is imperative. Fungal pigments could be a boon to the food industry.

Author Contributions: Writing—original draft, conceptualization (R.P.); writing—review and editing, supervision, resources (S.P.); supervision, conceptualization, writing—review and editing, validation (L.D.); visualization, supervision, conceptualization, resources (J.K.). All authors ha read and agreed to the published version of the manuscript.

Funding: This research received no external funding.

Institutional Review Board Statement: Not applicable.

Informed Consent Statement: Not applicable.

Data Availability Statement: Not applicable.

Conflicts of Interest: The authors declare no conflict of interest. The funders had no role in the desi of the study; in the collection, analyses, or interpretation of data; in the writing of the manuscript, in the decision to publish the results.

References

1. Kalra, R.; Conlan, X.A.; Goel, M. Fungi as a Potential Source of Pigments: Harnessing Filamentous Fungi. *Front. Chem.* **2020**, *8*, 3 [CrossRef] [PubMed]
2. Yusuf, M.; Shabbir, M.; Mohammad, F. Natural Colorants: Historical, Processing and Sustainable Prospects. *Nat. Prod. Bioprospect.* **2017**, *7*, 123–145. [CrossRef]
3. Martins, N.; Roriz, C.L.; Morales, P.; Barros, L.; Ferreira, I.C.F.R. Food colorants: Challenges, opportunities and current desires agro-industries to ensure consumer expectations and regulatory practices. *Trends Food Sci. Technol.* **2016**, *52*, 1–15. [CrossRef]
4. Ntrallou, K.; Gika, H.; Tsochatzis, E. Analytical and Sample Preparation Techniques for the Determination of Food Colorants Food Matrices. *Foods* **2020**, *9*, 58. [CrossRef]
5. Sigurdson, G.T.; Tang, P.; Giusti, M.M. Natural Colorants: Food Colorants from Natural Sources. *Annu. Rev. Food Sci. Techn* **2017**, *8*, 261–280. [CrossRef]
6. Rodriguez-Amaya, D.B. Natural food pigments and colorants. *Curr. Opin. Food Sci.* **2016**, *7*, 20–26. [CrossRef]
7. Gmoser, R.; Ferreira, J.A.; Lennartsson, P.R.; Taherzadeh, M.J. Filamentous ascomycetes fungi as a source of natural pigmer *Fungal Biol. Biotechnol.* **2017**, *4*, 4. [CrossRef]
8. Narsing Rao, M.P.; Xiao, M.; Li, W.-J. Fungal and Bacterial Pigments: Secondary Metabolites with Wide Applications. *Frc Microbiol.* **2017**, *8*, 1113. [CrossRef] [PubMed]
9. Rao, P.; Bhat, R.V. A comparative study on the synthetic food colours usage in foods procured from urban and rural areas Hyderabad. *Nutr. Food Sci.* **2003**, *33*, 230–234. [CrossRef]
10. Mittal, J. Permissible Synthetic Food Dyes in India. *Resonance* **2020**, *25*, 567–577. [CrossRef]
11. Oplatowska-Stachowiak, M.; Elliott, C.T. Food colors: Existing and emerging food safety concerns. *Crit. Rev. Food Sci. Nutr.* **20** *57*, 524–548. [CrossRef] [PubMed]
12. Feng, Y.; Shao, Y.; Chen, F. Monascus pigments. *Appl. Microbiol. Biotechnol.* **2012**, *96*, 1421–1440. [CrossRef] [PubMed]
13. WHO. *Toxicological Evaluation of Certain Food Additives and Contaminants*; WHO Food Additives Series; World Health Organizati Geneva, Switzerland, 1991; Volume 28.
14. Rao, P.; Bhat, R.V.; Sudershan, R.V.; Prasanna Krishna, T. Consumption of synthetic food colours during festivals in Hyderab India. *Br. Food J.* **2005**, *107*, 276–284. [CrossRef]
15. Rowe, K.S.; Rowe, K.J. Synthetic food coloring and behavior: A dose response effect in a double-blind, placebo-controlle repeated-measures study. *J. Pediatr.* **1994**, *125*, 691–698. [CrossRef]
16. Dufossé, L.; Fouillaud, M.; Caro, Y.; Mapari, S.A.; Sutthiwong, N. Filamentous fungi are large-scale producers of pigments ar colorants for the food industry. *Curr. Opin. Biotechnol.* **2014**, *26*, 56–61. [CrossRef]
17. Gupta, N.; Poddar, K.; Sarkar, D.; Kumari, N.; Padhan, B.; Sarkar, A. Fruit waste management by pigment production ar utilization of residual as bioadsorbent. *J. Environ. Manag.* **2019**, *244*, 138–143. [CrossRef]
18. Aruldass, C.A.; Dufossé, L.; Ahmad, W.A. Current perspective of yellowish-orange pigments from microorganisms—A revie *J. Clean. Prod.* **2018**, *180*, 168–182. [CrossRef]
19. Panesar, R.; Kaur, S.; Panesar, P. Production of microbial pigments utilizing agro-industrial waste: A review. *Curr. Opin. Food S* **2015**, *1*, 70–76. [CrossRef]
20. Tirumale, S.; Wani, N.A. *Biopigments: Fungal Pigments*; Springer: Singapore, 2018.
21. Nigam, P.S.; Luke, J.S. Food additives: Production of microbial pigments and their antioxidant properties. *Curr. Opin. Food S* **2016**, *7*, 93–100. [CrossRef]
22. Venil, C.K.; Zakaria, Z.A.; Ahmad, W.A. Bacterial pigments and their applications. *Process. Biochem.* **2013**, *48*, 1065–10 [CrossRef]
23. Joshi, V.A.; Bala, A.; Bhushan, S. Microbial Pigments. *Indian J. Biotechnol.* **2003**, *2*, 362–369.
24. Chattopadhyay, P.; Chatterjee, S.; Sen, S.K. Biotechnological potential of nature food grade biocolourants. *Afr. J. Biotechnol.* **20** *7*, 2972–2985.
25. Lin, L.; Xu, J. Fungal Pigments and Their Roles Associated with Human Health. *J. Fungi* **2020**, *6*, 280. [CrossRef]

Tuli, H.S.; Chaudhary, P.; Beniwal, V.; Sharma, A.K. Microbial pigments as natural color sources: Current trends and future perspectives. *J. Food Sci. Technol.* **2015**, *52*, 4669–4678. [CrossRef] [PubMed]

Velmurugan, P.; Kim, M.-J.; Park, J.-S.; Karthikeyan, K.; Lakshmanaperumalsamy, P.; Lee, K.-J.; Park, Y.-J.; Oh, B.-T. Dyeing of cotton yarn with five water soluble fungal pigments obtained from five fungi. *Fibers Polym.* **2010**, *11*, 598–605. [CrossRef]

Nirlane da Costa Souza, P.; Luiza Bim Grigoletto, T.; Alberto Beraldo de Moraes, L.; Abreu, L.M.; Henrique Souza Guimarães, L.; Santos, C.; Ribeiro Galvão, L.; Gomes Cardoso, P. Production and chemical characterization of pigments in filamentous fungi. *Microbiology* **2016**, *162*, 12–22. [CrossRef]

Heo, Y.M.; Kim, K.; Kwon, S.; Na, J.; Lee, H.; Jang, S.; Kim, C.-H.; Jung, J.; Kim, J.-J. Investigation of Filamentous Fungi Producing Safe, Functional Water-Soluble Pigments. *Mycobiology* **2018**, *46*, 269–277. [CrossRef]

Mostafa, M.; Abbady, M. Secondary Metabolites and Bioactivity of the Monascus Pigments—Review Article. *Glob. J. Biotechnol. Biochem.* **2014**, *9*, 1–13.

Saravanan, A.; Jayasree, R.; Kumar, P.S.; Varjani, S.; Hemavathy, R.; Jeevanantham, S.; Yaashikaa, P.R. Production of pigment using Aspergillus tamarii: New potentials for synthesizing natural metabolites. *Environ. Technol. Innov.* **2020**, *19*, 100967. [CrossRef]

Kongsak Boonyapranai, R.T. Sorasak Lhieochaiphant and Suree Phutrakul. Optimization of Submerged Culture for the Production of Naphthoquinones Pigment by Fusarium verticillioides. *Chiang Mai J. Sci.* **2008**, *35*, 457–466.

Dufossé, L. Microbial Production of Food Grade Pigments. *Food Technol. Biotechnol.* **2006**, *44*, 313–321.

Lagashetti, A.C.; Dufossé, L.; Singh, S.K.; Singh, P.N. Fungal Pigments and Their Prospects in Different Industries. *Microorganisms* **2019**, *7*, 604. [CrossRef]

Carvalho, J.C.; Oishi, B.O.; Pandey, A.; Soccol, C. Biopigments from Monascus: Strains selection, citrinin production and color stability. *Braz. Arch. Biol. Technol.* **2005**, *48*, 885–894. [CrossRef]

Dufossé, L. 16—Current and Potential Natural Pigments From Microorganisms (Bacteria, Yeasts, Fungi, Microalgae). In *Handbook on Natural Pigments in Food and Beverages*; Carle, R., Schweiggert, R.M., Eds.; Woodhead Publishing: Sawston, UK, 2016; pp. 337–354. [CrossRef]

Bezirhan Arikan, E.; Canli, O.; Caro, Y.; Dufossé, L.; Dizge, N. Production of Bio-Based Pigments from Food Processing Industry By-Products (Apple, Pomegranate, Black Carrot, Red Beet Pulps) Using Aspergillus carbonarius. *J. Fungi* **2020**, *6*, 240. [CrossRef]

The Joint FAO/WHO Expert Committee on Food Additives. *Evaluation of Certain Food Additives and Contaminants. FAO/WHO Expert Committee on Food Additive*; World Health Organization: Geneva, Switzerland, 1996; pp. 1–100.

Woutersen, R.A.; Wolterbeek, A.P.; Appel, M.J.; van den Berg, H.; Goldbohm, R.A.; Feron, V.J. Safety evaluation of synthetic beta-carotene. *Crit. Rev. Toxicol.* **1999**, *29*, 515–542. [CrossRef]

Cambaza, E. Comprehensive Description of Fusarium graminearum Pigments and Related Compounds. *Foods* **2018**, *7*, 165. [CrossRef] [PubMed]

Díaz-Sánchez, V.; Avalos, J.; Limón, M.C. Identification and regulation of fusA, the polyketide synthase gene responsible for fusarin production in Fusarium fujikuroi. *Appl. Environ. Microbiol.* **2012**, *78*, 7258–7266. [CrossRef] [PubMed]

Soumya, K.; Narasimha Murthy, K.; Sreelatha, G.L.; Tirumale, S. Characterization of a red pigment from Fusarium chlamydosporum exhibiting selective cytotoxicity against human breast cancer MCF-7 cell lines. *J. Appl. Microbiol.* **2018**, *125*, 148–158. [CrossRef] [PubMed]

Menezes, B.S.; Solidade, L.S.; Conceição, A.A.; Santos Junior, M.N.; Leal, P.L.; de Brito, E.S.; Canuto, K.M.; Mendonça, S.; de Siqueira, F.G.; Marques, L.M. Pigment production by Fusarium solani BRM054066 and determination of antioxidant and anti-inflammatory properties. *AMB Express* **2020**, *10*, 117. [CrossRef] [PubMed]

Santos, M.C.d.; Mendonça, M.d.L.; Bicas, J.L. Modeling bikaverin production by Fusarium oxysporum CCT7620 in shake flask cultures. *Bioresour. Bioprocess.* **2020**, *7*, 13. [CrossRef]

Mohankumari, H.P.; Naidu, K.A.; Narasimhamurthy, K.; Vijayalakshmi, G. Bioactive Pigments of Monascus purpureus Attributed to Antioxidant, HMG-CoA Reductase Inhibition and Anti-atherogenic Functions. *Front. Sustain. Food Syst.* **2021**, *5*, 590427. [CrossRef]

Shi, K.; Tang, R.; Huang, T.; Wang, L.; Wu, Z. Pigment fingerprint profile during extractive fermentation with Monascus anka GIM 3.592. *BMC Biotechnol.* **2017**, *17*, 46. [CrossRef] [PubMed]

Darwesh, O.M.; Matter, I.A.; Almoallim, H.S.; Alharbi, S.A.; Oh, Y.-K. Isolation and Optimization of Monascus ruber OMNRC45 for Red Pigment Production and Evaluation of the Pigment as a Food Colorant. *Appl. Sci.* **2020**, *10*, 8867. [CrossRef]

Mapari, S.A.S.; Meyer, A.S.; Thrane, U.; Frisvad, J.C. Identification of potentially safe promising fungal cell factories for the production of polyketide natural food colorants using chemotaxonomic rationale. *Microb. Cell Factories* **2009**, *8*, 24. [CrossRef] [PubMed]

Mapari, S.A.S.; Meyer, A.S.; Thrane, U.; Frisvad, J. Production of Monascus-like Pigments. European Patent EP2262862B1, 28 March 2012.

Khan, A.A.; Iqubal, S.S.; Shaikh, I.A.; Niyongabo Niyonzima, F.; More, V.S.; Muddapur, U.M.; Bennur, R.S.; More, S.S. Biotransformation of longifolene by *Penicillium europium*. *Biocatal. Biotransform.* **2021**, *39*, 41–47. [CrossRef]

Sopandi, T.; Wardah, W. Sub-Acute Toxicity of Pigment Derived from Penicillium resticulosum in Mice. *Microbiol. Indones.* **2012**, *6*, 6. [CrossRef]

Krishnamurthy, S.; Narasimha Murthy, K.; Thirumale, S. Characterization of ankaflavin from Penicillium aculeatum and its cytotoxic properties. *Nat. Prod. Res.* **2020**, *34*, 1630–1635. [CrossRef]

53. Gopal Pandit, S.; Honganoor Puttananjaiah, M.; Serva Peddha, M.; Appasaheb Dhale, M. Corrigendum to 'Safety efficacy a chemical profiling of water-soluble Talaromyces purpureogenus CFRM02 pigment' [Food Chem. 310 (2020) 125869]. *Food Ch* **2020**, *317*, 126403. [CrossRef]
54. Poorniammal, R.; Gunasekaran, S.; Sriharasivakumar, H. Toxicity evaluation of funfal food colourant from Thermomyces sp albino mice. *J. Sci. Ind. Res.* **2011**, *70*, 773–777.
55. Almurshidi, B.H.; Van Court, R.C.; Vega Gutierrez, S.M.; Harper, S.; Harper, B.; Robinson, S.C. Preliminary Examination of Toxicity of Spalting Fungal Pigments: A Comparison between Extraction Methods. *J. Fungi* **2021**, *7*, 155. [CrossRef] [PubMed]
56. Latha, B.V.; Jeevaratanm, K. Thirteen-week oral toxicity study of carotenoid pigment from Rhodotorula glutinis DFR-PDY in r *Indian J. Exp. Biol.* **2012**, *50*, 645–651.
57. Naidu, K.A.; Venkateswaran, G.; Vijayalakshmi, G.; Manjula, K.; Viswanatha, S.; Murthy, K.N.; Srinivas, L.; Joseph, R. Toxicolog assessment of the yeast Rhodotorula gracilis in experimental animals. *Z. Für Lebensm. Und Forsch. A* **1999**, *208*, 444–448. [CrossR
58. Kot, A.M.; Błażejak, S.; Gientka, I.; Kieliszek, M.; Bryś, J. Torulene and torularhodin: "new" fungal carotenoids for indust *Microb. Cell Fact.* **2018**, *17*, 49. [CrossRef] [PubMed]
59. Groenewald, M.; Boekhout, T.; Neuvéglise, C.; Gaillardin, C.; van Dijck, P.W.; Wyss, M. Yarrowia lipolytica: Safety assessmen an oleaginous yeast with a great industrial potential. *Crit. Rev. Microbiol.* **2014**, *40*, 187–206. [CrossRef] [PubMed]
60. Venkatesh, K.S. Strain and Process Improvement for Polygalacturonase Production by *Aspergillus carbonarius*. Ph.D. The University of Mysore, Mysore, India, 2004.
61. Narendrababu, B.N.; Shishupala, S. Spectrophotometric detection of pigments from Aspergillus and Penicillium isolates. *J. A Biol. Biotechnol.* **2017**, *5*, 53–58. [CrossRef]
62. Sanjay, K.R.; Kumaresan, N.; Akhilender Naidu, K.; Viswanatha, S.; Narasimhamurthy, K.; Umesh Kumar, S.; Vijayalakshmi Safety evaluation of pigment containing Aspergillus carbonarius biomass in albino rats. *Food Chem. Toxicol.* **2007**, *45*, 431– [CrossRef]
63. Mantzouridou, F.; Tsimidou, M.Z. On the monitoring of carotenogenesis by Blakeslea trispora using HPLC. *Food Chem.* **2007**, 439–444. [CrossRef]
64. Sen, T.; Barrow, C.J.; Deshmukh, S.K. Microbial Pigments in the Food Industry-Challenges and the Way Forward. *Front. N* **2019**, *6*, 7. [CrossRef]
65. Papaioannou, E.H.; Liakopoulou-Kyriakides, M. Substrate contribution on carotenoids production in Blakeslea trispora culti tions. *Food Bioprod. Process.* **2010**, *88*, 305–311. [CrossRef]
66. Papadaki, E.; Mantzouridou, F. Natural [2]-Carotene Production by Blakeslea trispora Cultivated in Spanish-Style Green Ol Processing Wastewaters. *Foods* **2021**, *10*, 327. [CrossRef]
67. Finkelstein, M.; Huang, C.C.; Byng, G.S.; Tsau, B.R.; Leach, J. Blakeslea Trispora Mated Culture Capable of Increased Beta-carot Production. U.S. Patent US5422247A, 6 June 1995.
68. Nabae, K.; Ichihara, T.; Hagiwara, A.; Hirota, T.; Toda, Y.; Tamano, S.; Nishino, M.; Ogasawara, T.; Sasaki, Y.; Nakamura, M.; e A 90-day oral toxicity study of beta-carotene derived from Blakeslea trispora, a natural food colorant, in F344 rats. *Food Ch Toxicol.* **2005**, *43*, 1127–1133. [CrossRef] [PubMed]
69. Tudzynski, B. Gibberellin biosynthesis in fungi: Genes, enzymes, evolution, and impact on biotechnology. *Appl. Microb Biotechnol.* **2005**, *66*, 597–611. [CrossRef]
70. Avalos, J.; Prado-Cabrero, A.; Estrada, A.F. Neurosporaxanthin production by Neurospora and Fusarium. *Methods Mol. Biol.* **20** *898*, 263–274. [CrossRef]
71. Ma, J.; Li, Y.; Ye, Q.; Li, J.; Hua, Y.; Ju, D.; Zhang, D.; Cooper, R.; Chang, M. Constituents of red yeast rice, a traditional Chin food and medicine. *J. Agric. Food Chem.* **2000**, *48*, 5220–5225. [CrossRef] [PubMed]
72. Chen, W.; He, Y.; Zhou, Y.; Shao, Y.; Feng, Y.; Li, M.; Chen, F. Edible Filamentous Fungi from the Species Monascus: Ea Traditional Fermentations, Modern Molecular Biology, and Future Genomics. *Compr. Rev. Food Sci. Food Saf.* **2015**, *14*, 555–5 [CrossRef]
73. Kallscheuer, N. Engineered Microorganisms for the Production of Food Additives Approved by the European Union— Systematic Analysis. *Front. Microbiol.* **2018**, *9*, 1746. [CrossRef]
74. Liu, J.; Zhou, Y.; Yi, T.; Zhao, M.; Xie, N.; Lei, M.; Liu, Q.; Shao, Y.; Chen, F. Identification and role analysis of an intermedi produced by a polygenic mutant of Monascus pigments cluster in Monascus ruber M7. *Appl. Microbiol. Biotechnol.* **2016**, 7037–7049. [CrossRef]
75. Blanc, P.J.; Loret, M.O.; Goma, G. Production of citrinin by various species ofMonascus. *Biotechnol. Lett.* **1995**, *17*, 291–2 [CrossRef]
76. Gunasekaran Sanjeevi, P.R. Optimization of fermentation conditions for red pigment production from Penicillium sp. und submerged cultivation. *Afr. J. Biotechnol.* **2008**, *7*, 1894–1898. [CrossRef]
77. Sardaryan, E.; Zihlova, H.; Strnad, R.; Cermakova, Z. Arpink red—meet a new natural red food colourant of microbial origin. *Pigments in Food, More Than Colours*; Elsevier: Amsterdam, TheNetherlands, 2004; pp. 207–208.
78. Kojima, R.; Arai, T.; Matsufuji, H.; Kasumi, T.; Watanabe, T.; Ogihara, J. The relationship between the violet pigment PF production and intracellular ammonium level in Penicillium purpurogenum. *AMB Express* **2016**, *6*, 43. [CrossRef]
79. Corrêia Gomes, D.; Takahashi, J.A. Sequential fungal fermentation-biotransformation process to produce a red pigment fro sclerotiorin. *Food Chem.* **2016**, *210*, 355–361. [CrossRef] [PubMed]

Visagie, C.M.; Houbraken, J.; Frisvad, J.C.; Hong, S.B.; Klaassen, C.H.W.; Perrone, G.; Seifert, K.A.; Varga, J.; Yaguchi, T.; Samson, R.A. Identification and nomenclature of the genus Penicillium. *Stud. Mycol.* **2014**, *78*, 343–371. [CrossRef]

Khan, A.A.; Alshabi, A.M.; Alqahtani, Y.S.; Alqahtani, A.M.; Bennur, R.S.; Shaikh, I.A.; Muddapur, U.M.; Iqubal, S.M.S.; Mohammed, T.; Dawoud, A.; et al. Extraction and identification of fungal pigment from Penicillium europium using different spectral studies. *J. King Saud Univ. Sci.* **2021**, *33*, 101437. [CrossRef]

Sethi, B.K.P.; Parida, P.; Sahoo, S.L.; Dikshit, B.; Pradhan, C.; Sena, S.; Behera, B.C. Extracellular production and characterization of red pigment from Penicillium purpurogenum BKS9. *Alger. J. Nat. Prod.* **2016**, *4*, 379–392.

Pandit, S.G.; Puttananjaih, M.H.; Harohally, N.V.; Dhale, M.A. Functional attributes of a new molecule-2-hydroxymethyl-benzoic acid 2′-hydroxy-tetradecyl ester isolated from Talaromyces purpureogenus CFRM02. *Food Chem.* **2018**, *255*, 89–96. [CrossRef]

Poorniammal, R.; Gunasekaran, S.; Gnanasambandam, A.V.; Murugesan, R. Process of Extracting Yellow Pigment from *Thermomyces* sp. Indian Patent 304979, 15 December 2018.

Somasundaram, T.; Rao, S.S.R.; Maheswari, R. Pigments in Thermophilic fungi. *Curr. Sci.* **1986**, *55*, 957–960.

Poorniammal, R.; Balachandar, D.; Gunasekaran, S. Evaluation of antioxidant property of some fungal pigments by DNA protection assay. *Ann. Phytomedicine* **2018**, *7*, 106–111. [CrossRef]

Poorniammal, R.; Gunasekaran, S.; Murugesan, R. In Vivo Antioxidant Activities of Thermomyces sp Pigment in Albino Mice. *Int. J. Agric. Environ. Biotechnol.* **2014**, *7*, 355–360. [CrossRef]

Poorniammal, R.; Prabhu, S.; Sakthi, A.R. Evaluation of In Vitro antioxidant activity of fungal pigments. *Pharma Innov.* **2019**, *8*, 326–330.

Chitale, A.; Jadhav, D.V.; Waghmare, S.R.; Sahoo, A.K.; Ranveer, R.C. Production and characterization of brown coloured pigment from *Trichoderma viride*. *Electron. J. Environ. Agric. Food Chem.* **2012**, *11*, 529–537.

Vega Gutierrez, S.M.; Stone, D.W.; He, R.; Vega Gutierrez, P.T.; Walsh, Z.M.; Robinson, S.C. Potential Use of the Pigments from Scytalidium cuboideum and Chlorociboria aeruginosa to Prevent 'Greying' Decking and Other Outdoor Wood Products. *Coatings* **2021**, *11*, 511. [CrossRef]

Perkins, D.D.; Davis, R.H. Evidence for safety of Neurospora species for academic and commercial uses. *Appl. Environ. Microbiol.* **2000**, *66*, 5107–5109. [CrossRef] [PubMed]

Priatni, S. Review: Potential production of carotenoids from Neurospora. *Nusant. Biosci.* **2016**, *6*, 63–68.

Gaffoor, I.; Brown, D.W.; Plattner, R.; Proctor, R.H.; Qi, W.; Trail, F. Functional analysis of the polyketide synthase genes in the filamentous fungus Gibberella zeae (anamorph Fusarium graminearum). *Eukaryot. Cell* **2005**, *4*, 1926–1933. [CrossRef] [PubMed]

Poorniammal, R.; Sarathambal, C.; Sakthi, A.R.; Arun, S. Yeast carotenoids importance in food and feed industries and its health benefits. *Agric. Rev.* **2013**, *34*, 307–312. [CrossRef]

Bogacz-Radomska, L.; Harasym, J. β-Carotene—properties and production methods. *Food Qual. Saf.* **2018**, *2*, 69–74. [CrossRef]

Bhosale, P.B.; Gadre, R.V. Production of beta-carotene by a mutant of Rhodotorula glutinis. *Appl. Microbiol. Biotechnol.* **2001**, *55*, 423–427. [CrossRef] [PubMed]

Zhao, Y.; Guo, L.; Xia, Y.; Zhuang, X.; Chu, W. Isolation, Identification of Carotenoid-Producing Rhodotorula sp. from Marine Environment and Optimization for Carotenoid Production. *Mar. Drugs* **2019**, *17*, 161. [CrossRef] [PubMed]

Bhosale, P.; Motiwale, L.; Ingle, A.D.; Gadre, R.V.; Rao, K.V.K. Protective effect of Rhodotorulaglutinis NCIM 3353 on the development of hepatic preneoplastic lesions. *Curr. Sci.* **2002**, *83*, 303–308.

Ben Tahar, I.; Kus-Liśkiewicz, M.; Lara, Y.; Javaux, E.; Fickers, P. Characterization of a nontoxic pyomelanin pigment produced by the yeast Yarrowia lipolytica. *Biotechnol. Prog.* **2020**, *36*, e2912. [CrossRef]

). Grenfell-Lee, D.; Zeller, S.; Cardoso, R.; Pucaj, K. The safety of β-carotene from Yarrowia lipolytica. *Food Chem. Toxicol.* **2014**, *65*, 1–11. [CrossRef] [PubMed]

Journal of
Fungi

icle

:eliminary Examination of the Toxicity of Spalting Fungal .gments: A Comparison between Extraction Methods

lria H. Almurshidi [1], R.C. Van Court [1], Sarath M. Vega Gutierrez [1], Stacey Harper [2], Bryan Harper [2] and i C. Robinson [1,*]

[1] Department of Wood Science, Oregon State University, Corvallis, OR 97333, USA; badria@email.sc.edu (B.H.A.); ray.vancourt@oregonstate.edu (R.C.V.C.); sarathth@yahoo.co.uk (S.M.V.G.)
[2] Department of Toxicology, Oregon State University, Corvallis, OR 97331, USA; Stacey.Harper@oregonstate.edu (S.H.); Bryan.Harper@oregonstate.edu (B.H.)
* Correspondence: seri.robinson@oregonstate.edu

Abstract: Spalting fungal pigments have shown potential in technologies ranging from green energy generation to natural colorants. However, their unknown toxicity has been a barrier to industrial adoption. In order to gain an understanding of the safety of the pigments, zebrafish embryos were exposed to multiple forms of liquid media and solvent-extracted pigments with concentrations of purified pigment ranging from 0 to 50 mM from *Chlorociboria aeruginosa, Chlorociboria aeruginascens,* and *Scytalidium cuboideum.* Purified xylindein from *Chlorociboria sp.* did not show toxicity at any tested concentration, while the red pigment dramada from *S. cuboideum* was only associated with significant toxicity above 23.2 uM. However, liquid cultures and pigment extracted into dichloromethane (DCM) showed toxicity, suggesting the co-production of bioactive secondary metabolites. Future research on purification and the bioavailability of the red dramada pigment will be important to identify appropriate use; however, purified forms of the blue-green pigment xylindein are likely safe for use across industries. This opens the door to the adoption of green technologies based on these pigments, with potential to replace synthetic colorants and less stable natural pigments.

Keywords: spalting; fungal pigment; xylindein; dramada; *Chlorociboria aeruginosa; Chlorociboria aeruginascens; Scytalidium cuboideum;* natural pigment; natural colorant

tion: Almurshidi, B.H.; Van
:t, R.C.; Vega Gutierrez, S.M.;
per, S.; Harper, B.; Robinson, S.C.
minary Examination of the
:ity of Spalting Fungal Pigments:
mparison between Extraction
nods. *J. Fungi* **2021**, *7*, 155.
s://doi.org/10.3390/jof7020155

lemic Editor: Laurent Dufossé

ived: 10 December 2020
pted: 15 February 2021
ished: 22 February 2021

1. Introduction

Spalting fungi are a specific group of wood decay fungi that have the ability to internally color wood [1]. The coloration that they cause can be classified into three types: bleaching, zone lines, and pigmentation. The first two types are produced mostly by white-rotting fungi. Pigmentation is caused by ascomycete fungi through the generation of secondary metabolites that cause coloration in wood. Known colors produced by these fungi include blue-green produced by *Chlorociboria* spp. [2–4], red from *Scytalidium cuboideum* (Sacc. and Ellis) Sigler and Kang [5], and yellow from *Scytalidium ganodermophthorum* Sigler and Kang [6], among others.

Beginning in the 15th century, wood stained blue-green by fungi from the genus *Chlorociboria* was a prized commodity in fine woodworking [7–9]. Artworks containing the blue-green pigment, named xylindein [10] (Figure 1), retain their coloration today, attesting to its stability. The structure of this pigment has since been established [11–13], and its impressive UV stability and thermal stability have been the subject of research [14]. The properties of other pigments have also been the subject of recent investigation, including the identification of the napthoquinonic crystal produced by *Scytalidium cuboideum*, called dramada [15]. This red compound has also been isolated from an actinomycete [16].

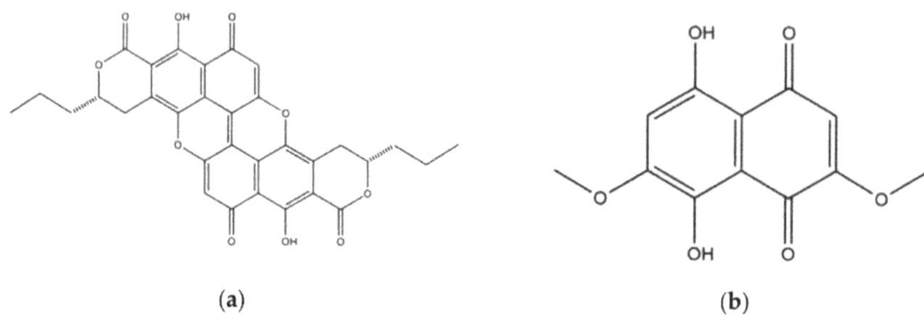

Figure 1. Structure of spalting fungal pigments. (**a**) Blue/green pigment xylindein produced by *Chlorociboria spp.*; (**b**) red pigment "dramada" (5,8-Dihydroxy-2,7-dimethoxy-1,4-naphthoquinone) produced by *Scytalidium cuboideum*.

Pigments produced by these fungi have been investigated for their use in a variety fields, for example, as a coloring agent for wood stain [17], in paint [18], and as a text dye [19–21]. The use of fungal pigments in these applications could replace current uns tainable industrial practices. For example, conventional textile dyeing practices produ toxic wastewater associated with negative environmental and health consequences [22– The use of sustainably produced fungal pigments instead would allow for the prod tion of desired colored cloth without the toxic tradeoff. In addition to use as a colora the pigments have also been the subject of investigation into use as organic semic ductors [14,26], and may allow for fully sustainable energy generation through orga photovoltaic systems. However, before the adoption of these green technologies is possi especially for those associated with extended human contact, the toxicity of the pigme must be understood.

Many filamentous fungi produce secondary metabolites with bioactive effects. So have a beneficial effect, with pharmacological uses like antibiotics [27,28]. However, fu also produce a range of mycotoxins such as aflatoxins and rhizonin, which are severe hea hazards [29]. A number of filamentous fungi also produce pigments that are themsel toxic [30]. For example, pigments from *Monascus* spp., used for coloring food and pharmacological applications, are restricted due to fungal co-production of citrinin [which is nephrotic [32]. This has led to research attempting to reduce citrinin producti through growth condition variation or strains [33–37].

Spalting fungi have received attention for their potential associated health risks, m notably by woodturners who have spread fear about their supposed toxicity. Theor spread have included spalted wood causing allergic reactions, releasing "carloads" spores, and implications of brain infection, which have likely been driven through limi understanding of fungal biology [38]. These and more urban myths around any potent threats that spalted wood might pose compared to non-decayed wood have been debunk as spalted wood is not inherently more dangerous or toxic than non-decayed wood [39

To determine the potential hazard of toxins in humans, a broadly used method volves the testing of compounds in zebrafish (*Danio rerio*) embryos. Zebrafish are a tropi freshwater fish that have been highly studied and used as a model organism for rapid a low-cost research in the fields of toxicology, genetics and developmental biology [40– Zebrafish assays are used to indicate bioactive drugs and therapeutic compounds for ph maceutical applications and to understand effects on developmental mechanisms [44– The use of the zebrafish embryo model for toxicology research is accepted internationa as an alternate animal model on testing hazard and risk assessment [48,49]. The zebrafi model has also been used to evaluate the toxicity of natural products from plants and ot organisms such as secondary metabolite extracts [50–52], making it an appropriate mo for preliminary examination of the toxicity of spalting fungal pigments.

This study sought to characterize the potential toxic effects of spalting fungal pigme and identify whether they came from the pigments themselves or due to the presence other compounds in the extract. Understanding the toxicity of these pigments, as well the cause (whether the pigment *itself* is toxic or the accompanying secondary metabolites are toxic), will enable the determination of what products they are appropriate for use in, and what level of processing after fungal production they require to be safely used. This information will inform future research and industrial adoption of these unique, sustainably sourced pigments.

2. Materials and Methods

2.1. Preparation of Pigment Samples

2.1.1. Solvent-Extracted Pigment

Chlorociboria aeruginosa UAMH 11657 (isolated from a decaying hardwood log in Haliburton, ON, Canada), *Chlorociboria aeruginascens* UAMH 7615 (isolated in Lake District, UK), and Scytalidium *cuboideum* (Sacc. and Ellis) Sigler and Kang UAMH 11517 (isolated from *Quercus* sp. in Memphis, TN, USA) were used to inoculate petri dishes containing 2% malt extract agar (MEA) (20 g of bacteriological malt extract (VWR, Radnor, PA, USA), 15 g of agar (VWR), 1 l of deionized water) amended with sterile white rotted wood chips from either *Acer saccharum* or *Populus grandidentata*, following the protocol set by Robinson et al. [53]. Cultures were harvested once plates were completely pigmented, with times ranging from four (*Scytalidium cuboideum*) to twelve weeks (*Chlorociboria* spp.). Plates were opened and left to dry for 48 h, then ground using a blender (Oster Precise Blend, Boca Raton, FL, USA) until reaching a maximum size of ~5 mm. The resulting powder and 45 mL of dichloromethane (DCM) (VWR, Radnor, PA, USA) were combined in a 250 mL Erlenmeyer flask with a 2 mm × 5 mm VWR Spinbar magnetic stir bar. The flask was closed with a rubber cap and a stirred at 220 rpm for 30 min on a VWR Dylastir stir plate. The resulting solution was then filtered through VWR 415 Whatman Filter Paper to remove the wood chip particles. The extract was collected in a borosilicate glass vial (Acc glass, Vineland, NJ, USA) and sealed with non-evaporative polyseal-cone-lined caps.

2.1.2. Pigments from Liquid Media

Liquid media were prepared following methods in Weber et al. [54]. Sterilized and cooled 150 mL mason jars containing 50 mL of 2% malt broth (20 g of VWR bacteriological malt extract, 1 l of deionized water) were inoculated with active fungal cultures of either *C. aeruginosa* or *S. cuboideum* using one plug of approximately 2 mm in diameter. Jars were then incubated at room temperature (21 °C) for 28 days on an open shelf.

Pigment from liquid media was tested in three ways. First, liquid media were used directly in zebrafish assays. Second, liquid media were autoclaved at 121 °C for 30 min. Finally, media from fungal liquid malt cultures were also cleaned independently using Strata SPE 2 g/12 mL columns (Phenomenex). The column was conditioned by adding 4 mL of HPLC acetonitrile (CAN) solvent to remove trapped air and activate the SPE particles, before the solvent was removed and 4 mL HPLC grade water was added to maximize the sorbent interaction with target analytes. For all species, liquid media culture was filtered through 415 Whatman filter paper (VWR) twice before 10 mL was loaded onto activated column, where a visible band of pigment was formed. Contaminants were removed from the column through the addition of 10 mL 50% acetonitrile (ACN) in HPLC grade water. Pigment was eluted using 2 to 4 mL of 100% of HPLC-grade chloroform (EMD Millipore, Burlington, MA, USA). About 10 mL of the pigment mixture sample was used to achieve less than 0.5 mL of each purified pigment in two to three hours. This method was used to obtain pigment with a reduced amount of contaminants to reduce and identify potential effects of the extracts on the zebrafish embryos.

2.1.3. Solid Pigments

Scytalidium cuboideum pigment (dramada) crystals were precipitated by applying 200 mL of liquid nitrogen to a solution of 100 mL of concentrated acetone extract from *S. cuboideum* following the method stated by [15]. The acetone carrier differed from the traditional DCM as its melting point (−95 ° C) was preferred for the crystallization by precipitation method. After the crystals were formed in the cold solvent, they were filter with the use of 415 Whatman filter paper (VWR). The crystals were then air-dried a placed in a glass vial. This method has been shown to result in crystals of high purity [15,5 with samples tested for purity in previous work [55].

To obtain solid xylindein from *Chlorociboria* spp., 15 mL of standardized *Chlorocibo* spp. extract in DCM was placed in a 30 mL ACE borosilicate glass vial and left uncover to fully evaporate. Then, another 15 mL of extract was added. This process was repeat 15 times to form a solid pigment layer attached to the glass of the vial. Once the last fill DCM extract was completely evaporated, 10 mL of acetone was added to the vial and was closed with a non-evaporative polyseal-cone-lined cap. The vial was then shaken one minute by hand before the mix was filtered using a VWR glass funnel equipped wit 415 Whatman filter paper to collect solid pigment. The acetone wash was repeated un the solid pigment was completely removed from the glass vial. After finishing this proce the filter paper was left to dry overnight. The resulting solid pigment was removed fr the filter paper with the use of forceps and stored in a borosilicate glass vial. Solidifi xylindein collected via this method does not yield a pure compound [56]; however, standardized purification methodology has not yet been developed, though methods a in development [57].

2.1.4. Solvent-Extracted Pigment Concentration

Standards corresponding to the extracts with DCM have been previously used determine pigment concentration based on CIE L*a*b* values, not dry weight, wh working with pigmenting spalting fungi [17,53,55,58]. The color values were used inste of the dry weight due to the simplicity of a direct extraction from the dry plates with DC The standard CIEL*a*b values utilized were established by [59] with a range of +/− 2 for each fungal species, including: *C. aeruginosa* L* = 82.28, a* = − 11.06, b* = −5.40 a *S. cuboideum* L* = 82.32, a* = 26.84, b* = 13.19. Three milliliters of each of the extract solutions was added to a VWR glass cuvette for analysis on a Konica Minolta Chron Meter CR−5 colorimeter, and concentration adjusted through the addition of solvent evaporation to match these color values. After the read, the dye solution was returned the vial and stored for future use.

2.2. *Zebrafish Preparation and Exposure for Pigment Extracts*

All experiments were performed in compliance with national care and use guidelin and were approved by the Institutional Animal Care and Use Committee (IACUC) Oregon State University (ACUP 5113). Adult wild-type D5 zebrafish (*Danio rerio*) e bryos were raised at the Sinnhuber Aquatic Research Laboratory (SARL) at Oregon Sta University (Corvallis, Oregon, USA). Fish were maintained in fish water, consisting reverse osmosis water supplemented with 0.3 g/l Instant Ocean salts (Aquatic Ecosysten Apopka, FL) with pH adjusted with sodium bicarbonate to pH 7 ± 0.2, with a temperatu of 28 ° C and a 14 h light to 10 h dark photoperiod. After group spawn and egg collectio an Olympus-SZ51 stereomicroscope was used to select and remove the abnormal ar non-fertilized eggs. Six hours post-fertilization (hpf), all normal embryos were dechoric ated to ensure contact with test materials [60]. Embryos were placed into a 60 mm gla petri dish with 25 mL fish water and exposed to 50 µL of 50 mg/mL pronase enzyn (Sigma-Aldrich, cat # 81750, St. Louis, MO, USA) to degrade the outer chorionic lay After chorion deflation (~7 min) solution was diluted with fresh fish water and recover in a petri dish at room temperature until 8 hpf, when waterborne exposure testing w carried out. At this time, each embryo was placed in its own well in a prepared 96-w plate containing fish water and tested pigment condition. The embryos were incubated 28 °C for 24 hfp for the first assessment.

2.2.1. Pigment from Solvent Extraction

The pigment bioactivity testing procedure using zebrafish embryos followed methods laid out in Truong, Harper and Tanguay [60]. Standardized pigments in DCM were placed in Zinser 96-well glass petri dishes and the solvent was allowed to evaporate under a fume hood for 24 h or until the DCM had evaporated completely, with 100 μL of each pigment extract used per well. At 8 hpf, 200 μL of fish water containing 0.3 g/L of aquatic salt was transferred with a VWR disposable wide-bore glass pipette in the Zinser 96-well glass petri dish, and one dechorionated embryo was added per well. These were then incubated at 28 °C until 24 hpf, then the appropriate assessments were performed as described below.

Extracted pigment toxicity was compared across multiple conditions for pigments from all tested fungal species. First, embryos were exposed to standardized pigments from fungi grown on aspen wood chip amended malt agar plates in one 96-well plate at 100% concentration ($n = 12$ for *Chlorociboria* species and $n = 24$ for *Scytalidium cuboideum*). Next, extracted pigment from fungi grown in maple amended wood chip plates at standard concentration was carried out in individual plates ($n = 72$ per pigment). Testing across concentrations was then carried out using only pigment extracts from amended maple wood chip plates. Each pigment extract was tested in separate 96-well plates using 72 embryos ($n = 72$ per pigment), with separate tests for 100%, 200%, and 400% concentrations. This range of concentrations was used to allow for comparison with previous publications using standardized pigment extract.

2.2.2. Pigment from Liquid Culture

Three liquid culture solutions (live, autoclaved, and filtered, as described above) from each fungal species and SPE column purified pigment from *Chlorociboria* spp. and *S. cuboideum* were tested. These forms were tested in order to compare the toxicity of the extracted pigment solution to the full panel of compounds present in fungal cultures used to produce the pigments. For each, 100 μL of pigment solution was transferred into a Falcon sterile 96-well plate and fish water was added to give a final working volume of 250 μL, with controls of fresh fish water and sterile liquid malt extract. At 8 hpf, a dechorionated embryo was transferred into each individual well of the 96-well plate using a VWR disposable wide-bore glass pipette. The plates were incubated at 28 °C until 24 hpf, then assessed as described below.

First, pigmented liquid media taken directly from fungal cultures were applied to 12 embryos per pigment. This was then repeated using media autoclaved at 120 °C for 30 min. Next, pigment solutions were filtered to remove fungal cells from the medium using 0.2 μm EMD Millipore filters. This experiment was run twice, with 24 embryos per test condition each time. Finally, green pigment from *C. aeruginosa* and *C. aeruginascens* and red pigment from *S. cuboideum* collected from liquid media samples and purified using SPE columns (as described in Section 2.1.2) were tested on 72 embryos each.

2.2.3. Assessment Protocol

The assessment method was modified from Truong, Harper and Tanguay [60]. At 24 hpf, developmental stages and spontaneous kinetics were observed over a two-minute period. At 120 hpf, embryo morphology, including body axis, ocular perceiver, snout, jaw, notochord, heart, brain, somite, fin, yolk sac, trunk, circulation, pigment, swim bladder and behavioral endpoints (motility, tactile replication), was observed and recorded. Mortality rate was also recorded at 24 and 120 hpf. The assessments were conducted in a binary form, as present or not present.

2.3. Zebrafish Preparation and Exposure for Solid Purified Pigments

2.3.1. Preparation of Zebrafish

Adult tropical 5D strain zebrafish (*Danio rerio*) embryos were collected and staged. Chorion was enzymatically removed using pronase (63.3 mg/mL) using a custom automated dechorionator [61] at 4 h past fertilization.

2.3.2. Exposure Protocol

Six concentrations of solidified xylindein from *C. aeruginosa* and dramada fr *S. cuboideum* (as described in Section 2.1.3) were compared to aniline (≥99.5% ACS gra Sigma-Aldrich). Tested concentrations ranged from 50 to 2.32 mM for dramada and anil Roughly the same concentrations were used for xylindein; however, as yet no meth exists that results in a pure compound, so tested concentrations ranged from 28.42 mg/ (~50 mM) to 1.27 mg/mL (~2.32 mM). Compounds were exposed in 96-well plates, w one embryo per well loaded at 6 hpf and 100 ul of exposure solution. There were a tal of 32 embryos per concentration across multiple plates, with 8 embryos exposed concentration per plate.

2.3.3. Embryo Photomotor Response (EPR) Behavior

Embryos were assessed at 24 hpf using a custom photomotor response analy tool [62] in plate. The light cycle consisted of 30 s of dark background, a short light pu followed by a second light pulse nine seconds later and 10 more seconds of dark. Ev exposure plate had 850 frames of digital video recorded from below at 17 frames second, with white LED and infrared lights above. Recorded periods were truncated at beginning and end of the experiment to ensure the same recorded period was compare

2.3.4. Larval Photomotor Response Behavior (LPR)

Embryos were assessed at 120 hpf using a Zebrabox behavior chamber (ViewPo Life Sciences, Montreal, CA, USA) with an infrared backlight stage. Total movement response to three light cycles (3 min of light to 3 min dark) was tracked in 96 wells dur a 24-min assay. HD video was recorded at 15 frames/second.

2.3.5. Mortality and Morphology Response

Embryos were exposed statically and assessed at 24 hpf for four developmen toxicity endpoints (mortality, developmental progression, spontaneous movement, a notochord distortion), and again at 120 hpf for 18 developmental endpoints [60] by Zebrafish Acquisition and Analysis Program (ZAAP) custom program, with evaluati conducted by evaluators from Sinnhuber Aquatic Research Laboratory.

2.4. Statistical Analysis

2.4.1. Zebrafish Exposure to Pigment Extracts

Statistical analysis followed methods in Truong et al. [60]. Binary data from zebrafi assessments were compared using the Fisher's exact test. This statistically compared mortality of fish embryos exposed to pigments versus a baseline five percent mortality r of control embryos, using the *proc freq* function in SAS 9.4.

2.4.2. Zebrafish Exposure to Solid Pigments

Zebrafish exposure to solid pigments analysis followed analysis in Troung et al. [6 Embryo Photomotor Response (EPR) behavior was analyzed by comparing the ba ground, excitatory, and refractory intervals to the negative control (0 µg/mL of co pound) activity using a combination of percent change and a Kolmogorov–Smirnov te (Bonferroni-corrected p-value threshold). Dead or deformed fish were excluded fr behavioral datasets.

Larval photomotor response movement data from the behavior chamber were in grated into 6 s bins, and the area under the curve was compared to control movement t-test for each exposure concentration. LPR was considered valid when percent chan in area under the curve was greater than or equal to 40% above the control group a statistical significance (at $p < 0.05$) was met. Dead or deformed fish were excluded frc behavioral datasets.

Statistical analysis of mortality and morphology endpoints was performed in R [6 based on binary indices for each endpoint ($n = 32$). A significance threshold was comput for each chemical and endpoint combination in comparison to the control incidence rate, and Fisher's exact test was used to compare treatment groups to control groups to account for low category counts. Control data were used to check for confounding plate, well, and chemical effects. Slight differences in chemical effects lead to multiple comparisons used to control the family wise error rate.

Concentration response modeling based on mortality and morphology data was carried out on mortality data at 24 hpf and at 120 hpf for tested compounds showing significant responses compared to the control. R was used to fit a Hill model to the average of all individuals at each exposure concentration following methods in Truong et al. [63], using the four parameters of lower limit, upper limit, the median effective concentration (EC50) curve inflection point, and the "Hill" slope. Curves were fit with the *drm()* function in *drc* package in R, using least squares estimation. The strength of each curve was assessed for goodness of fit using Normalized Root Mean Square Error and Akaike Information Criterion.

3. Results and Discussion

3.1. Solvent-Extracted Pigment and Liquid Culture Testing

3.1.1. Pigment from Solvent Extraction (No Purification)

Dichloromethane-extracted pigment resulted in significantly more mortality than the control across a number of test conditions (Figure 2), including tested pigments, exposure concentrations, and fungal growth substrate. Mortality results were also reflected in the number of embryos with sublethal effects, presented in Figure S1. Deformations seen included pericardial edema, yolk sac edema, reduced total length, and axis problems, with deflated swim bladder and bent tail/trunk seen in 400× standard concentration.

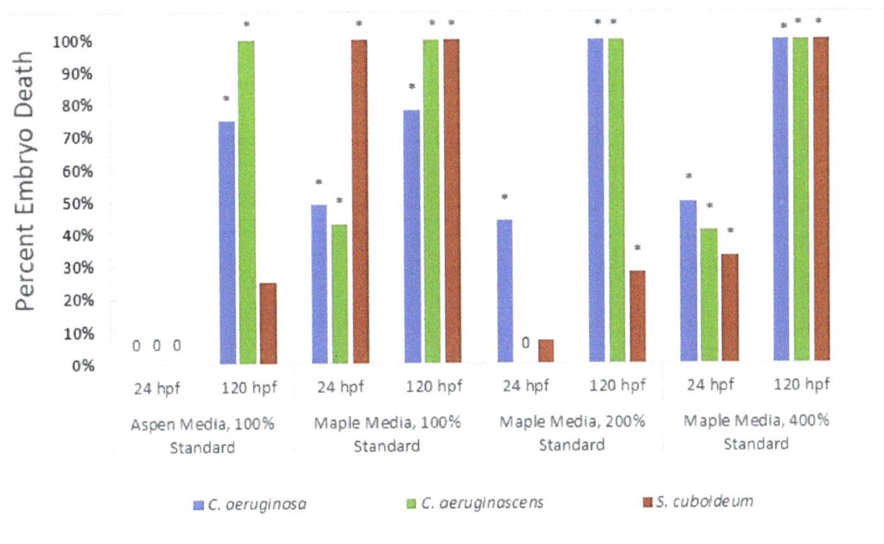

Figure 2. Percent Embryo Death After Exposure to Dichloromethane Extracted Pigments from Fungi at 24 and 120 hpf. Asterisks denote significant difference from control, with all significant values having a *p*-value of <0.0001 as determined by Fisher's exact test. Zeros indicate there was no fish mortality in that condition. DCM-extracted pigment is shown to result in significantly more mortality than control across a number of test conditions, with higher embryo death rates after longer exposure. Aspen media resulted in less mortality than maple media, and higher concentrations of maple media resulted in more deaths for pigments from most fungal species.

The toxicity of pigments varied based on if fungi were grown in wood chips amended with either maple or aspen. Pigments from fungi grown in maple media resulted in significantly more deaths at 24 hpf than the control for all fungal species, most notably the case of *S. cuboideum* where exposure resulted in 100% mortality. In contrast, no dea or deformity was seen in extracts from any fungi grown on aspen at 24 hpf, though 120 hpf both species of *Chlorociboria* showed significant deaths and high levels of deform (Figures 2 and 3). The two wood varieties have differing extractive profiles [65], with map containing bioactive compounds such as resorcinol [66], which likely accounts for the differences. *Chlorociboria aeruginascens* resulted in complete mortality by 120 hpf in all test conditions, with *C. aeruginosa* showing similar toxicity. Both species also showed hi rates of sublethal effects, with deformations seen in all or nearly all embryos by 120 h (Supplemental Figure S1). This suggests that other bioactive secondary metabolites a likely produced by *Chlorociboria* species in amended plate cultures. Finally, *S. cuboide* extract showed total mortality by 24 h at 100% concentration; however, it did not ha significantly different mortality than the control at the same time under 200% exposu ($p = 1.000$) and did not show complete mortality by 120 hpf. This suggests that there m be variation in embryo response, leading to inconsistency in the lethality of pigment extra though the eventual significance of embryo mortality indicates overall toxicity.

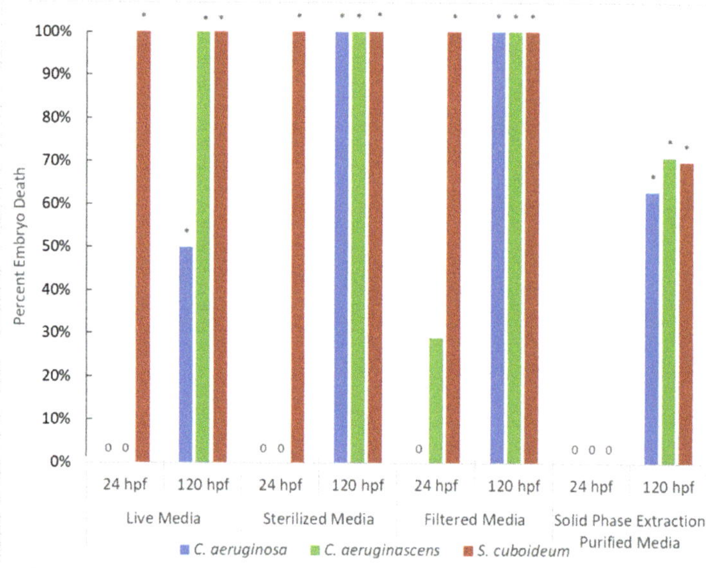

Figure 3. Percent Embryo Death After Exposure to Pigments from Fungi Grown in Liquid Media 24 and 120 hpf. Asterisks denote significant differences from control at the $\alpha = 0.05$ level, with significant values having a p-value of <0.0001, apart from live media *C. aeruginosa* at 120 hpf whi had a p-value of 0.0028. Zeros indicate there was no embryo mortality in that condition.

3.1.2. Pigment from Liquid Culture

Pigments from liquid cultures were associated with high rates of mortality, with a pigments across test conditions resulting in mortality significantly different than the contr by 120 hpf (Figure 2). *Scytalidium cuboideum* liquid culture pigment was especially leth with all embryos dead at 24 h in every test condition except SPE purified pigment (Figure This was a contrast compared to DCM-extracted pigments, which did not show the san consistent toxicity. The differences in toxicity seen between the liquid cultures and the sol cultures are likely explained by variation in differential metabolite production. Variatic in growth conditions is known to alter the secondary metabolite production of fun generally [67–69], and variation in metabolite production in spalting fungi specifically [7 Increased co-production of bioactive compounds in the liquid cultures compared to th in the wood chip plates is likely, especially as pigmented metabolite production is often seen earlier in the lab in liquid compared to solid cultures. In addition to this effect, differences between liquid cultures and DCM extracted pigment were likely influenced by the differences in pigment concentration, the presence of growth media in liquid cultures, and the likelihood that DCM extract contains fewer products of fungal metabolism due to its polarity limiting transfer of compounds.

At 24 hpf there were differences between fungal species, with *Chlorociboria* species showing no significant toxicity in live, sterilized, or filtered media. However, *C. aeruginascens* showed relatively high sublethal effects at 24 hpf despite the lack of significant mortality (Figure S2). This variation in the production of secondary metabolites in addition to target pigment was also observed in DCM-extracted cultures, while both *Chlorociboria* spp. showed high levels of toxicity in DCM- and liquid media solutions. This suggests that other fungal metabolites may have been responsible for the effects seen, especially as toxicity was also lower in SPE-purified samples compared to liquid media and there appeared to be variation in toxicity between species. *Chlorociboria aeruginosa* showed lower percent mortality than *C. aeruginascens* in multiple tests. Notably, in live media testing at 120 hpf, *C. aeruginosa* showed only half the mortality of *C. aeruginascens*, and in sterilized media at 24 hpf it showed no toxicity while *C. aeruginascens* had 100% deformities. Variation in pigment production between the two species has been seen in other studies [70,71], including variation in the production of a yellow pigment in addition to differential production of xylindein [54].

SPE purification resulted in no mortality seen at 24 hpf, though by 120 hpf there was significant mortality for all tested pigments (Figure 2) and high levels of total sublethal effects. Deformations observed included pericardial edema, yolk sac edema, trunk, axis, and craniofacial malformations.

3.2. *Solid Pigments and Behavior Response Testing*

3.2.1. Embryo Photomotor Response

The red pigment crystal dramada produced by *S. cuboideum* was the only compound shown to have significant bioactivity. Exposure to concentrations of 10.7 and 23.2 μM was associated with hyperactive tail bending in the excitatory phase ($p = 0.043$ and 0.022, respectively), and in the case of 23.2 μM also in the baseline phase ($p = 0.007$). In contrast, at 50 μM an absence of any activity was observed in both the excitatory and baseline phases, with p values of 0.024 and 0.022, respectively.

3.2.2. Larval Photomotor Response

Dramada was also the only tested compound that showed any significant effect on larval photomotor response, though all tested compounds were modestly bioactive. Many were associated with effects during the light phase of testing, with the exception of one concentration of the red pigment dramada. Aniline was shown to have a significant hyper effect on photomotor response behaviors in the light interval at 2.32 μM ($n = 27$, $p < 0.01$), at 5 μM ($n = 30$, $p < 0.01$), at 10.7 μM ($n = 26$, $p < 0.01$), and at 23.2 μM ($n = 27$, $p < 0.01$). Xylindein from *C. aeruginosa* was associated with a hyperactive response on LPR in light at ~5 mM or 2.8 mg/mL ($n = 15$, $p < 0.01$), and a hypoactive response at ~10.7 mM or 5.7 mg/mL ($n = 14$, $p < 0.01$). Finally, dramada from *S. cuboideum* at 2.32 mM showed a hypoactive effect in both dark ($n = 26$, $p < 0.01$) and light ($n = 26$, $p < 0.01$) conditions, and a hypoactive effect in light at 5 mM ($n = 23$, $p < 0.01$) and 10 mM ($n = 23$, $p < 0.01$). Dramada also did not have enough animals remaining at 120 hpf to allow for analysis, unlike xylindein and aniline which showed no significant effects at the highest concentration. These abnormalities seen in dramada and impure xylindein from *C. aeruginosa* were modest in comparison to other known compounds that have a major effect on LPR in the lighted phase [42]. In addition, the lack of association with differences in the dark phase is unusual, suggesting that the effect may be at the level of detection and spurious. The control compound used for comparison, aniline, showed similar moderate levels of abnormalities.

3.2.3. Mortality and Morphology

Dramada was the only tested compound that exhibited significant morbidity in tes endpoints, with 50 μM concentrations associated with near 100% mortality by 24 h (Figure 4). Deformations found to be significantly associated with exposure incluc caudal fin deformity at 10.7 and 23.2 μM, abnormal pigmentation at 23.2 μM, and mod body length shortening at 10.7 and 23.2 μM. Concentration response modeling for drama showed that the LC_{50} for mortality at 24 hpf was estimated to be 38.2 μM (±1.4), a mortality (LD_{50}) at 120 hpf was estimated to be 25.5 μM (±3.4). This higher bioactiv is not surprising, as many other naphthoquinones from various sources are known have bioactive properties [72], and the red pigment has been previously described having modest bioactivity against Gram-negative bacteria and fungi [16]. Other naphthoquinonic pigments extracted from filamentous fungi have shown cytotoxic against cell lines [73,74], and one of these compounds, erythreostominone, has also be shown to induce malformations and impair locomotor activity [75,76].

In contrast, xylindein and aniline showed non-significant toxicity responses with a clear dose-response relationship (Figure 4). Aniline is associated with known toxic a potential mutagenic effects [77], and has been shown to have developmental and sublet effects in zebrafish [78] with reported LD_{50} values at 96 h of 618 (±43.0) μmol/L [higher than the concentrations tested here.

Based on the ratio of LD_{50} value to body mass, it would take 510.4 kg (2040 M 250.20 g/mol) of dramada for an 80 kg (~176 lbs) human to experience equivalent expos to the 25.5 μM experienced by a 1 mg embryo – while also being fully immersed in pigm For reference, the LD_{50} of table salt taken orally is 3 g/kg for a rat [80], equivalent to 24 for an 80 kg human. Previous work on the toxicity of the red pigment when isolated fr an actinomycete showed it to be non-toxic at 600 mg/kg administered intraperitoneally mice [16]. These factors suggest that while dramada may present some toxicity, it may present a human health risk except at extreme quantities.

The applications developed based on dramada from *Scytalidium cuboideum* have cused on its use as a coloring agent for textiles and bamboo [20,21,58], using a standa concentration of extracted pigment. Estimates of the pigment concentration in this "st dard" pigment solution range from 0.73 to 2.4 mM [55], making the standard used mu higher than the calculated LD_{50} of 25.5 μM (±3.4) at 120 hpf. This suggests that solutic used in current methods likely contain quantities of pigment that may pose a risk developing embryos, if the pigments were transferred to water to allow for exposure.

However, while dramada has demonstrated a toxic effect on zebrafish, this res does not necessarily prevent its use in industry. Many dyes currently in use in the text industry have been shown to have toxic effects [81], and the development of natural dy to replace toxic synthetics has been a focus of research [82]. Additionally, as the process currently used for dyeing with dramada do not produce wastewater [59,83,84], there a limited likelihood of effluent reaching watersheds and impacting fish development. addition, dramada has limited solubility in water and strongly binds to materials su as textiles, wood, and glass, making transfer and high levels of exposure unlikely. Wh any toxicity suggests that caution should be taken in regard to the use of the pigment applications where consistent exposure to humans is expected, further testing of prepar textiles and other materials should be carried out to determine bioavailability and likelihood of exposure.

While the red pigment produced by *S. cuboideum* demonstrates some toxicity, the gre pigment produced by *Chlorociboria* spp. in solid form has limited bioactivity. Howev pigmented fungal media solutions and, in some cases, extraction using DCM were asso ated with significant mortality. This suggests the co-production of other toxic seconda metabolites, while xylindein itself may be non-toxic. Purer *Chlorociboria* extract has a been shown to have improved semiconductive capabilities [56], further supporting t need for improved methods of xylindein purification. As xylindein is under investigati for use in solar cells, an industry associated with toxic byproducts and materials [85–8 the adoption of xylindein as an alternative could improve the environmental friendliness of energy generation in addition to its sustainability. Future work should focus on the identification and removal of possible mycotoxins and the use of purified pigments. This research supports previous research into the use of these pigments for a variety of applications, allowing for future industrial adoption. The use of these sustainably sourced pigments with limited toxicity has the potential to improve sustainability and replace toxic byproducts across multiple industries.

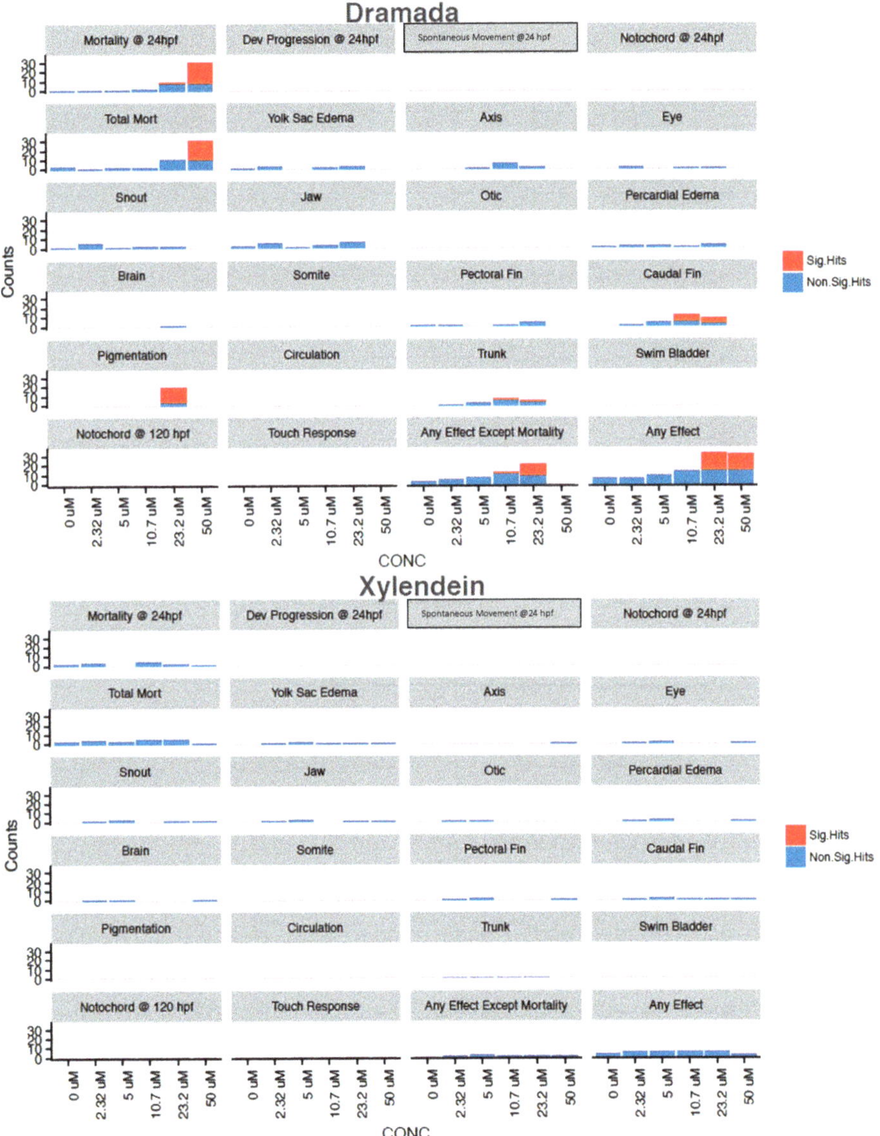

Figure 4. Mortality and Morphology Endpoint Counts for Dramada and Xylindein. Presence/absence data across 32 replicates with points above and at threshold for binomial significance in red. Concentration for xylindein is approximate due to lack of pure compound for testing. Control incidence of all morphological and touch response endpoints below 20% cutoff for biological validity.

4. Conclusions

Pigments from spalting fungi show varying levels of toxicity to zebrafish embry across species and growth/extraction methodology. The red pigment dramada fro *S. cuboideum* was the only solidified and pure pigment associated with significant to city in zebrafish, though with a relatively high LD_{50} value, making it unlikely to aff humans. Solidified and washed xylindein did not demonstrate toxicity. However, s nificant mortality was associated with impure solutions of fungal metabolites contain pigments from all tested species. This suggests the co-production of mycotoxins or toxic related to media components. The development of improved purification methodolog especially for xylindein from *Chlorociboria* species, is therefore of paramount importan for future industrial adoption. However, the low levels of toxicity seen in the solidifi xylindein are sufficient to suggest that future technologies are likely to be both sustainab and environmentally safe. The adoption of these sustainably produced pigments has t potential to replace conventional technologies currently associated with toxicity, allowi for a greener future.

Supplementary Materials: The following are available online at https://www.mdpi.com/2309-6 X/7/2/155/s1, Figure S1: Percent of Embryos Showing Sublethal Effects After Exposure to DC Extracted Pigments from Fungi at 24 and 120 hpf, Figure S2: Percent of Embryos Showing Subleth Effects After Exposure to Pigments from Fungi Grown in Liquid Media at 24 and 120 hpf.

Author Contributions: Conceptualization, B.H.A. and S.C.R.; methodology, B.H.A., S.M.V.G., S and B.H.; validation, S.H., B.H.; formal analysis, B.H.A.; investigation, B.H.A. and S.M.V.G.; resourc S.H. and S.C.R.; writing—original draft preparation, R.C.V.C.; writing—review and editing, S.C. S.H.; visualization, R.C.V.C.; supervision, S.C.R.; project administration, S.C.R.; funding acquisiti S.C.R. All authors have read and agreed to the published version of the manuscript.

Funding: This research was funded by the Oregon BEST Innovation Grant (not Vertue Lab) partial funding of this research. This work was also supported by the USDA National Institute Food and Agriculture, McIntire Stennis project number 1009811 (SR) and partially supported by N grants ES017552 and ES016896 (SH).

Institutional Review Board Statement: The study was conducted according to the guidelines the Declaration of Helsinki, and approved by approved by the Institutional Animal Care and U Committee (IACUC) at Oregon State University (ACUP 5113).

Informed Consent Statement: Not applicable.

Data Availability Statement: Not applicable.

Acknowledgments: The authors would like to thank the Tanguay Lab at Sinnhuber Aquatic Resear Laboratory at Oregon State University for carrying out the experiment and analysis described in t Solid Pigments and Behavior Response Testing. The authors would also like to thank Ariel Muldo (College of Forestry, Oregon State University) for help with statistical analysis review.

Conflicts of Interest: The authors declare no conflict of interest.

References

1. Robinson, S.C.; Michaelsen, H.; Robinson, J.C. *Spalted wood. The History, Science and Art of a Unique Material*, 1st ed.; Schiff Publishing, Ltd.: Atglen, PA, USA, 2016; p. 287.
2. Seaver, F.G. Photographs and descriptions of cup fungi –XXIV. Chlorociboria. *Mycologia* **1936**, *28*, 309–394.
3. Dixon, J.R. Chlorosplenium and its segregates. The genera chlorociboria and chlorencoelia. *Mycotaxon* **1975**, *1*, 193–237.
4. Oeder, G.C. *Flora Danica. Verlegts Heineck, Mumme und Faber*; Nicolaus Möller: Kopenhagen, Denmark, 1770.
5. Chidester, M.S. A pink stain of wood caused by a species of geotrichum. *Phytopathology* **1940**, *30*, 530–533.
6. Kang, H.; Sigler, L.; Lee, J.; Gibas, C.; Yun, S.; Lee, Y. *Xylogone ganodermophthora* sp. nov., an ascomycetous pathogen causi yellow rot on cultivated mushroom *Ganoderma lucidum* in Korea. *Mycologia* **2010**, *102*, 1167–1184. [CrossRef] [PubMed]
7. Blanchette, R.A.; Wilmering, A.M.; Baumeister, M. The use of green-stained wood caused by the fungus *Chlorociboria* in Intars masterpieces from the 15th century. *Holzforschung* **1992**, *46*, 225–232. [CrossRef]
8. Vega Gutierrez, T.P.; Robinson, C.S. Determining the Presence of Spalted Wood in Spanish Marquetry Woodworks of the 150 through the 1800s. *Coatings* **2017**, *7*, 188. [CrossRef]

Otterstedt, A. Investigating green Marquetry on bowed-string instruments. The leaves be greene. *Galpin Soc. J.* **2001**, 330–338. [CrossRef]
Rommier, P.T.A. Sur un nouvelle matière colorante appelée xylindeine et extraite de certains bois morts. *Comptes Rendus Hebd. Des Séances De L'académie Des Sci.* **1868**, *66*, 108–109.
Blackburn, G.M.; Ekong, D.E.; Nielson, A.H.; Todd, L. Xylindein. *Chimia* **1965**, *19*, 208–212.
Edwards, R.L.; Kale, N. The structure of xylindein. *Tetrahedron* **1965**, *21*, 2095–2107. [CrossRef]
Saikawa, Y.; Watanabe, T.; Hashimoto, K.; Nakata, A. Absolute configuration and tautomeric structure of xylindein, a blue-green pigment of *Chlorociboria* species. *Phytochemistry* **2000**, *55*, 237–240. [CrossRef]
Harrison, R.; Quinn, A.; Weber, G.; Johnson, B.; Rath, J.; Remcho, V.; Robinson, S.; Ostroverkhovaa, O. Fungi-Derived Pigments as Sustainable Organic (Opto)electronic Materials. Available online: https://www.spiedigitallibrary.org/conference-proceedings-of-spie/10101/1/Fungi-derived-pigments-as-sustainable-organic-optoelectronic-materials/10.1117/12.2251265.short?SSO=1 (accessed on 20 February 2021).
Vega Gutierrez, M.S.; Hazell, K.K.; Simonsen, J.; Robinson, C.S. Description of a Naphthoquinonic Crystal Produced by the Fungus Scytalidium cuboideum. *Molecules* **2018**, *23*, 1905. [CrossRef]
Gerber, N.N.; Wieclawek, B. The Structures of Two Naphthoquinone Pigments from an Actinomycete1. *J. Org. Chem.* **1966**, *31*, 1496–1498. [CrossRef]
Robinson, S.C.; Hinsch, E.; Weber, G.; Leipus, K.; Cerney, D. Wood Colorization through Pressure Treating: The Potential of Extracted Colorants from Spalting Fungi as a Replacement for Woodworkers' Aniline Dyes. *Materials* **2014**, *7*, 5427–5437. [CrossRef]
Robinson, S.C.; Vega Gutierrez, S.M.; Garcia, R.A.C.; Iroume, N.; Vorland, N.R.; Andersen, C.; de Oliveira Xaxa, I.D.; Kramer, O.E.; Huber, M.E. Potential for fungal dyes as colorants in oil and acrylic paints. *J. Coat. Technol. Res.* **2018**, *15*, 845–849. [CrossRef]
Hinsch, E.M. A Comparative Analysis of Extracted Fungal Pigments and Commercially Available Dyes for Colorizing Textiles. Master's Thesis, Oregon State University, Corvallis, OR, USA, 2015.
Hinsch, E.; Robinson, S. Comparing Colorfastness to Light of Wood-Staining Fungal Pigments and Commercial Dyes: An Alternative Light Test Method for Color Fastness. *Coatings* **2018**, *8*, 189. [CrossRef]
Weber, G.; Chen, H.-L.; Hinsch, E.; Freitas, S.; Robinson, S. Pigments extracted from the wood-staining fungi Chlorociboria aeruginosa, Scytalidium cuboideum, and S. ganodermophthorum show potential for use as textile dyes. *Coloration Technol.* **2014**, *130*, 445–452. [CrossRef]
Khan, S.; Malik, A. Environmental and Health Effects of Textile Industry Wastewater. In *Environmental Deterioration and Human Health: Natural and anthropogenic determinants*; Malik, A., Grohmann, E., Akhtar, R., Eds.; Springer Netherlands: Dordrecht, The Netherlands, 2014; pp. 55–71. [CrossRef]
Markandeya, S.P.; Shukla, S.P.; Mohan, D. Toxicity of Disperse Dyes and its Removal from Wastewater Using Various Adsorbents: A Review. *Res. J. Environ. Toxicol.* **2017**, *11*, 72–89.
Sharma, K.P.; Sharma, S.; Sharma, S.; Singh, P.K.; Kumar, S.; Grover, R.; Sharma, P.K. A comparative study on characterization of textile wastewaters (untreated and treated) toxicity by chemical and biological tests. *Chemosphere* **2007**, *69*, 48–54. [CrossRef]
Khan, R.; Bhawana, P.; Fulekar, M.H. Microbial decolorization and degradation of synthetic dyes: A review. *Rev. Environ. Sci. Bio/Technol.* **2013**, *12*, 75–97. [CrossRef]
Giesbers, G.; Van Schenck, J.; Vega Gutierrez, M.S.; Robinson, S.; Ostroverkhovaa, O. Fungi-Derived Pigments for Sustainable Organic (Opto)Electronics. *Mrs Adv.* **2018**, *3*. [CrossRef]
Radić, N.; Štrukelj, B. Endophytic fungi—The treasure chest of antibacterial substances. *Phytomedicine* **2012**, *19*, 1270–1284. [CrossRef] [PubMed]
Beekman, A.M.; Barrow, R.A. Fungal Metabolites as Pharmaceuticals. *Aust. J. Chem.* **2014**, *67*, 827–843. [CrossRef]
Moss, M.O. Mycotoxins of Aspergillus and other filamentous fungi. *J. Appl. Bacteriol.* **1989**, *67*, 69s–81s. [CrossRef]
Narsing Rao, M.P.; Xiao, M.; Li, W.-J. Fungal and Bacterial Pigments: Secondary Metabolites with Wide Applications. *Front. Microbiol.* **2017**, *8*, 1113. [CrossRef]
Dufossé, L.; Fouillaud, M.; Caro, Y.; Mapari, S.A.S.; Sutthiwong, N. Filamentous fungi are large-scale producers of pigments and colorants for the food industry. *Curr. Opin. Biotechnol.* **2014**, *26*, 56–61. [CrossRef]
Dubravka, F.; Maja, P. Toxicological Properties of Citrinin. *Arch. Ind. Hyg. Toxicol.* **2009**, *60*, 457–464. [CrossRef]
Carvalho, J.C.d.; Oishi, B.O.; Pandey, A.; Soccol, C.R. Biopigments from Monascus: Strains selection, citrinin production and color stability. *Braz. Arch. Biol. Technol.* **2005**, *48*, 885–894. [CrossRef]
Liang, B.; Du, X.-J.; Li, P.; Sun, C.-C.; Wang, S. Investigation of Citrinin and Pigment Biosynthesis Mechanisms in Monascus purpureus by Transcriptomic Analysis. *Front. Microbiol.* **2018**, *9*. [CrossRef]
Marič, A.; Skočaj, M.; Likar, M.; Sepčić, K.; Cigić, I.K.; Grundner, M.; Gregori, A. Comparison of lovastatin, citrinin and pigment production of different Monascus purpureus strains grown on rice and millet. *J. Food Sci. Technol.* **2019**, *56*, 3364–3373. [CrossRef] [PubMed]
Panda, B.P.; Ali, M. Reduction of citrinin biosynthesis by fatty acids in Monascus fermented food. *World Mycotoxin J.* **2012**, *5*, 163–167. [CrossRef]
Zhen, Z.; Xiong, X.; Liu, Y.; Zhang, J.; Wang, S.; Li, L.; Gao, M. NaCl Inhibits Citrinin and Stimulates Monascus Pigments and Monacolin K Production. *Toxins (Basel)* **2019**, *11*, 118. [CrossRef]

38. Gibson, S. A Warning on Spalted Wood. *Fine Woodworking* **1995**, *118*, 110–111.
39. Robinson, S.C. Spalted Wood: Health and Safety. *Am. Woodturn.* 2011. Available online: https://www.google.com.hk/url?t&rct=j&q=&esrc=s&source=web&cd=&ved=2ahUKEwjMmK_YpPzuAhVCQKwKHUbJAJ0QFjAAegQIBxAD&url=https3A%2F%2Fwww.peaceriverwoodturners.org%2Fresources%2FDocuments%2FUNIQUE%2520TECHNIQUES%2FSpalted%20Wood%2520Safety.pdf&usg=AOvVaw0XAz6Tj2Ow1izwSMtkPQmL (accessed on 21 February 2021).
40. Ali, S.; Champagne, D.L.; Spaink, H.P.; Richardson, M.K. Zebrafish embryos and larvae: A new generation of disease models a drug screens. *Birth Defects Res. Part C: Embryo Today: Rev.* **2011**, *93*, 115–133. [CrossRef]
41. Kari, G.; Rodeck, U.; Dicker, A.P. Zebrafish: An Emerging Model System for Human Disease and Drug Discovery. *Clin. Pharma Ther.* **2007**, *82*, 70–80. [CrossRef]
42. Hagstrom, D.; Truong, L.; Zhang, S.; Tanguay, R.; Collins, E.-M.S. Comparative Analysis of Zebrafish and Planarian Mo Systems for Developmental Neurotoxicity Screens Using an 87-Compound Library. *Toxicol. Sci.* **2018**, *167*, 15–25. [CrossRef]
43. Detrich, H.W.; Westerfield, M.; Zon, L.I. Chapter 1 Overview of the Zebrafish System. In *Methods in Cell Biology*; Detrich, H. Westerfield, M., Zon, L.I., Eds.; Academic Press: Cambridge, MA, USA, 1998; Volume 59, pp. 3–10.
44. Zon, L.I.; Peterson, R.T. In vivo drug discovery in the zebrafish. *Nat. Rev. Drug Discov.* **2005**, *4*, 35–44. [CrossRef]
45. MacRae, C.A.; Peterson, R.T. Zebrafish as tools for drug discovery. *Nat. Rev. Drug Discov.* **2015**, *14*, 721. [CrossRef]
46. Gehrig, J.; Pandey, G.; Westhoff, J.H. Zebrafish as a Model for Drug Screening in Genetic Kidney Diseases. *Front. Pediatrics* **201** [CrossRef] [PubMed]
47. Zhang, C.; Willett, C.; Fremgen, T. Zebrafish: An Animal Model for Toxicological Studies. *Curr. Protoc. Toxicol.* **2003**, 1.7.1–1.7.18. [CrossRef]
48. Embry, M.R.; Belanger, S.E.; Braunbeck, T.A.; Galay-Burgos, M.; Halder, M.; Hinton, D.E.; Léonard, M.A.; Lillicrap, A.; Norbe King, T.; Whale, G. The fish embryo toxicity test as an animal alternative method in hazard and risk assessment and scient research. *Aquat. Toxicol.* **2010**, *97*, 79–87. [CrossRef] [PubMed]
49. Scholz, S. Zebrafish embryos as an alternative model for screening of drug-induced organ toxicity. *Arch. Toxicol.* **2013**, *87*, 767– [CrossRef]
50. Challal, S.; Bohni, N.; Buenafe, O.E.; Esguerra, C.V.; de Witte, P.A.M.; Wolfender, J.-L.; Crawford, A.D. Zebrafish Bioassay-guic Microfractionation for the Rapid in vivo Identification of Pharmacologically Active Natural Products. *Chim. Int. J. Chem.* **2012**, 229–232. [CrossRef]
51. Crawford, A.D.; Esguerra, C.V.; de Witte, P.A.M. Fishing for Drugs from Nature: Zebrafish as a Technology Platform for Natu Product Discovery. *Planta Med* **2008**, *74*, 624–632. [CrossRef]
52. Zuberi, Z.; Eeza, M.N.H.; Matysik, J.; Berry, J.P.; Alia, A. NMR-Based Metabolic Profiles of Intact Zebrafish Embryos Exposed Aflatoxin B1 Recapitulates Hepatotoxicity and Supports Possible Neurotoxicity. *Toxins (Basel)* **2019**, *11*, 258. [CrossRef]
53. Robinson, S.C.; Hinsch, E.; Weber, G.; Freitas, S. Method of extraction and resolubilisation of pigments from Chlorocibo aeruginosa and Scytalidium cuboideum, two prolific spalting fungi. *Coloration Technol.* **2014**, *130*, 221–225. [CrossRef]
54. Weber, G.; Boonloed, A.; Naas, K.M.; Koesdjojo, M.T.; Remcho, V.T.; Robinson, S.C. A method to stimulate production extracellular pigments from wood-degrading fungi using a water carrier. *Curr. Res. Environ. Appl. Mycol.* **2016**, *6*, 218–2 [CrossRef]
55. Vega Gutierrez, M.S.; Van Court, R.C.; Stone, D.W.; Konkler, M.J.; Groth, E.N.; Robinson, C.S. Relationship between Molarity a Color in the Crystal (Dramada') produced by *Scytalidium cuboideum*, in Two Solvents. *Molecules* **2018**, *23*, 2581. [CrossRef]
56. Giesbers, G.; Krueger, T.; Schenck, J.V.; Court, R.V.; Moore, J.; Fang, C.; Robinson, S.; Ostroverkhova, O. Fungi-derived xylind Effect of purity on optical and electronic properties. *Mrs Adv.* **2019**, *4*, 1769–1777. [CrossRef]
57. Boonloed, A.; Weber, G.L.; Ramzy, K.M.; Dias, V.R.; Remcho, V.T. Centrifugal partition chromatography: A preparative tool isolation and purification of xylindein from Chlorociboria aeruginosa. *J. Chromatogr. A* **2016**, *1478*, 19–25. [CrossRef]
58. Vega Gutierrez, S.; Vega Gutierrez, P.; Godinez, A.; Pittis, L.; Huber, M.; Stanton, S.; Robinson, S. Feasibility of Coloring Bamb with the Application of Natural and Extracted Fungal Pigments. *Coatings* **2016**, *6*, 37. [CrossRef]
59. Hinsch, E.M.; Weber, G.; Chen, H.-L.; Robinson, S.C. Colorfastness of Extracted Wood-staining Fungal Pigments on Fabrics new potential for textile dyes. *J. Text. Appar. Technol. Manag.* **2015**, *9*. [CrossRef]
60. Truong, L.; Harper, S.L.; Tanguay, R.L. Evaluation of embryotoxicity using the zebrafish model. *Methods Mol. Biol.* **2011**, 6 271–279. [CrossRef] [PubMed]
61. Mandrell, D.; Truong, L.; Jephson, C.; Sarker, M.R.; Moore, A.; Lang, C.; Simonich, M.T.; Tanguay, R.L. Automated Zebraf Chorion Removal and Single Embryo Placement: Optimizing Throughput of Zebrafish Developmental Toxicity Screens. *J. I Autom.* **2012**, *17*, 66–74. [CrossRef]
62. Reif, D.M.; Truong, L.; Mandrell, D.; Marvel, S.; Zhang, G.; Tanguay, R.L. High-throughput characterization of chemical-associa embryonic behavioral changes predicts teratogenic outcomes. *Arch. Toxicol.* **2016**, *90*, 1459–1470. [CrossRef] [PubMed]
63. Truong, L.; Zaikova, T.; Baldock, B.L.; Balik-Meisner, M.; To, K.; Reif, D.M.; Kennedy, Z.C.; Hutchison, J.E.; Tanguay, R Systematic determination of the relationship between nanoparticle core diameter and toxicity for a series of structurally analog gold nanoparticles in zebrafish. *Nanotoxicology* **2019**, 1–15. [CrossRef] [PubMed]
64. R Core Team. *R: A Language and Environment for Statistical Computing*; R Foundation for Statistical Computing: Vienna, Aust 2016.
65. Forest Products Lab. *Extractives in Eastern Hardwoods—A Review*; US Department of Agriculture: Masidon, WI, USA, 1979.

He, Z.; Sleighter, R.L.; Hatcher, P.G.; Liu, S.; Wu, F.; Zou, H.; Olanya, O.M. Molecular level comparison of water extractives of maple and oak with negative and positive ion ESI FT-ICR mass spectrometry. *J. Mass Spectrom.* **2019**, *54*, 655–666. [CrossRef]
VanderMolen, K.M.; Raja, H.A.; El-Elimat, T.; Oberlies, N.H. Evaluation of culture media for the production of secondary metabolites in a natural products screening program. *Amb Express* **2013**, *3*, 71. [CrossRef]
Son, S.Y.; Lee, S.; Singh, D.; Lee, N.-R.; Lee, D.-Y.; Lee, C.H. Comprehensive Secondary Metabolite Profiling Toward Delineating the Solid and Submerged-State Fermentation of Aspergillus oryzae KCCM 12698. *Front. Microbiol.* **2018**, *9*, 1076. [CrossRef]
Bode, H.; Bethe, B.; Hofs, R.; Zeeck, A. Big effects from small changes: Possible ways to explore nature's chemical diversity. *Chembiochem* **2002**, *3*, 619–627. [CrossRef]
Van Court, R.C.; Giesbers, G.; Ostroverkhova, O.; Robinson, C.S. Optimizing Xylindein from Chlorociboria spp. for (Opto)electronic Applications. *Processes* **2020**, *8*. [CrossRef]
Tudor, D.; Margaritescu, S.; Sánchez-Ramírez, S.; Robinson, S.C.; Cooper, P.A.; Moncalvo, J.M. Morphological and molecular characterization of the two known North American Chlorociboria species and their anamorphs. *Fungal Biol.* **2014**, *118*, 732–742. [CrossRef]
Petr, B.; Vojtech, A.; Ladislav, H.; Rene, K. Noteworthy Secondary Metabolites Naphthoquinones—Their Occurrence, Pharmacological Properties and Analysis. *Curr. Pharm. Anal.* **2009**, *5*, 47–68. [CrossRef]
Kittakoop, P.; Punya, J.; Kongsaeree, P.; Lertwerawat, Y.; Jintasirikul, A.; Tanticharoen, M.; Thebtaranonth, Y. Bioactive naphthoquinones from Cordyceps unilateralis. *Phytochemistry* **1999**, *52*, 453–457. [CrossRef]
Abe, F.R.; de Oliveira, D.P. Evaluation of apoptotic and necrotic cells of the natural dye erythrostominone. *Toxicol. Lett.* **2014**, *229*, S114. [CrossRef]
Abe, F.R.; Soares, A.M.V.M.; Oliveira, D.P.d.; Gravato, C. Toxicity of dyes to zebrafish at the biochemical level: Cellular energy allocation and neurotoxicity. *Environ. Pollut.* **2018**, *235*, 255–262. [CrossRef] [PubMed]
Abe, F.R.; Mendonça, J.N.; Moraes, L.A.B.; Oliveira, G.A.R.d.; Gravato, C.; Soares, A.M.V.M.; Oliveira, D.P.d. Toxicological and behavioral responses as a tool to assess the effects of natural and synthetic dyes on zebrafish early life. *Chemosphere* **2017**, *178*, 282–290. [CrossRef] [PubMed]
EPA (Ed.) *Aniline Fact Sheet: Support Document (CAS No. 62-53-3)*; United States Evnironmental Protection Agency: Washington, DC, USA, 1994.
Horie, Y.; Yamagishi, T.; Koshio, M.; Iguchi, T.; Tatarazako, N. Lethal and sublethal effects of aniline and chlorinated anilines on zebrafish embryos and larvae. *J. Appl. Toxicol.* **2017**, *37*, 836–841. [CrossRef]
Zok, S.; Görge, G.; Kalsch, W.; Nagel, R. Bioconcentration, metabolism and toxicity of substituted anilines in the zebrafish (Brachydanio rerio). *Sci. Total Environ.* **1991**, *109–110*, 411–421. [CrossRef]
Scientific, F. Material Safety Data Sheet Sodium chloride. Available online: https://fscimage.fishersci.com/msds/21105.htm (accessed on 17 February 2021).
Verma, Y. Acute toxicity assessment of textile dyes and textile and dye industrial effluents using Daphnia magna bioassay. *Toxicol. Ind. Health* **2008**, *24*, 491–500. [CrossRef]
Lopes, F.C.; Tichota, D.M.; Pereira, J.Q.; Segalin, J.; de Oliveira Rios, A.; Brandelli, A. Pigment Production by Filamentous Fungi on Agro-Industrial Byproducts: An Eco-Friendly Alternative. *Appl. Biochem. Biotechnol.* **2013**, *171*, 616–625. [CrossRef] [PubMed]
Palomino Agurto, E.M.; Vega Gutierrez, M.S.; Chen, H.-L.; Robinson, C.S. Wood-Rotting Fungal Pigments as Colorant Coatings on Oil-Based Textile Dyes. *Coatings* **2017**, *7*, 152. [CrossRef]
Palomino Agurto, M.; Vega Gutierrez, S.; Van Court, R.; Chen, H.; Robinson, S. Oil-Based Fungal Pigment from *Scytalidium cuboideum* as a Textile Dye. *J. Fungi* **2020**, *6*. [CrossRef] [PubMed]
Brun, N.R.; Wehrli, B.; Fent, K. Ecotoxicological assessment of solar cell leachates: Copper indium gallium selenide (CIGS) cells show higher activity than organic photovoltaic (OPV) cells. *Sci. Total Environ.* **2016**, *543*, 703–714. [CrossRef] [PubMed]
Cyrs, W.D.; Avens, H.J.; Capshaw, Z.A.; Kingsbury, R.A.; Sahmel, J.; Tvermoes, B.E. Landfill waste and recycling: Use of a screening-level risk assessment tool for end-of-life cadmium telluride (CdTe) thin-film photovoltaic (PV) panels. *Energy Policy* **2014**, *68*, 524–533. [CrossRef]
Tsoutsos, T.; Frantzeskaki, N.; Gekas, V. Environmental impacts from the solar energy technologies. *Energy Policy* **2005**, *33*, 289–296. [CrossRef]

Article

Production of Bio-Based Pigments from Food Processing Industry By-Products (Apple, Pomegranate, Black Carrot, Red Beet Pulps) Using *Aspergillus carbonarius*

Ezgi Bezirhan Arikan [1,*], Oltan Canli [2], Yanis Caro [3,4], Laurent Dufossé [4,*] and Nadir Dizge [1]

1 Department of Environmental Engineering, Mersin University, Mersin 33343, Turkey; ndizge@mersin.edu.tr
2 Environment and Clean Production Institute, The Scientific and Technological Research Council of Turkey, Marmara Research Center, Kocaeli 41470, Turkey; oltan.canli@tubitak.gov.tr
3 Département Hygiène Sécurité Environnement (HSE), IUT La Réunion, Université de La Réunion, 40 avenue de Soweto, BP 373, F-97455 Saint-Pierre, Réunion, France; yanis.caro@univ-reunion.fr
4 Laboratoire de Chimie et de Biotechnologie des Produits Naturels, Chemistry and Biotechnology of Natural Products (CHEMBIOPRO), Université de La Réunion, ESIROI Agroalimentaire, 15 Avenue René Cassin, CS 92003, F-97744 Saint-Denis, Réunion, France
* Correspondence: ezgibezirhan@gmail.com (E.B.A.); laurent.dufosse@univ-reunion.fr (L.D.)

Received: 21 September 2020; Accepted: 20 October 2020; Published: 22 October 2020

Abstract: Food processing industry by-products (apple, pomegranate, black carrot, and red beet pulps) were evaluated as raw materials in pigment production by the filamentous fungi *Aspergillus carbonarius*. The effect of fermentation conditions (solid and submerged-state), incubation period (3, 6, 9, 12, and 15 d), initial substrate pH (4.5, 5.5, 6.5, 7.5, and 8.5), and pulp particle size (<1.4, 1.4–2.0, 2–4, and >4 mm) on fungal pigment production were tested to optimize the conditions. Pigment extraction analysis carried out under solid-state fermentation conditions showed that the maximum pigment production was determined as 9.21 ± 0.59 absorbance unit at the corresponding wavelength per gram (AU/g) dry fermented mass (dfm) for pomegranate pulp (PP) by *A. carbonarius* for 5 d. Moreover, the highest pigment production was obtained as 61.84 ± 2.16 AU/g dfm as yellowish brown at initial pH 6.5 with < 1.4 mm of substrate particle size for 15-d incubation period. GC×GC-TOFMS results indicate that melanin could be one of the main products as a pigment. SEM images showed that melanin could localize on the conidia of *A. carbonarius*.

Keywords: *Aspergillus carbonarius*; bioconversion; food processing industry by-product valorization; filamentous fungi; bio-based pigment

1. Introduction

Color has always been the basis for the evaluation of both aesthetics and quality for humanity [1]. Pigments that can be defined as colorant compounds are used in many industries such as textiles, cosmetics, dyes, pharmaceuticals, food etc. [2,3]. Before the discovery of synthetic colorants in the mid-19th century, pigments were obtained from natural sources such as animals, plants, and rocks [4]. The discovery of a synthetic pigment named mauveine in 1856 [5] triggered the industries' usage of synthetic pigments [6]. However, recent studies have shown that some synthetic pigments may have carcinogenic, teratogenic, and allergenic effects [2,7]. For this reason, utilization of some synthetic pigments in food, pharmaceutical, and cosmetic products is limited or prohibited by organizations such as the World Health Organization (WHO), the Food and Agriculture Organization of the United Nations

(FAO), and the US Food and Drug Administration (FDA) [8,9]. The increasing legal requirements and consumer awareness in recent years have encouraged industries to use a larger amount of natural pigments. Hence, recent studies have focused on production of cost-effective natural pigments [10–13], also called bio-based pigments.

Natural pigments that can be obtained from plants, animals, and microorganisms are mostly biocompatible, biodegradable, environmental-friendly, and they have low toxicity [11,14,15]. The usage of plants and animals for natural pigment production has many disadvantages such as the non-stability and high solubility of pigments, dependency on the season, and the loss of certain species for large scale production [16]. However, microbial pigment production is considered more advantageous due to their higher growth rate, the fact that they are unaffected by the seasonal changes, and the high stability of produced pigments [3,7]. A literature survey showed that among the microorganisms, fungi are mostly preferred for industrial-scale production of natural pigments because algae requires sunlight and bacteria are more vulnerable to environmental conditions.

On the one hand, it is estimated that the market of natural pigments presents the highest growth rate of around 7% per year [17]. On the other hand, the production of natural pigments at industrial-scale has depended on the design of a cost-effective production process [18–20]. It is known that the cost of microbial bio-pigment production has been affected by 38–73% of raw material selection [7]. Thus, the raw materials selected as a substrate in the production of bio-pigments using fungi should be both inexpensive and rich in carbon and nitrogen sources. In this context, agricultural or food by-products originated from industries have recently gained great attention due to their applicability for obtaining new valuable products with a zero-waste strategy [21]. Therefore, food by-products were chosen as a low-cost substrate for pigment production in this study.

Filamentous fungi are known to be producers of many types of bio-pigments, such as carotenoids, melanins, flavins, phenazines, quinones, monacins, and indigo [22,23]. Hence, filamentous fungi such as *Monascus*, *Aspergillus*, *Penicillium*, *Neurospora*, *Eurotium*, *Drechslera*, and *Trichoderma* have been found to be the subject of many studies as a potential producer of bio-pigments [19,24–28]. Among the filamentous fungi, *Aspergillus* is the fungal genus most commonly found on foods [29] and they are able to produce various pigments that contain hydroxyanthraquinoid [22]. Recently, many studies have focused on potential pigment production from *Aspergillus spp.* and the optimizing of production [28,30–32]. However, to the best of our knowledge, pigment production from food by-products as substrates using *Aspergillus carbonarius* filamentous fungus has not yet been studied.

Based on the aspects cited above, the main objective of this study is to evaluate the potential ability of pigment production by the filamentous fungi *Aspergillus carbonarius*, using food processing industry by-products including apple, pomegranate, black carrot, and red beet pulp. Furthermore, the effect of fermentation conditions, incubation time, initial pH, and pulp particle size to optimize pigment production were evaluated.

2. Materials and Methods

2.1. Fungal Species and Inoculum

Filamentous fungi *Aspergillus carbonarius* M333 were obtained from the Department of Environmental Engineering Laboratory, Mersin University, Mersin, Turkey. This fungus was maintained on a potato dextrose agar (PDA, Merck, Darmstadt, Germany) slant at 4 °C and sub-cultured monthly. *A. carbonarius* was transferred to PDA medium in Petri dishes in a UV laminar chamber (Faster, UCS2–4, Ferrara, Italy) and incubated at 25 ± 1 °C for 7 d. After incubation, spores of fungus were individually harvested from the surface of the Petri dishes. A suspension of spores was prepared in sterile distilled water containing Tween 80 (0.1%, Fisher). The concentration of fungus was then adjusted to 1×10^6 colony-forming unit/mL by using a Thoma cell counting chamber [33]. The spore suspension of *A. carbonarius* was used for inoculation for further studies.

2.2. Substrate

In this study, apple, pomegranate, black carrot, and red beet pulp were selected as substrates for pigment production by fungi. Pulps were kindly provided by Anadolu Etap Agriculture and Fruit Products Industry and Trade Inc. in Mersin City, Turkey. To evaluate the effect of the initial carbon to nitrogen (C:N) ratio, elemental analysis of each type of pulp was performed at this stage. For further studies, pulps were dried at 60 ± 1 °C for 24 h in an oven (Figure S1), followed by particle size reduction using a kitchen blender and sieved (Tyler mesh 10–12) [34] to obtain particles of 1.4 mm diameter. All dried pulp particles were stored at 4 °C prior to use.

2.3. Selection of Pulp Type

The experimental flow chart during this study is illustrated in Figure S2. In the first stage of the experiment, pigment production capacities of *A. carbonarius* on each pulp (apple, pomegranate, black carrot, and red beet) were evaluated by solid-state fermentation (Figure S2A) based on previous studies [35]. The particle size of all pulps was chosen to be under 1.4 mm in order to increase the penetration of fungal hypha. All types of pulp particles (5 g) were added in Erlenmeyer flasks (250 mL) individually and all flasks subsequently were autoclaved (Sanyo, MLS-3781L, Moriguchi, Japan) at 121 °C for 20 min. After cooling to room temperature, sterilized distilled water was added to each flask to adjust the initial moisture content of the substrate to 50% (*w/w*, on a dry basis) at aseptic conditions [36,37]. Then, spore suspension of fungus (1 mL) was added into the flasks. Some flasks were not inoculated, and these flasks were used as blanks. All flasks were incubated at 25 ± 1 °C in a static incubator (Sanyo, MIR152, Japan) for 5 d. After incubation, the complete solid mass (including fungal biomass and pulp mass) in each flask was harvested, dried at 60 ± 1 °C in an oven for 24 h, and used for pigment extraction and analysis (Figure S2A). Additionally, the harvested wet mass was used directly for pigment extraction in order to test extraction efficiency. For the rest of the study, the pulp particle type that has the highest pigment production was named as the optimum pulp. Further experiments were performed with the optimum pulp for evaluating the impact of fermentation strategy, incubation period, particle size of the substrate, and initial pH of the substrate on pigment production.

2.4. Effect of Fermentation Strategy and Incubation Period

After the optimum pulp type was determined, the effect of fermentation strategy (solid-state and submerged-state) and incubation period (3, 6, 9, 12, and 15 d) were investigated for pigment production capacities of fungus. For this purpose, 5 g of pulp (particle size < 1.4 mm) was transferred into Erlenmeyer flasks (250 mL). All flasks were sterilized at 121 °C for 20 min and cooled to room temperature. Solid-state fermentation (SSF) was carried out by adding sterilized distilled water to these flasks in order to obtain an initial moisture of 50%. For submerged-state fermentation (SmF), 50 mL of sterilized distilled water was added to flasks to reach the final working volume. Then, each flask was inoculated with 1 mL spore suspension of *A. carbonarius*. The flasks were incubated in a shaking incubator at 100 rpm and 25 ± 1 °C for SmF and in a static incubator at 25 ± 1 °C for SSF for 15 d. During the incubation period, the contents of flasks were harvested every 3 d and dried at 60 ± 1 °C for 24 h. Dried masses were utilized for pigment extraction and analysis (Figure S2B).

2.5. Effect of Substrate Particle Size

To determine the effect of particle size on pigment production, the optimum dried pulp type was sieved and classified as < 1.4 mm, 1.4–2.0 mm, 2–4 mm, and > 4 mm. Each particle size was evaluated for pigment production. Optimum pulp type at different particle sizes was weighed as 5 g and transferred to flasks. Following the sterilization at 121 °C for 20 min, sterile distilled water was added to the flasks, according to the optimum fermentation condition determined in the previous stage (solid-state or submerged-state). After inoculation of 1 mL of the spore suspension of *A. carbonarius*, all flasks were incubated at 25 ± 1 °C. During the incubation period, pigment extraction and estimation analysis were conducted to dried mass every 3 d (Figure S2C).

2.6. Effect of Initial Substrate pH

After determining the optimum pulp type, fermentation condition, incubation period, and pulp particle size for pigment production, the initial pH of the substrate was tested. For this purpose, 5 g of pulp particles and a known volume of distilled water according to the optimum fermentation condition were added to flasks. Then, the pH value of each flask was individually adjusted to 4.5 (original pH), 5.5, 6.5, 7.5, and 8.5 using a pH/Cond 340i Handheld Multimeter. The solution of sodium hydroxide (0.1 N) (Merck, Germany) and hydrogen chloride (0.1 N) (37%, Merck, Germany) was used for pH adjustment. After sterilization of all flasks that contain pulp particles at different initial pH levels, each flask was inoculated with spore suspension (1 mL) and incubated at 25 ± 1 °C for the optimum incubation period (Figure S2D). During the incubation period, the mass was separated from flasks, dried, and used for pigment extraction.

In the last stage of the experiment, pigment production was carried out at optimal conditions during the optimum incubation period (Figure S2E). The obtained pigment supernatant was used for GC×GC-TOFMS analysis, and dyeing tests. The obtained fungal biomass and pulp after fermentation were used for scanning electron microscopy (SEM) analysis and elemental analysis, respectively.

2.7. Instrumental Analysis

The morphological characterization of dried fermented mass obtained from optimal conditions was observed by SEM (Zeiss Supra 55, Oberkochen, Germany). Images were taken by applying an electron beam with an acceleration voltage of 5 kV. In addition, SEM analyses were performed on the unfermented optimum type of pulp and *A. carbonarius* growing in PDB to compare morphological changes.

The elemental analysis of each type of pulp was performed before fermentation due to evaluating C:N ratio utilization in the selection of pulp type experiments. Moreover, it was performed both before and after fermentation for the optimum pulp type, which provides higher pigment extraction. Elemental analysis was performed using a TruSpec Micro (LECO, St. Joseph, MI, USA) elementary analyzer (for C, H, N, and S weight percentages).

Two-dimensional gas chromatography (GC×GC) was used for the qualitative analysis of the extract. Approximately 10 mg of pulp, without any pre-treatment, was weighed and dissolved with 1.0 mL of hexane:acetone (1:1) solvent mixture. Liquid nitrogen used for cold pulses was automatically filled. The Agilent 7890 B (Palo Alto, CA, USA) Gas Chromatograph System was equipped with a LECO Pegasus® BT 4D mass spectrometer (Leco, St. Joseph, MI, USA) dual-stage, quad jet thermal modulator and with a split/splitless injector. The GC primary column had 30 m × 0.25 mm id. × 0.25 μm film thickness Rxi®-17Sil MS (Restek Corp., Bellefonte, PA, USA). The GC secondary column had 0.75 m × 0.25 mm id. × 0.25 μm film thickness Rxi®-5Sil MS (Restek Corp, Bellefonte, PA, USA) mounted in a separate oven installed within the main GC oven. The carrier gas was helium and set 1 mL/min. Injection speed was 3 μL/s and the inlet purge time was 60 s. A 1 μL injection was made in splitless mode with an inlet temperature of 250 °C. The temperature program of the first column was as follows: 40 °C kept for 4 min, then raised at 8 °C/min up to 310 °C kept for 20 min. The temperature of the second oven was programmed with an offset of 10 °C and the modulator temperature offset was 25 °C relative to the first GC oven temperature. The second-dimension separation time (modulation time) was 5 s divided into a hot pulse time of 1.50 s and a cold pulse time between the stages of 1 s. The transfer line from the secondary oven into the mass spectrometer was maintained at 280 °C. The ion source was operated at 250 °C. The electron energy was −70 eV. The data acquisition rate was 200-scans/s, covering a mass range of 50–550 m/z.

2.8. Analytical Techniques

During the pulp type selection experiments, different pigment extraction protocols were performed to evaluate the effect of wet or dry fermented solids. During the following experiments, pigment extraction protocol was performed on a dry basis. Pigment extraction was performed according to the method reported by Kantifedaki et al. [34] with slight modifications. After the fermentation of every case, solid material was weighed as 0.5 g (on wet or dry basis), transferred into a tube and mixed with 5 mL of ethanol (95%). The mixture was placed in an ultrasonic bath at 60 Hz (Witeg, Wertheim, Germany) for 30 min at 25 °C. Then, the mixture was mixed on a rotary shaker at 180 rpm for 1 h at 30 °C followed by centrifugation at 6000 rpm for 20 min. After centrifugation, the supernatant was used for pigment estimation by measuring the absorbance with a UV–vis spectrophotometer (DR3900, Hach Co, Loveland, CO, USA) along with utilizing a quartz cuvette. Yellow, orange, and red pigments were determined by measuring the absorbance in three different wavelengths, 400 nm, 475 nm, and 500 nm, respectively, taking into consideration the dilution factor of the sample [38]. In every step, the unfermented substrate was subjected to sterilization, but inoculation was not carried out. These unfermented pulp particles were subjected to pigment extraction. The results were expressed as absorbance unit at the corresponding wavelength per gram (AU/g). In this study, intracellular pigment production was assessed due to the application of pigment extraction to only fermented solids including fungal biomass and pulp mass.

2.9. Dyeing Tests

The wool fabric sample obtained from a local firm was cut into a size of 10 cm × 10 cm (3 g) and it was immersed into a glass beaker containing 50 mL extracted dye solution (10 mg/L). Then, this glass beaker that contained a dye solution liquor volume to fabric weight ratio of 50 mL/3 g was placed in a water bath and left at 93 °C for 1 h. For the uniform dyeing, the sample was stirred regularly. The dyed wool fabric piece was firstly washed with non-ionic detergent and then washed with tap water to remove the detergent [39]. Thereafter, the dyed fabric piece was dried at 60 °C for 24 h.

Furthermore, the dyeing test was also conducted with the mordanting process. For this purpose, the wool fabric was supplied by BOSSA Trade and Industry Enterprises Turkısh Inc. The fabrics (6 g) were mordanted at 50 °C for 60 min with iron sulfate (100 mL, 25 mg/L) [40]. After mordanting, the dyeing process of the mordanted fabric was carried out with an extracted dye solution (10 mg/L), so that the final dye solution liquor volume to fabric weight ratio was 100 mL/6 g. The dye bath was run for 60 min at 85 °C. After dyeing, color fastness to wash was assessed by obeying ISO 105-C10:2006 method.

3. Results and Discussion

3.1. Optimal Fungal Species and Pulp Type

Elemental analysis of pulps performed before fermentation are shown in Table 1. Results show that each type of pulp had essential nutrient content for fungal growth [41]. In addition to this, after the inoculation of *A. carbonarius* onto different pulps, it was observed that fungus was grown on each different pulp type and covered the surface of the pulps within 3 d. This observation (Figure S3) supports the results of elemental analysis.

Figure 1 shows the results of pigment extraction obtained from wet and dry fermented mass after SSF using *A. carbonarius* cultivated on the pulp of black carrot (Figure 1A), apple (Figure 1B), pomegranate (Figure 1C), and red beet (Figure 1D). The literature describes that nitrogen and carbon sources have effects on secondary metabolite regulation [42]. Therefore, it is possible that different levels and types of pigments could be produced by the fungus grown on the different types of pulps, which have different contents of C and N sources.

Table 1. Elemental analysis of unfermented pulps.

Type of Pulp	C (%)	H (%)	N (%)	S (%)	C:N
Black carrot	29.79	4.27	1.32	0.00	22.57/1
Red beet	37.54	5.43	1.82	0.00	20.63/1
Pomegranate	50.94	6.08	3.02	0.02	16.87/1
Apple	52.12	7.64	4.52	0.11	11.53/1

C: Carbon, H: Hydrogen, N: Nitrogen, S: Sulphur, C:N: carbon to nitrogen ratio.

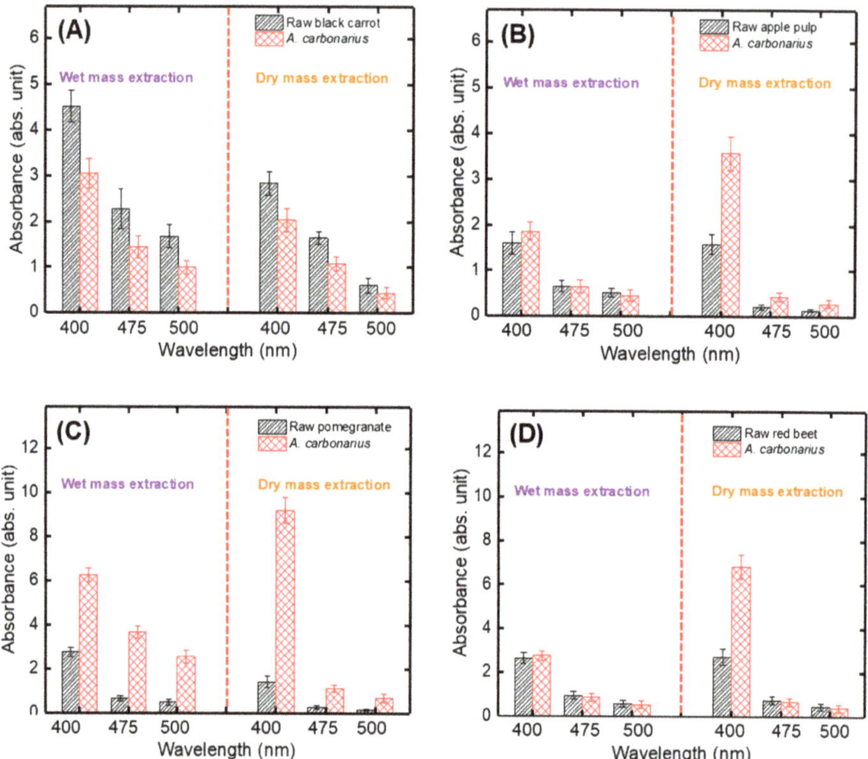

Figure 1. Evaluation of wet and dry mass pigment extraction methods during solid-state fermentation using *A. carbonarius* cultivated on (**A**) black carrot, (**B**) apple, (**C**) pomegranate, (**D**) red beet pulps (experimental conditions: incubation time, 5 d).

Generally, a high C:N ratio has been reported to induce pigment production for filamentous fungi [43]. For example, Palacio-Barrera et al. [44] studied the effect of C:N ratio on pigment production by *Aspergillus chevalieri* and they found that a high C:N ratio (20/1, glucose/yeast extract) was optimum for inducing pigment production. However, it was not the case in this work for all type of pulps. The highest pigment production by *A. carbonarius* was on pomegranate, red beet and apple pulp, respectively. Nevertheless, the highest initial C:N ratio was red beet (20.63/1), pomegranate (16.87/1) and apple (11.53/1), respectively. This could be attributed to the decreasing pH levels during the fermentation of red beet pulps [45]. In this study, a more complex substrate such as a real industrial pulp was used for pigment production. It is known that pigment production is affected not only by the C:N ratio, but also by other factors such as types of C and N sources and the presence of organic acids and minerals [46]. For this reason, it is possible to obtain different results in pigment production from synthetic media and from more complex substrates.

On the other hand, the highest pigment extraction was mostly achieved with dry fermented mass compared to wet fermented mass, for apple (Figure 1B), pomegranate (Figure 1C), and red beet (Figure 1D) pulp. This can be explained as the wet mass (0.5 g on wet basis) used in the extraction method contained more water, whereas the dry mass (0.5 g on dry basis) contained less water due to drying. Hence, it can be concluded that the dry mass, as a result of dehydration, may contain more concentrated pigment due to the lack of water. Pigment extraction analysis carried out under SSF conditions showed that the maximum pigment production was determined as 9.21 ± 0.59 AU/g dry fermented mass (dfm) in pomegranate pulp by *A. carbonarius* for 5 d (Figure 1C) at 400 nm. It was also found that the highest absorbance and color were determined at 400 nm as yellow hue for all pulp types. It is known that yellow hydroxyanthraquinone (HAQ) pigments are produced by many species of *Aspergillus* [47]. Hence, the following studies were performed with pomegranate pulp (PP) for optimal pigment production.

3.2. Optimal Fermentation Strategy and Incubation Period

The cost and yield of bio-pigment production depend on the fermentation strategy used in production [35]. To date, different incubation periods have been tested for the production of pigment from different fungal species and wastes/pulps. For example, Babitha et al. [36] investigated pigment production by *Monascus purpureus* from jack fruit seed and they found that the maximum pigment production was on the sixth day. Gmoser et al. [48] found that the highest pigment (0.7 mg carotene/g waste) was produced by *Neurospora intermedia* from waste bread on the sixth day. In the study conducted by Padmavathi and Prabhudessai [49], the highest pigment production by *Monascus sanguineus* from potato peel was obtained on the 15th day.

Therefore, pigment production by *A. carbonarius* on PP was evaluated with SSF and SmF for 3, 6, 9, 12, and 15 d. In addition, the effect of incubation time on pigment production was limited to 15 d in this study due to the fact that a short fermentation time is desired in industrial production to obtain a competitive production [14]. Figure 2 illustrates the effect of two fermentation strategies (SSF and SmF) on pigment production over 15 d. SSF exhibited higher yellow pigment production (22.9 ± 1.34 AU/g dfm) than SmF. In addition, it was determined that pigment production mostly increased with increasing incubation time for both fermentation strategies.

Pigments are synthesized as secondary metabolites by the fungus [50] and these metabolites often produce at the stationary phase of fungal growth as a result of nutrient limitations and/or under stress conditions [48]. Furthermore, secondary metabolism is commonly associated with sporulation processes for microorganisms, including fungi [51]. It is known that pigment production can be related to the formation of both sexual and asexual spores for some fungal species [51]. On the other hand, *Aspergillus* species can produce asexual spores, which are called conidia [52], from their conidiophores as a result of differentiation of their aerial hyphae. For example, Teertstra et al. [53] showed melanin pigments being produced in the conidia of *Aspergillus niger*. For these reasons, two possible reasons can explain the increasing of pigment production when the incubation day increases: (i) for both fermentation strategies, the stationary phase began after 6 d (Figure 2) and pigment production increased with increasing incubation day because of decreasing nutrients or stress conditions, (ii) conidia that contain pigments were produced after 6 d. In addition, it is known that conidia are only formed in the air [48,54]. Due to the presence of water in SmF, conidia were not produced by *A. carbonarius* until day 12 (Figure 2). Therefore, SSF was found to be more successful in pigment production than SmF. Consequently, further studies were conducted with SSF for maximum pigment production.

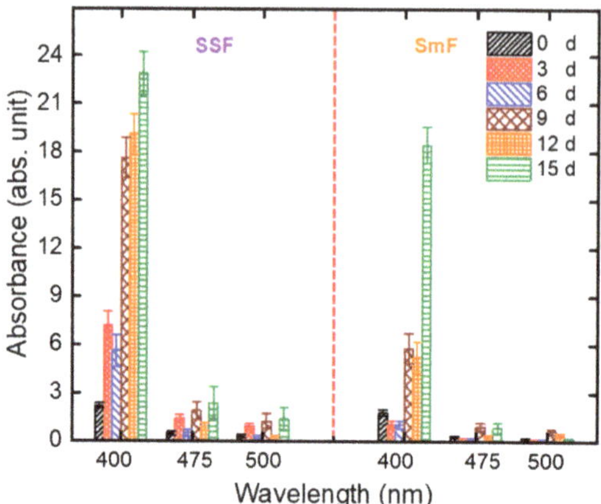

Figure 2. Evaluation of pigment production of *A. carbonarius* on pomegranate pulp (PP) at solid-state fermentation (SSF) and submerged-state fermentation (SmF) conditions (experimental conditions: pH: 4.5; shaking speed for SmF: 100 rpm).

3.3. Optimal Substrate Particle Size

Figure 3 illustrates the effect of PP particle sizes (<1.4, 1.4–2, 2–4, and >4 mm) on pigment production during SSF with *A. carbonarius*. The smallest particle size (<1.4 mm) supplied higher pigment production, particularly in the case of the yellow pigment (25.38 ± 1.60 AU/g dfm for 15 d) (Figure 3A). Moreover, 2.12 ± 0.11 AU/g dfm (Figure 3B) and 1.19 ± 0.15 AU/g dfm (Figure 3C) were measured for orange and red pigments for 12 and 9 d, respectively. The same trend was observed by Kantifedaki et al. [34] and they reported that the smallest particle size (<2 mm) exhibited higher pigment synthesis (9 AU/g dfm for 16 d). This could be attributed to the fact that smaller substrate particles provided a larger surface area for the fungal attack to the substrate [55]. Due to obtaining higher pigment production with the smallest particle size, < 1.4 mm of PP was used for the production of pigments in the subsequent experiments.

3.4. Optimal Initial Substrate pH

The optimization of initial pH is an important parameter to increase pigment production. Figure 4 shows the effect of initial pH (4.5, 5.5, 6.5, 7.5, and 8.5) of PP on pigment production during SSF with *A. carbonarius*. During the incubation period, the highest pigment production was determined as 61.84 ± 2.16 AU/g dfm at 400 nm for 15 d for initial pH 6.5 (Figure 4A). Moreover, 11.52 ± 1.01 AU/g dfm (Figure 4B) and 7.56 ± 1.03 AU/g dfm (Figure 4C) were measured for orange and red pigments for 15 d, respectively. It was found that as pH increased between pH 4.5–6.5, pigment production increased. However, pigment production decreased after pH 6.5. Therefore, optimum pH was determined as pH 6.5 for the pigment production at the maximum yield in this study. A similar result was obtained, so that the production of pigment by *Aspergillus nidulans* increased when the initial pH of the substrate was at 6.8 compared to pH 8.0 [56]. Furthermore, Afshari et al. [57] studied pigment production from another filamentous fungal species *Penicillium aculeatum* and they found that the best production of yellow pigment was obtained with a pH value of 6.5. The effects of pH on fungal pigment production are connected with changes in enzyme activity [56]. It is known that the optimal pH range is 5.0–6.5 for most fungi [58]. Therefore, it is thought that the growth of fungi on PP and its enzyme activity for pigment production could affect pigment production.

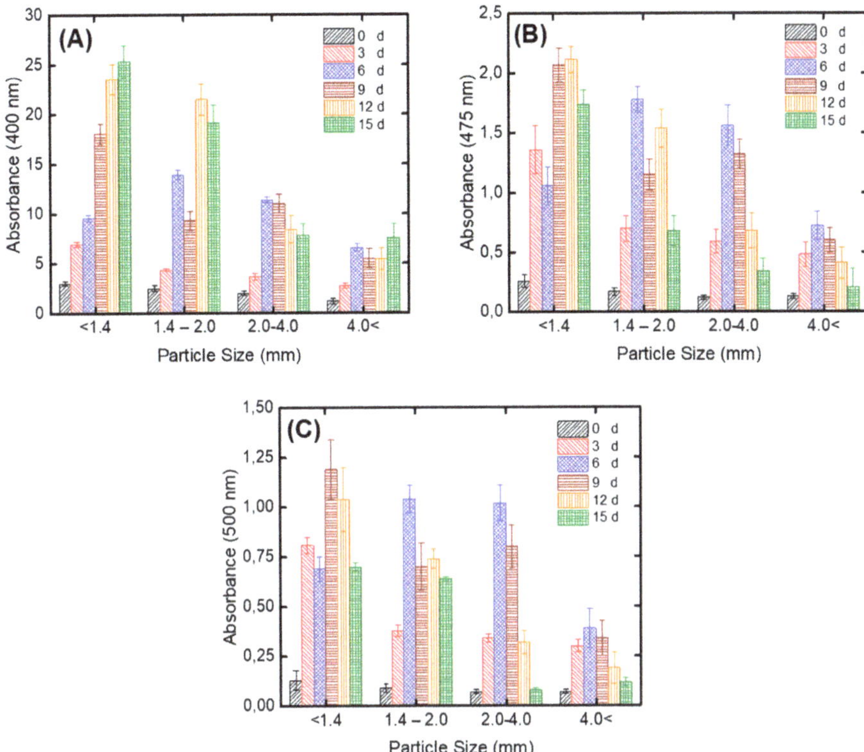

Figure 3. The effect of PP particle sizes on pigment production at (**A**) 400 nm, (**B**) 475 nm, (**C**) 500 nm during SSF with *A. carbonarius* (experimental conditions: pH: 4.5).

3.5. C:N Ratio Utilization

Carbon and nitrogen are essential for fungal growth and pigment production. Elemental analyses of the unfermented and fermented pomegranate pulp are summarized in Table 2. The elemental analysis of the PP showed that their composition changed after fermentation. PP lost a percentage of nitrogen (from 3.02% to 1.60%), hydrogen (from 6.08% to 5.09%), sulfur (from 0.02 % to 0.00 %), and carbon (from 50.94% to 42.46%) after fermentation. It is known that the C:N ratio affects the biosynthesis of many metabolites, such as pigments, in fungi [59]. It was detected that the C:N ratio significantly increased from 16.87/1 to 26.54/1 after fermentation in this study. These results demonstrate that the stress condition for the pigment production in this experiment could be the nitrogen decrease [43].

Table 2. Elemental analysis of unfermented and fermented pomegranate pulp

Sample	C (%)	H (%)	N (%)	S (%)	C:N
Unfermented PP	50.94	6.08	3.02	0.02	16.87/1
Fermented PP	42.46	5.09	1.60	0.00	26.54/1

C: Carbon, H: Hydrogen, N: Nitrogen, S: Sulphur, C:N: carbon to nitrogen ratio.

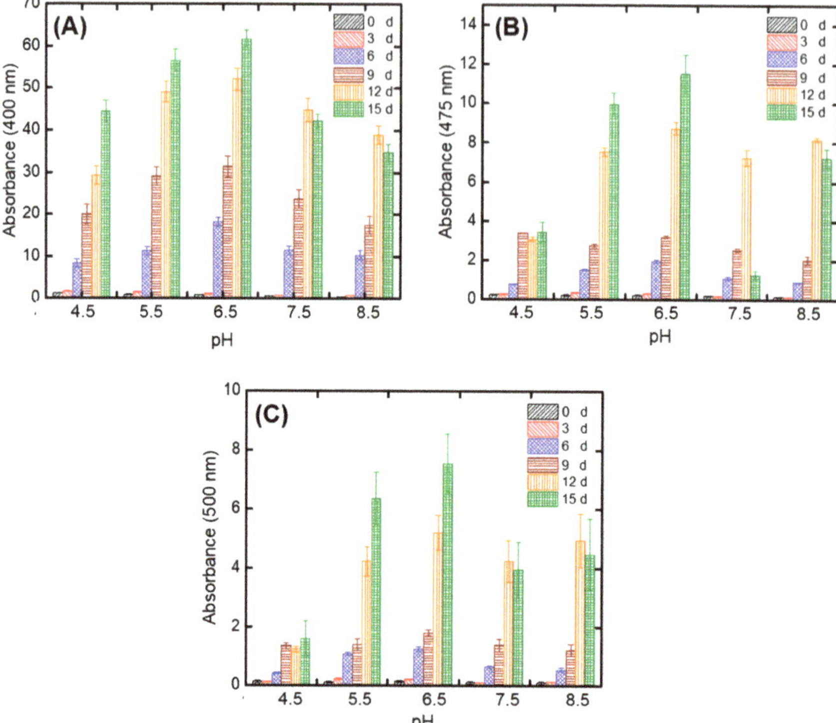

Figure 4. The effect of initial substrate pH on pigment production at (**A**) 400 nm, (**B**) 475 nm, (**C**) 500 nm during SSF with *A. carbonarius* (experimental conditions: particle size: < 1.4 mm).

3.6. Characterization of Fungal Biomass

To characterize fungal biomass, SEM is the one of most appropriate techniques. Figure 5 shows the SEM image of the fermented *A. carbonarius* on PP under optimal conditions. Fermented *A. carbonarius* on PP (Figure 5A) has more conidia whereas *A. carbonarius* grown on YMB has more hyphae (Figure 5C). This result supports our hypothesis that pigments were produced in conidia. Furthermore, Figure 5B shows the pigmented conidia of fermented *A. carbonarius* on PP. Unfermented PP (Figure 5D) shows a perforated structure.

It is known from the literature that *Aspergillus* species can produce melanin [30]. However, Pihet et al. [60] studied melanin production in *Aspergillus fumigatus* and SEM analysis results show that conidia that do not have smooth walls contain melanin. Therefore, this could support our second hypothesis that *Aspergillus carbonarius* could produce melanin from pomegranate pulp.

GC×GC-TOFMS is a reliable method for the detection of organic compounds in complex matrices [61], and its results are given in Table 3 for the extracted solution from fermented *A. carbonarius* on PP under optimal conditions. It is known that pomegranate contains sugars such as xylose, arabinose, and glucose [62]. GC×GC-TOFMS results show that fermentation was conducted successfully due to occurring alcohols (xylitol, sorbitol, and arabinitol) [63] and fatty acids (linoleic, oleic, and erucic acids) (Table 3) [64]. Ergosterol, which is a secondary metabolite of *Aspergillus spp.*, might come from its fungal cell membrane [65,66]. Furthermore, squalene and β-amyrin detected by GC×GC-TOFMS could be in the pathway of ergosterol synthesis [67].

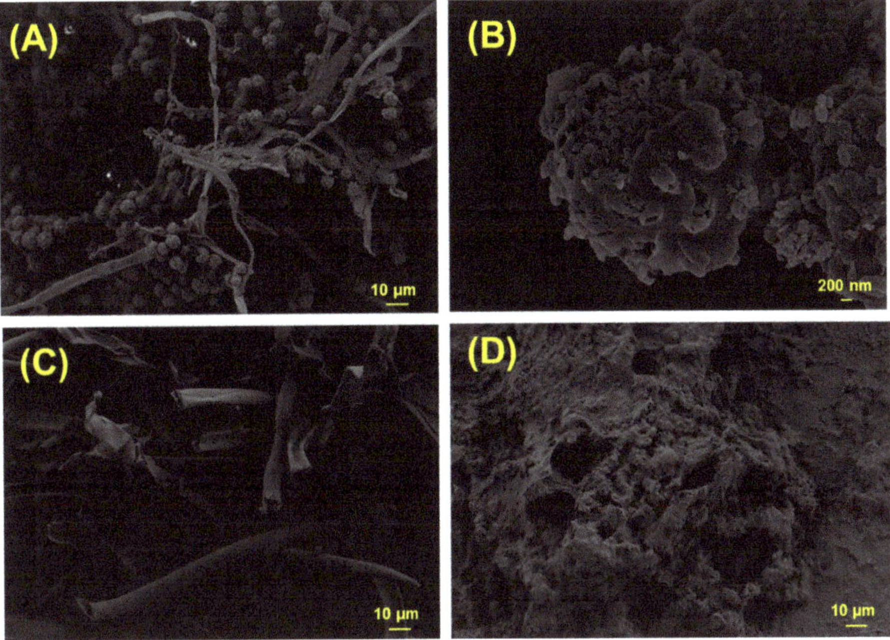

Figure 5. SEM images of (**A**) fermented *A. carbonarius* on PP under optimal conditions, (**B**) pigmented conidia of fermented *A. carbonarius* on PP under optimal conditions, (**C**) *A. carbonarius* grown on YMB, (**D**) unfermented PP.

Furthermore, 2,6-diisopropylnaphthalene detected by GC×GC-TOFMS (Table 3) could be caused by the usage of pesticides in pomegranate growth [68]. In a similar way, it is though that γ-tocopherol, which is an antioxidant, may be caused by the composition of pomegranate [69].

Due to the detection of 1,8-dihydroxynaphthalene by GC×GC-TOFMS (Table 3), our findings suggest that *A. carbonarius* might produce melanin from the polymerization of 1,8- dihydroxynaphthalene (DHN) from PP through the DHN pathway [30]. Furthermore, phthalic acid, mequinol, and galangin are known to be potent inhibitors of the DHN pathway [70–72]. Therefore, it is concluded that inhibitors could not surpass the melanin production. A potential metabolic pathway of melanin production by *Aspergillus carbonarius* from pomegranate pulp is illustrated in Figure S3.

To date, several species of *Aspergillus* such as *A. niger*, *A. flavus*, *A. tamarii*, *A. terreus*, *A. tubingensis*, *A. sydowii* and *A. fumigatus* were investigated for melanin production [30,31,60]. However, there is very little information about melanin production by *A. carbonarius*. Babitskaya et al. [55] studied melanin production by *A. carbonarius* on Czapek's medium and they determined that melanin belongs to the DHN pathway. Another study on antioxidant properties of melanin produced by *A. carbonarius* was conducted by Shcherba et al. [73]. In accordance with these studies, it was thought that *A. carbonarius* could produce melanin from pomegranate pulp, which is an economical way, and that the produced melanin contains antioxidant. When evaluated together with SEM results, it was determined that the localization of this melanin was conidia.

Table 3. Chemical compounds determined by GC×GC-TOFMS

Name of Compounds	Chemical Structure	Formula	Classification	S
Xylitol	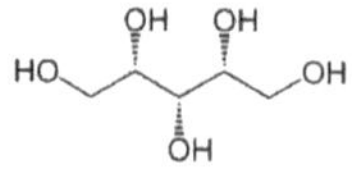	$C_5H_{12}O_5$	Alcohol	951
L-Arabinitol	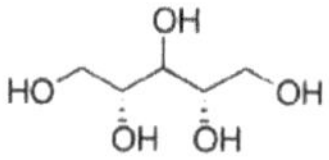	$C_5H_{12}O_5$	Alcohol	946
Sorbitol	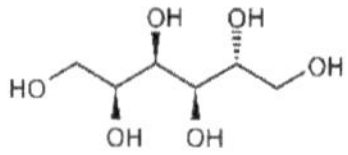	$C_6H_{14}O_6$	Alcohol	796
Ethanol	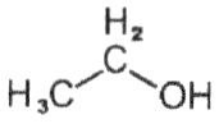	C_2H_5OH	Alcohol	949
Linoleic acid	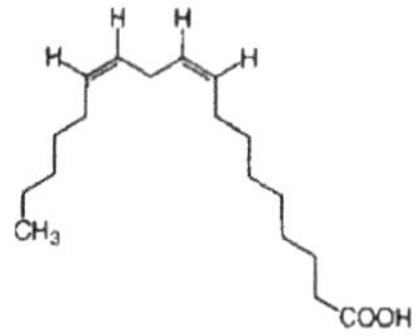	$C_{18}H_{32}O_2$	Fatty acid	851
Oleic acid	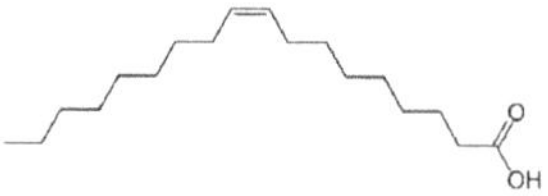	$C_{18}H_{34}O_2$	Fatty acid	829
Erucic acid	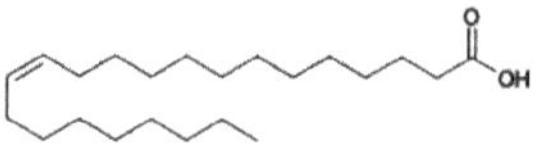	$C_{22}H_{42}O_2$	Fatty acid	753
Ergosterol	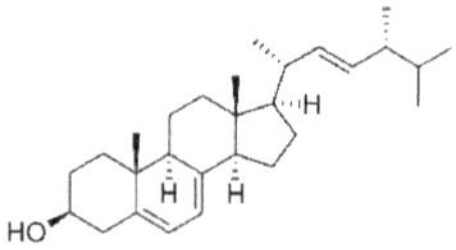	$C_{28}H_{44}O$	Sterol	794
Squalene	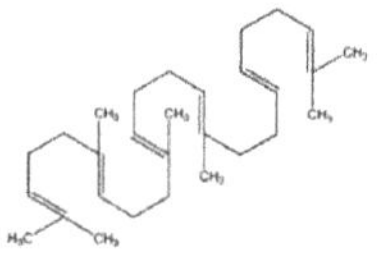	$C_{30}H_{50}$	Triterpene	778
β-Amyrin	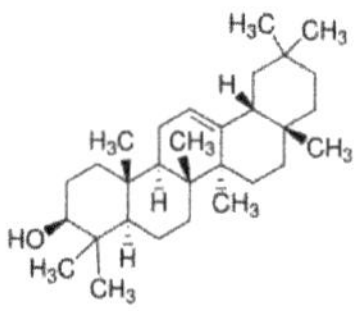	$C_{30}H_{50}O$	Triterpene	887

Table 3. Cont.

Name of Compounds	Chemical Structure	Formula	Classification	S
2,6-diisopropylnaphthalene	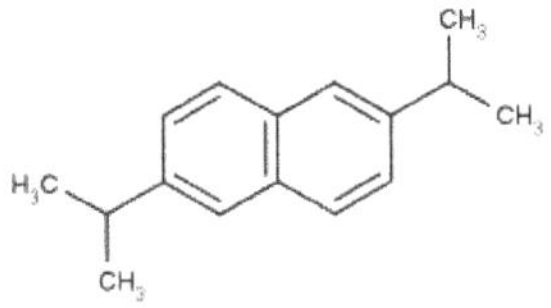	$C_{16}H_{26}$	Naphthalene	796
γ-Tocopherol	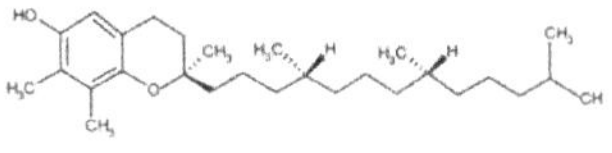	$C_{28}H_{48}O_2$	Vitamin E	899
Galangin	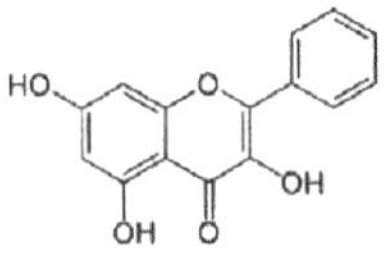	$C_{15}H_{10}O_5$	Flavonoid	754
Phthalic acid	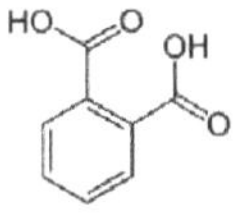	$C_6H_4(CO_2)$	Dicarboxylic acid	884
Mequinol (4-methoxyphenol)	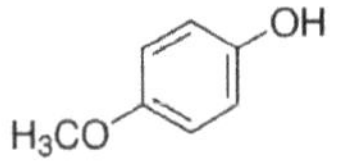	$C_7H_8O_2$	Methoxyphenol	723
1,8-dihydroxynaphthalene	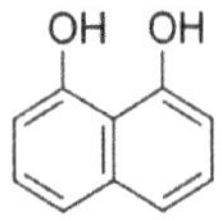	$C_{10}H_8O_2$	Naphthalene	941

S: similarity.

Fungal melanin has attracted interest due to its potential usage in many industries including nanotechnology, biomedicine, dermo-cosmetics, and materials science [56]. However, *Aspergillus carbonarius* is also known as a producer of ochratoxin A (OTA) [74]. For this reason, further research on melanin production, OTA level and physical OTA removal methods such as filtration and pressure needs to be undertaken. It is known that OTA biosynthesis in *A. carbonarius* is driven by polyketide synthase (pks) and nonribosomal peptide synthetase genes (NRPS) [75]. The inactivation of these genes could be another way to eliminate the ability of this fungus to produce OTA [76] for a techno-economic way of producing pigment from pomegranate pulp.

In addition, GC×GC-TOFMS results show that the production of fatty acids such as linoleic, erucic and oleic acids by *A. carbonarius* could be another option for by-product valorization for the food processing industry [7].

3.7. Potential Usage of Fungal Pigment in Textile Industry

Images of dyed and undyed fabric pieces without the mordanting process are shown in Figure S4A,B, respectively. The dyeing test result shows that pigment obtained from this study could serve as a source for the natural dyeing of wool textiles. Dyed fabric was washed in the washing machine at a temperature of 60 °C five times and it was observed that the color did not change (Figure S4B). Dyed and undyed fabric pieces with mordant processing are shown in Figure S4C,D, respectively. Color fastness to washing was good to excellent (grade 3–4). On the one hand, the observation hue for the pieces was yellowish brown and yellowish pink for without mordanting and with mordanting process, respectively. However, it is known that the hue of dyes can be changed by using mordants such as iron sulfate [77]. Furthermore, colorimetric parameters such as dyeing rate constant, half-time of dyeing, and values of pigment uptake are worth further investigation.

On the other hand, the thermal stability of pigment can affect the coloring. However, it is known that thermal degradation of melanin occurs at very high temperatures (above 150 °C) [78]. Therefore, it is thought that the temperatures throughout our experimental studies (60 and 93 °C) after fermentation could not affect the melanin itself.

4. Conclusions

The results of our work reveal that pomegranate pulps derived from the food processing industry have a high potential for pigment production by *A. carbonarius*. The higher production of yellow pigment (61.84 ± 2.16 AU/g dfm) was obtained for 15 d at initial substrate pH of 6.5 under solid-state fermentation conditions. Moreover, our findings suggest that *A. carbonarius* might produce melanin due to the detection of 1,8-dihydroxynaphthalene by GC×GC-TOFMS and this will be investigated in the future by LC-MS-ToF. Optimal conditions such as incubation time, fermentation strategy, initial pH, and particle size of substrate findings in the current study can give basic information for the scaled-up production of bio-pigment production by filamentous fungus *Aspergillus carbonarius*. However, further studies are needed in order to detect potential utilization industrial areas of this bio-pigment.

Supplementary Materials: The following are available online at http://www.mdpi.com/2309-608X/6/4/240/s1, Figure S1: Dried (A) apple pulp, (B) black carrot pulp, (C) pomegranate pulp, (D) red beet pulp, Figure S2: Flow chart of experiments (A) Selection of fungal species and pulp type, (B) Effect of fermentation strategy and incubation period, (C) Effect of substrate particle size, (D) Effect of initial substrate pH, (E) Pigment production at optimal conditions, Figure S3: Potential metabolic pathway of melanin production by *Aspergillus carbonarius* from pomegranate pulp, Figure S4: Images of (A) dyed and (B) undyed fabric pieces without mordanting, (C) dyed fabric pieces with mordanting, (D) undyed fabric pieces with mordanting (Dyeing of cotton fabric made with a fungal pigmented extract obtained by cultivation of *Aspergillus carbonarius* during 15 days on pomegranate pulp).

Author Contributions: Conceptualization, N.D.; methodology, E.B.A.; validation, O.C. and Y.C.; investigation, N.D.; resources, N.D.; writing—original draft preparation, N.D.; writing—review and editing, L.D.; project administration, N.D. and L.D. All authors have read and agreed to the published version of the manuscript.

Funding: This research received no external funding.

Conflicts of Interest: The authors declare no conflict of interest.

References

1. Vendruscolo, F.; Luise Müller, B.; Esteves Moritz, D.; de Oliveira, D.; Schmidell, W.; Luiz Ninow, J. Thermal stability of natural pigments produced by *Monascus ruber* in submerged fermentation. *Biocatal. Agric. Biotechnol.* **2013**, *2*, 278–284. [CrossRef]
2. Shahid, M.; ul-Islam, S.; Mohammad, F. Recent advancements in natural dye applications: A review. *J. Clean. Prod.* **2013**, *53*, 310–331. [CrossRef]
3. Nigam, P.S.; Luke, J.S. Food additives: Production of microbial pigments and their antioxidant properties. Curr. *Opin. Food Sci.* **2016**, *7*, 93–100. [CrossRef]
4. Karger-Kocsis, J. Paints, coatings and solvents. *Compos. Sci. Technol.* **1994**, *51*, 613–614. [CrossRef]
5. Barnett, J.R.; Miller, S.; Pearce, E. Colour and art: A brief history of pigments. *Opt. Laser Technol.* **2006**, *38*, 445–453. [CrossRef]
6. Downham, A.; Collins, P. Colouring our foods in the last and next millennium. *Int. J. Food Sci. Technol.* **2000**, *35*, 5–22. [CrossRef]
7. Panesar, R.; Kaur, S.; Panesar, P.S. Production of microbial pigments utilizing agro-industrial waste: A review. *Curr. Opin. Food Sci.* **2015**, *1*, 70–76. [CrossRef]
8. US FDA. Laws & Regulations-FDA Authority over Cosmetics: How Cosmetics Are not FDA-Approved, but Are FDA-Regulated. 2016. Available online: https://www.fda.gov/cosmetics/cosmetics-laws-regulations/fda-authority-over-cosmetics-how-cosmetics-are-not-fda-approved-are-fda-regulated (accessed on 14 October 2020).

9. FAO. WHO. Safety Evaluation of Certain Food Additives. 2017. Available online: https://apps.who.int/iris/bitstream/handle/10665/258934/9789241660730-eng.pdf;jsessionid=CEC17CE6D1D291C848E8289CFCD2A713?sequence=1 (accessed on 14 October 2020).
10. Embaby, A.M.; Hussein, M.N.; Hussein, A. Monascus orange and red pigments production by *Monascus purpureus* ATCC16436 through co-solid state fermentation of corn cob and glycerol: An eco-friendly environmental low cost approach. *PLoS ONE* **2018**, *13*, e0207755. [CrossRef]
11. Gupta, N.; Poddar, K.; Sarkar, D.; Kumari, N.; Padhan, B.; Sarkar, A. Fruit waste management by pigment production and utilization of residual as bioadsorbent. *J. Environ. Manag.* **2019**, *244*, 138–143. [CrossRef]
12. Liu, J.; Luo, Y.; Guo, T.; Tang, C.; Chai, X.; Zhao, W.; Bai, J.; Lin, Q. Cost-effective pigment production by *Monascus purpureus* using rice straw hydrolysate as substrate in submerged fermentation. *J. Biosci. Bioeng.* **2020**, *129*, 229–236. [CrossRef]
13. Thejus, P.K.; Krishnapriya, K.V.; Nishanth, K.G. A cost-effective intense blue colour inorganic pigment for multifunctional cool roof and anticorrosive coatings. *Sol. Energy Mater. Sol. Cells* **2021**, *219*, 110778. [CrossRef]
14. Aruldass, C.A.; Dufossé, L.; Ahmad, W.A. Current perspective of yellowish-orange pigments from microorganisms—A review. *J. Clean. Prod.* **2018**, *180*, 168–182. [CrossRef]
15. Tirumale, S.; Wani, N.A. Biopigments: Fungal Pigments. In *Fungi and Their Role in Sustainable Development: Current Perspectives*, 1st ed.; Gehlot, P., Singh, J., Eds.; Springer: Singapore, 2018; Chapter 23; pp. 413–426. [CrossRef]
16. Narsing Rao, M.P.; Xiao, M.; Li, W.-J. Fungal and bacterial pigments: Secondary metabolites with wide applications. *Front. Microbiol.* **2017**, *8*, 1113. [CrossRef] [PubMed]
17. Sen, T.; Barrow, C.J.; Deshmukh, S.K. Microbial pigments in the food industry-Challenges and the way forward. *Front. Nutr.* **2019**, *6*, 1–14. [CrossRef]
18. Venil, C.K.; Zakaria, Z.A.; Ahmad, W.A. Bacterial pigments and their applications. *Process Biochem.* **2013**, *48*, 1065–1079. [CrossRef]
19. Nirlane da Costa Souza, P.; Luiza Bim Grigoletto, T.; Alberto Beraldo de Moraes, L.; Abreu, L.M.; Henrique Souza Guimarães, L.; Santos, C.; Ribeiro Galvão, L.; Gomes Cardoso, P. Production and chemical characterization of pigments in filamentous fungi. *Microbiology* **2016**, *162*, 12–22. [CrossRef] [PubMed]
20. Lopes, F.C.; Tichota, D.M.; Pereira, J.Q.; Segalin, J.; de Oliveira Rios, A.; Brandelli, A. Pigment Production by Filamentous Fungi on Agro-Industrial Byproducts: An Eco-Friendly Alternative. *Appl. Biochem. Biotechnol.* **2013**, *171*, 616–625. [CrossRef] [PubMed]
21. Karimi, S.; Mahboobi Soofiani, N.; Mahboubi, A.; Taherzadeh, M. Use of Organic Wastes and Industrial By-Products to Produce Filamentous Fungi with Potential as Aqua-Feed Ingredients. *Sustainability* **2018**, *10*, 3296. [CrossRef]
22. Dufossé, L.; Fouillaud, M.; Caro, Y.; Mapari, S.A.S.; Sutthiwong, N. Filamentous fungi are large-scale producers of pigments and colorants for the food industry. *Curr. Opin. Biotechnol.* **2014**, *26*, 56–61. [CrossRef]
23. Caro, Y.; Venkatachalam, M.; Lebeau, J.; Fouillaud, M.; Dufossé, L. Pigments and Colorants from Filamentous Fungi. In *Fungal Metabolites*, 1st ed.; Merillon, J.M., Ramawat, K., Eds.; Springer International Publishing: Chambridge, UK, 2017; pp. 499–568. [CrossRef]
24. Takahashi, J.A.; Carvalho, S.A. Nutritional potential of biomass and metabolites from filamentous fungi. In *Current Research, Technology and Education Topics in Applied Microbiology and Microbial Biotechnology*; Méndez-Vilas, A., Ed.; FORMATEX: Badajoz, Spain, 2010; Volume 2, pp. 1126–1135.
25. Teixeira, M.F.S.; Martins, M.S.; da Silva, J.C.; Kirsch, L.S.; Fernandes, O.C.C.; Carneiro, A.L.B.; de Conti, R.; Durán, N. Amazonian biodiversity: Pigments from *Aspergillus* and *Penicillium*-characterizations, antibacterial activities and their toxicities. *Curr. Trends Biotechnol. Pharm.* **2012**, *6*, 2230–7303.
26. Mostafa, M.E.; Saad Abbady, M. Secondary Metabolites and Bioactivity of the *Monascus* Pigments Review Article. *Glob. J. Biotechnol. Biochem.* **2014**, *9*, 1–13. [CrossRef]
27. Heo, Y.M.; Kim, K.; Kwon, S.L.; Na, J.; Lee, H.; Jang, S.; Kim, C.H.; Jung, J.; Kim, J.-J. Investigation of Filamentous Fungi Producing Safe, Functional Water-Soluble Pigments. *Mycobiology* **2018**, *46*, 269–277. [CrossRef]
28. Saravanan, A.; Jayasree, R.; Senthil Kumar, P.; Varjani, S.; Hemavathy, R.V.; Jeevanantham, S.; Yaashikaa, P.R. Production of pigment using *Aspergillus tamarii*: New potentials for synthesizing natural metabolites. *Environ. Technol. Innov.* **2020**, *19*, 100967. [CrossRef]

29. Taniwaki, M.H.; Pitt, J.I.; Magan, N. *Aspergillus* species and mycotoxins: Occurrence and importance in major food commodities. *Curr. Opin. Food Sci.* **2018**, *23*, 38–43. [CrossRef]
30. Pal, A.K.; Gajjar, D.U.; Vasavada, A.R. DOPA and DHN pathway orchestrate melanin synthesis in *Aspergillus* species. *Med. Mycol.* **2014**, *52*, 10–18. [CrossRef]
31. Geib, E.; Gressler, M.; Viediernikova, I.; Hillmann, F.; Jacobsen, I.D.; Nietzsche, S.; Hertweck, C.; Brock, M. A non-canonical melanin biosynthesis pathway protects *Aspergillus terreus* conidia from environmental stress. *Cell Chem. Biol.* **2016**, *23*, 587–597. [CrossRef]
32. Narendrababu, B.N.; Shishupala, S. Spectrophotometric detection of Pigments from Aspergillus and Penicillium isolates. *J. Appl. Biol. Biotechnol.* **2017**, *5*, 53–58. [CrossRef]
33. Bouras, H.D.; Yeddou, A.R.; Bouras, N.; Hellel, D.; Holtz, M.D.; Sabaou, N.; Chergui, A.; Nadjemi, B. Biosorption of Congo red dye by *Aspergillus carbonarius* M333 and *Penicillium glabrum* Pg1: Kinetics, equilibrium and thermodynamic studies. *J. Taiwan Inst. Chem. Eng.* **2017**, *80*, 915–923. [CrossRef]
34. Kantifedaki, A.; Kachrimanidou, V.; Mallouchos, A.; Papanikolaou, S.; Koutinas, A.A. Orange processing waste valorisation for the production of bio-based pigments using the fungal speciess *Monascus purpureus* and *Penicillium purpurogenum*. *J. Clean. Prod.* **2018**, *185*, 882–890. [CrossRef]
35. Agboyibor, C.; Kong, W.-B.; Chen, D.; Zhang, A.-M.; Niu, S.-Q. *Monascus* pigments production, composition, bioactivity and its application: A review. *Biocatal. Agric. Biotechnol.* **2018**, *16*, 433–447. [CrossRef]
36. Babitha, S.; Soccol, C.R.; Pandey, A. Solid-state fermentation for the production of *Monascus* pigments from jackfruit seed. *Bioresour. Technol.* **2007**, *98*, 1554–1560. [CrossRef] [PubMed]
37. Johns, M.R.; Stuart, D.M. Production of pigments by *Monascus purpureus* in solid culture. *J. Ind. Microbiol.* **1991**, *8*, 23–28. [CrossRef]
38. De Carvalho, J.C.; Cardoso, L.C.; Ghiggi, V.; Woiciechowski, A.L.; de Souza Vandenberghe, L.P.; Soccol, C.R. Microbial Pigments. In *Biotransformation of Waste Biomass into High Value Biochemicals*, 1st ed.; Brar, S.K., Dhillon, G.S., Soccol, C.R., Eds.; Springer: New York, NY, USA, 2014; pp. 73–97. [CrossRef]
39. Khan, M.I.; Ahmad, A.; Khan, S.A.; Yusuf, M.; Shahid, M.; Manzoor, N.; Mohammad, F. Assessment of antimicrobial activity of Catechu and its dyed substrate. *J. Clean. Prod.* **2011**, *19*, 1385–1394. [CrossRef]
40. Shibila, S.D.; Nanthini, A.U.R. Extraction and characterization of red pigment from *Talaromyces australis* and its application in dyeing cotton yarn. *Int. Arch. App. Sci. Technol.* **2019**, *10*, 81–91. [CrossRef]
41. Sankaran, S.; Khanal, S.K.; Jasti, N.; Jin, B.; Pometto, A.L.; Van Leeuwen, J.H. Use of Filamentous Fungi for Wastewater Treatment and Production of High Value Fungal Byproducts: A Review. *Crit. Rev. Environ. Sci. Technol.* **2010**, *40*, 400–449. [CrossRef]
42. Akilandeswari, P.; Pradeep, B.V. *Aspergillus terreus* Kmbf1501 A Potential Pigment Producer Under Submerged Fermentation. *Int. J. Pharm. Pharm. Sci.* **2017**, *9*, 38–43. [CrossRef]
43. Raman, N.M.; Shah, P.H.; Mohan, M.; Ramasamy, S. Improved production of melanin from *Aspergillus fumigatus* AFGRD105 by optimization of media factors. *AMB Express* **2015**, *5*, 72. [CrossRef]
44. Palacio-Barrera, A.M.; Areiza, D.; Zapata, P.; Atehortúa, L.; Correa, C.; Peñuela-Vásquez, M. Induction of pigment production through media composition, abiotic and biotic factors in two filamentous fungi. *Biotechnol. Rep.* **2019**, *21*, e00308. [CrossRef]
45. Said, F.M.; Brooks, J.; Chisti, Y. Optimal C:N ratio for the production of red pigments by *Monascus ruber*. *World J. Microbiol. Biotechnol.* **2014**, *30*, 2471–2479. [CrossRef]
46. Dufossé, L. Pigments, Microbial. Reference Module in Life Sciences. 2016. Available online: https://hal.archives-ouvertes.fr/hal-01734750/document (accessed on 14 October 2020).
47. Caro, Y.; Anamale, L.; Fouillaud, M.; Laurent, P.; Petit, T.; Dufosse, L. Natural hydroxyanthraquinoid pigments as potent food grade colorants: An overview. *Nat. Prod. Bioprospect.* **2012**, *2*, 174–193. [CrossRef]
48. Gmoser, R.; Ferreira, J.A.; Taherzadeh, M.J.; Lennartsson, P.R. Post-treatment of Fungal Biomass to Enhance Pigment Production. *Appl. Biochem. Biotechnol.* **2019**, *189*, 160–174. [CrossRef] [PubMed]
49. Padmavathi, T.; Prabhudessai, T. A Solid Liquid State Culture Method to Stimulate *Monascus* Pigments by Intervention of Different Substrates. *Int. Res. J. Biol. Sci.* **2013**, *2*, 22–29.
50. Satyanarayana, T.; Deshmukh, S.K.; Johri, B.N. *Developments in Fungal Biology and Applied Mycology*; Springer: Singapore, 2017; pp. 525–541.
51. Calvo, A.M.; Wilson, R.A.; Bok, J.W.; Keller, N.P. Relationship between Secondary Metabolism and Fungal Development. *Microbiol. Mol. Biol. Rev.* **2002**, *66*, 447–459. [CrossRef]

52. Baker, S.E.; Bennett, J. An Overview of the Genus *Aspergillus*. In *The Aspergilli: Genomics, Medical Aspects, Biotechnology, and Research Methods*; Machida, M., Gomi, K., Eds.; Caister Academic Press: Poole, UK, 2007; Volume 26, pp. 3–13.
53. Teertstra, W.R.; Tegelaar, M.; Dijksterhuis, J.; Golovina, E.A.; Ohm, R.A.; Wösten, H.A.B. Maturation of conidia on conidiophores of *Aspergillus niger*. *Fungal Genet. Biol.* **2017**, *98*, 61–70. [CrossRef]
54. Zalokar, M. Studies on biosynthesis of carotenoids in *Neurospora crassa*. *Arch. Biochem. Biophys.* **1954**, *50*, 71–80. [CrossRef]
55. Babitskaya, V.G.; Shcherba, V.V.; Filimonova, T.V.; Grigorchyuk, E.A. Melanin pigments from the fungi *Paecilomyces variotii* and *Aspergillus carbonarius*. *Appl. Biochem. Microbiol.* **2000**, *36*, 128. [CrossRef]
56. Pombeiro-Sponchiado, S.R.; Sousa, G.S.; Andrade, J.C.R.; Lisboa, H.F.; Gonçalves, R.C.R. Production of Melanin Pigment by Fungi and Its Biotechnological Applications. In *Melanin*; Blumenber, M., Ed.; IntechOpen: London, UK, 2017; p. 3760. Available online: https://www.intechopen.com/books/melanin/production-of-melanin-pigment-by-fungi-and-its-biotechnological-applications (accessed on 14 October 2020).
57. Afshari, M.; Shahidi, F.; Mortazavi, S.A.; Tabatabai, F.; Es'haghi, Z. Investigating the influence of pH, temperature and agitation speed on yellow pigment production by *Penicillium aculeatum* ATCC 10409. *Nat. Prod. Res.* **2015**, *29*, 1300–1306. [CrossRef]
58. Arora, D.S.; Chandra, P. Antioxidant Activity of *Aspergillus fumigatus*. *ISRN Pharmacol.* **2011**, *2011*, 1–11. [CrossRef]
59. Cho, Y.J.; Park, J.P.; Hwang, H.J.; Kim, S.W.; Choi, J.W.; Yun, J.W. Production of red pigment by submerged culture of *Paecilomyces sinclairii*. *Lett. Appl. Microbiol.* **2002**, *35*, 195–202. [CrossRef]
60. Pihet, M.; Vandeputte, P.; Tronchin, G.; Renier, G.; Saulnier, P.; Georgeault, S.; Mallet, R.; Chabasse, D.; Symoens, F.; Bouchara, J.-P. Melanin is an essential component for the integrity of the cell wall of *Aspergillus fumigatus* conidia. *BMC Microbiol.* **2009**, *9*, 177. [CrossRef]
61. Hernández, F.; Ibáñez, M.; Portolés, T.; Cervera, M.I.; Sancho, J.V.; López, F.J. Advancing towards universal screening for organic pollutants in waters. *J. Hazard. Mater.* **2015**, *282*, 86–95. [CrossRef] [PubMed]
62. Hasnaoui, N.; Wathelet, B.; Jiménez-Araujo, A. Valorization of pomegranate peel from 12 cultivars: Dietary fibre composition, antioxidant capacity and functional properties. *Food Chem.* **2014**, *160*, 196–203. [CrossRef] [PubMed]
63. Schiweck, H.; Bär, A.; Vogel, R.; Schwarz, E.; Kunz, M.; Dusautois, C.; Clement, A.; Lefranc, C.; Lüssem, B.; Moser, M.; et al. Sugar Alcohols. In *Ullmann's Encyclopedia of Industrial Chemistry*; Wiley-VCH Verlag GmbH & Co. KGaA: Weinheim, Germany, 2012. [CrossRef]
64. Sinha, M.; Weyda, I.; Sørensen, A.; Bruno, K.S.; Ahring, B.K. Alkane biosynthesis by *Aspergillus carbonarius* ITEM 5010 through heterologous expression of Synechococcus elongatus acyl-ACP/CoA reductase and aldehyde deformylating oxygenase genes. *AMB Express* **2017**, *7*, 18. [CrossRef] [PubMed]
65. Alcazar-Fuoli, L.; Mellado, E.; Garcia-Effron, G.; Lopez, J.F.; Grimalt, J.O.; Cuenca-Estrella, J.M.; Rodriguez-Tudela, J.L. Ergosterol biosynthesis pathway in *Aspergillus fumigatus*. *Steroids* **2008**, *73*, 339–347. [CrossRef]
66. Vadlapudi, V.; Borah, N.; Yellusani, K.R.; Gade, S.; Reddy, P.; Rajamanikyam, M.; Vempati, L.N.S.; Gubbala, S.P.; Chopra, P.; Upadhyayula, S.M.; et al. Aspergillus Secondary Metabolite Database, a resource to understand the Secondary metabolome of *Aspergillus* genus. *Sci. Rep.* **2017**, *7*, 7325. [CrossRef]
67. Gealt, M.A. Isolation of p-Amyrin from the Fungus *Aspergillus nidulans*. *J. Gen. Microbiol.* **1983**, *129*, 543–546. [CrossRef] [PubMed]
68. Mohamed, E.M. Flavoring and medicinal values of the yellow pigment produced by *Monascus ruber* 4066 species cultivated on static malt agar medium. *Int. Res. J. Biochem. Biotechnol.* **2016**, *3*, 37–43.
69. Caligiani, A.; Bonzanini, F.; Palla, G.; Cirlini, M.; Bruni, R. Characterization of a Potential Nutraceutical Ingredient: Pomegranate (*Punica granatum L.*) Seed Oil Unsaponifiable Fraction. *Plant Foods Hum. Nutr.* **2010**, *65*, 277–283. [CrossRef]
70. Yin, S.J.; Si, Y.X.; Qian, G.Y. Inhibitory Effect of Phthalic Acid on Tyrosinase: The Mixed-Type Inhibition and Docking Simulations. *Enzym. Res.* **2011**, *294724*, 7. [CrossRef]
71. Bandyopadhyay, D. Topical treatment of melasma. *Indian J. Dermatol.* **2009**, *54*, 303–309. [CrossRef]
72. Lu, Y.H.; Tao, L.; Wang, Z.T.; Wei, D.Z.; Xiang, H.B. Mechanism and inhibitory effect of galangin and its flavonoid mixture from *Alpinia officinarum* on mushroom tyrosinase and B16 murine melanoma cells. *J. Enzym. Inhib. Med. Chem.* **2007**, *22*, 433–438. [CrossRef] [PubMed]

73. Shcherba, V.V.; Babitskaya, V.G.; Kurchenko, V.P.; Ikonnikova, N.V.; Kukulyanskaya, T.A. Antioxidant Properties of Fungal Melanin Pigments. *Appl. Biochem. Microbiol.* **2000**, *36*, 491–495. [CrossRef]
74. Zeidan, R.; Ul-Hassan, Z.; Al-Thani, R.; Migheli, Q.; Jaoua, S. In-Vitro Application of a *Qatari Burkholderia cepacia* species (QBC03) in the Biocontrol of Mycotoxigenic Fungi and in the Reduction of Ochratoxin A biosynthesis by Aspergillus carbonarius. *Toxins* **2019**, *11*, 700. [CrossRef] [PubMed]
75. Gallo, A.; Bruno, K.S.; Solfrizzo, M.; Perrone, G.; Mulè, G.; Visconti, A.; Baker, S.E. New Insight into the Ochratoxin A Biosynthetic Pathway through Deletion of a Nonribosomal Peptide Synthetase Gene in *Aspergillus carbonarius*. *Appl. Environ. Microbiol.* **2012**, *78*, 8208–8218. [CrossRef]
76. Castellá, G.; Bragulat, M.R.; Puig, L.; Sanseverino, W.; Cabañes, F.J. Genomic diversity in ochratoxigenic and non ochratoxigenic strains of *Aspergillus carbonarius*. *Sci. Rep.* **2018**, *8*, 1–11. [CrossRef]
77. Moghaddam, M.K.; Adivi, M.G.; Dehkord, M.T. Effect of Acids and Different Mordanting Procedures on Color Characteristics of Dyed Wool Fibers Using Eggplant Peel (*Solanum melongena L*). *Prog. Color Colorants Coat.* **2019**, *12*, 219–230.
78. Pralea, I.E.; Moldovan, R.C.; Petrache, A.M.; Ilieș, M.; Hegheș, S.C.; Ielciu, I.; Nicoară, R.; Moldovan, M.; Ene, M.; Radu, M.; et al. From Extraction to Advanced Analytical Methods: The Challenges of Melanin Analysis. *Int. J. Mol. Sci.* **2019**, *20*, 3943. [CrossRef]

Journal of
Fungi

mmunication

oes Structural Color Exist in True Fungi?

iet Brodie [1], Colin J. Ingham [2,*] and Silvia Vignolini [3]

1 Department of Life Sciences, Natural History Museum, London SW7 5BD, UK; j.brodie@nhm.ac.uk
2 Hoekmine BV, 3515 GJ Utrecht, The Netherlands
3 Department of Chemistry, University of Cambridge, Cambridge CB2 1EW, UK; sv319@cam.ac.uk
* Correspondence: colinutrecht@gmail.com; Tel.: +31-642-477-078

Abstract: Structural color occurs by the interaction of light with regular structures and so generates colors by completely different optical mechanisms to dyes and pigments. Structural color is found throughout the tree of life but has not, to date, been reported in the fungi. Here we give an overview of structural color across the tree of life and provide a brief guide aimed at stimulating the search for this phenomenon in fungi.

Keywords: Myxomycetes; iridescence; pigmentation; evolution of color; mycelia; cell organization; living photonics

ation: Brodie, J.; Ingham, C.J.; nolini, S. Does Structural Color st in True Fungi? *J. Fungi* **2021**, *7*, https://doi.org/10.3390/ 020141

demic Editor: Laurent Dufossé

eived: 20 January 2021
epted: 12 February 2021
lished: 16 February 2021

lisher's Note: MDPI stays neutral h regard to jurisdictional claims in lished maps and institutional affil- ons.

The rich diversity of the color palette throughout the tree of life is usually obtained via dyes and pigments, with the fungal kingdom (*Eumycetes*) alone producing a vast array of hues [1]. However, most vibrant colors and color effects found in nature are created using transparent materials with architecture on the same scale as the wavelength of visible light (a few hundred nanometers). Such coloration, so-called structural color, is based on light interference phenomena and is usually independent from the chemical composition of the material. This is in contrast to pigments, whose color depends on light absorption dictated by their molecular composition. [2]. Structural color has been known since Hooke and Newton, who in the 17th century, studied how the peacock and house fly generated intensely colored surfaces without extractable pigments [3]. Four centuries later, the optics of structural color in life is now a growing field, and it is clear that multiple types of structural color are found widely in the eukaryotes [4–6]. Examples of structural color are found in red, green, and brown macroalgae [7,8], but also in different plant tissues: from leaves to flowers to fruits (Figure 1) [9,10]. The functions of structural color in both plants and algae are still not fully understood, and depending on the system considered, can span from light management, including improving photosynthesis and photoprotection [11], to interspecies communication [9]. Similarly, structural color is also extremely widespread within the invertebrates, (such as insects, arachnids, and marine cephalopods [4–6]), and vertebrates, such as birds, and reptiles including mammals [2,4,12–15]. Demonstrable advantages of structural color for animals are largely connected with visual properties and include intra- and interspecies signaling and camouflage [2,4].

In all cases, for a living organism to create structural color a formidable capacity is required for biological organization on the nanoscale, posing questions as to the evolution of structural organization and color.

Structural color also exists in microorganisms and is well-known within the Myxomycetes (Figure 2), a phylogenetic group distinct from true fungi but with many similarities, including sporulation and saprotrophy [16,17]. Structural color in the Myxomycetes (often multihued and pointillistic) is due to thin-film interference caused by light interacting with the peridium, the thin transparent membrane which encloses clusters of asexual spores [16]. Within the prokaryotes, we also find examples of structural color, albeit not as individual cells, but in bacteria colonies. Gliding bacteria such as *Flavobacterium* IR1

(Figure 3) and *Cellulophaga lytica* are known to produce vivid and iridescent coloration d to the periodic organization of their cells [18–21].

Figure 1. Examples of structural color in eukaryotes. (**a**) Female jumping spider *Phidippus audax* (Photo: T. Shahan). (**b**) Dogbane beetle, *Chrysochus auratus* (Photo: P. Ganai). (**c**) *Hibiscus trionum* flower (Photo: E. Moyroud). (**d**) Male superb bird of paradise, *Paradisaea rudolphi*. (**e**) Male mandrill, *Mandrillus sphinx*. (**f**) Male Indian peafowl (peacock), *Pavo cristatus*. (**g**) Bigfin reef squid (or glitter squid), *Sepioteuthis lessoniana*. (**h**) Red macroalgae, *Iridaea cordata* (Photo: J. Brodie). (**i**) Red macroalga, *Trichogloea* spp. (Photo: G. Saunders).

Therefore, considering that such structural colors are so widespread in a great varie of living organisms, it is puzzling that structural colors have never been reported in t Eumycetes. The fungi are a structurally and chromatically sophisticated kingdom, ar yet structural color appears, to the best of our knowledge, to have never been report From an evolutionary point of view, it would be fascinating if structural color remains be discovered within the fungi but also notable if genuinely absent.

Structural colors can be subtle to observe and often the optical effects are poor described in papers and therefore are probably under-reported. Unfortunately, there is substantive genomics information that would enable a bioinformatics approach. The are almost no genes relating to structural color known in eukaryotes except for the *opt* gene which has been identified as relevant to both pigmentation and structural effects butterflies [22]. Even in a well-known organism such as the peacock, with a fully sequenc genome, the genes encoding the nanoscale architects responsible for the vivid coloratio displayed by the feathers remain obscure [23]. Further, whilst some genes specifying structural colors are known in bacteria [18], it is not clear if there is any direct relevance to the fungi.

Figure 2. Examples of structural color in the Myxomycetes. (**a–c**) *Diachea leucopoda* (Photo: M. Inchaussandague) with spores enclosed within the transparent peridium, displaying bright, iridescent, structural color. (**c**) A close up of the surface of the sporulating body, with individual spores visible. (**d**) *Lamproderma* sp. (Photo: S. Lloyd) with iridescent sporulating bodies enclosed within the peridium but also iridescence visible on the stalk and hypothallus (e.g., white arrows). Scale bars: (**a**) = 1.2, (**b**) = 2, (**c**) = 0.3, (**d**) = 0.2 mm.

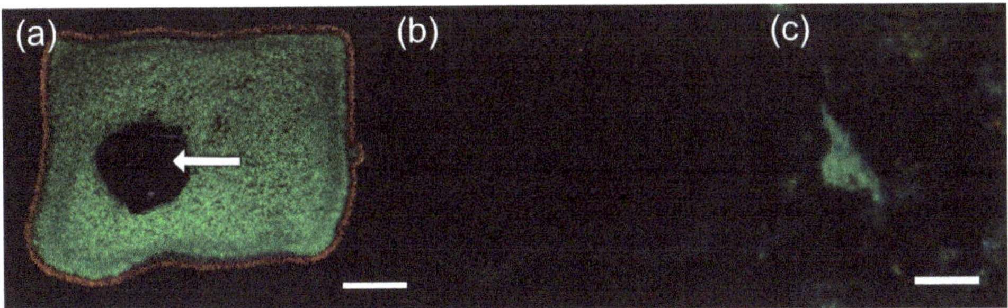

Figure 3. Structural color of *Flavobacterium* IR1 cultured on nutrient agar containing a black dye for contrast. (**a**) Structural color viewed at the optimal angle. Arrow indicates where the cells were mixed with an inoculation loop, disrupting the photonic crystal and so the coloration. (**b**) As panel (**a**) but viewed from a suboptimal angle for structural color. (**c**) Close up of structural color from panel (**a**) showing that what appears green to the eye is actually a pointillistic arrangement of multiple colors. Scale bars: (**a**) and (**b**) = 6 mm, (**c**) = 20 µm.

Discriminating between color produced by pigments and structural color can require a combination of spectroscopical and anatomical studies [24], often accompanied by modeling efforts. This is due to the fact that often structural colors are combined with coloration effects provided by pigments, making it harder to disentangle and differentiate these two distinct optical mechanisms. In practice, given most labs do not have these resources, there are a number of simple tests that could be useful to the mycologist in determining if an isolate might have structural colors. Firstly, many forms of structural color are best observed using a strong, directed white light, i.e., direct sunlight, a broad-spectrum LED, or an incandescent bulb. Viewing under diffuse illumination, such as fluorescent strip lighting, makes it more challenging the observation of structural colors as this can enha the coloration produced by the pigments. Secondly, the following checklist may be use in trying to decide if a fungus is structurally colored:

(1) Does the color shift with a change in viewing and/or illumination angle (Figure 3a
(2) If you disturb the structure by mixing or grinding, does the color vanish (Figure
(3) Are there punctuated, extremely saturated colors under direct white light (Figure 3a
(4) Does the color reversibly change (or appear/disappear) if the material is immersed water or other solvents;
(5) When viewed by electron microscopy, is there a degree of ordering into repetit units with each subunit on the hundreds of the scale of a few hundred nanomete

If the answer to one or more of these questions is "yes" then structural color may involved.

We hope that mycologists will be stimulated to search for colors beyond pigmentat and would be happy to assist in this quest.

Author Contributions: All authors contributed to the conceptualization, writing, providing ima for the article, and funding acquisition. All authors have read and agreed to the published versio the manuscript.

Funding: This project has received funding from the European Union's Horizon 2020 research a innovation program under the Marie Skłodowska–Curie grant agreement no. 860125 as par 'BEEP' Bio-inspired and bionic materials for enhanced photosynthesis. This material reflects only author's views and the union is not liable for any use that may be made of the information contai therein.

Institutional Review Board Statement: Not applicable.

Informed Consent Statement: Not applicable.

Data Availability Statement: Not applicable.

Acknowledgments: Figure 1e was taken from reference 15 with permission. We thank Sarah Ll for valuable discussions.

Conflicts of Interest: The authors declare no conflict of interest. The funders had no role in the des of the study; in the collection, analyses, or interpretation of data; in the writing of the manuscript in the decision to publish the results.

References

1. Venil, C.K.; Velmurugan, P.; Dufossé, L.; Devi, P.R.; Ravi, A.V. Fungal pigments: Potential coloring compounds for wide rang applications in textile dyeing. *J. Fungi* **2020**, *6*, 68. [CrossRef] [PubMed]
2. Burg, S.L.; Parnell, A.J. Self-assembling structural colour in nature. *J. Phys. Conden. Matt.* **2018**, *30*, 413001. [CrossRef] [PubM
3. Hooke, R. *Micrographia*; Martyn, J., Alleftry, J., Eds.; BoD–Books on Demand: London, UK, 1665.
4. Doucet, S.M.; Meadows, M.G. Iridescence: A functional perspective. *J. R. Soc. Interface* **2009**, *6*, S115–S132. [CrossRef] [PubM
5. Wong, V.L.; Marek, P.E. Structure and pigment make the eyed elater's eyespots black. *PeerJ* **2020**, *8*, e8161. [CrossRef]
6. Burresi, M.; Cortese, L.; Pattelli, L.; Kolle, M.; Vukusic, P.; Wiersma, D.S.; Steiner, U.; Vignolini, S. Bright-white beetle sca optimise multiple scattering of light. *Sci. Reps.* **2014**, *4*, 6075. [CrossRef] [PubMed]
7. Chandler, C.J.; Wilts, B.D.; Vignolini, S.; Brodie, J.; Steiner, U.; Rudall, P.J.; Glover, B.J.; Gregory, T.; Walker, R.H. Structural col in Irish moss, *Chondrus crispus*. *Sci. Rep.* **2015**, *5*, 11645. [CrossRef]
8. Chandler, C.J.; Wilts, B.D.; Brodie, J.; Vignolini, S. Structural color in marine algae. *Adv. Opt. Mat.* **2016**, *5*, 1600646. [CrossRe
9. Whitney, H.M.; Kolle, M.; Andrew, P.; Chittka, L.; Steiner, U.; Glover, B.J. Floral iridescence, produced by diffractive optics, act a cue for animal pollinators. *Science* **2009**, *323*, 130–133. [CrossRef]
10. Middleton, R.; Sinnott-Armstrong, M.; Ogawa, Y.; Jacucci, G.; Moyroud, E.; Rudall, P.J.; Prychid, C.; Conejero, M.; Glover, Donoghue, M.J.; et al. *Viburnum tinus* fruits use lipids to produce metallic blue structural color. *Curr. Biol.* **2020**, *30*, 3804–3 [CrossRef]
11. Steiner, L.M.; Ogawa, Y.; Johansen, V.E.; Lundquist, C.R.; Whitney, H.; Vignolini, S. Structural colours in the frond of *Microsor thailandicum*. *J. Roy. Soc. Interface Focus* **2019**, *9*, 20180055. [CrossRef]
12. Cai, C.; Tihelka, E.; Pan, Y.; Yin, Z.; Jiang, R.; Xia, F.; Huang, D. Structural colours in diverse Mesozoic insects. *Proc. Roy. Soc* **2020**, *287*, 20200301. [CrossRef]

Wilts, B.D.; Michielsen, K.; De Raedt, H.; Stavenga, D.G. Sparkling feather reflections of a bird-of-paradise explained by finite-difference time-domain modelling. *Proc. Natl. Acad. Sci. USA* **2014**, *111*, 4363–4368. [CrossRef]
Seago, A.E.; Brady, P.; Vigneron, J.P.; Schultz, T.D. Gold bugs and beyond: A review of iridescence and structural colour mechanisms in beetles (Coleoptera). *J. R. Soc. Interface* **2009**, *6*, S165–S184. [CrossRef]
Prum, R.O.; Torres, R.H. Structural colouration of mammalian skin: Convergent evolution of coherently scattering dermal collagen arrays. *J. Exp. Biol.* **2004**, *207*, 2157–2172. [CrossRef]
Dolinko, A.; Skigin, D.; Inchaussandague, M.; Carmaran, C. Photonic simulation method applied to the study of structural color in Myxomycetes. *Opt. Express* **2012**, *20*, 15139–15148. [CrossRef]
Lloyd, S.J. *Where the Slime Mould Creeps: The Fascinating World of Myxomycetes*; Tympanocryptis Press: Birralee, Australia, 2014.
Johansen, V.E.; Catón, L.; Hamidjaja, R.; Oosterink, E.; Wilts, B.D.; Rasmussen, T.S.; Sherlock, M.M.; Ingham, C.J.; Vignolini, S. Genetic manipulation of structural color in bacterial colonies. *Proc. Natl. Acad. Sci. USA* **2018**, *115*, 2652–2657. [CrossRef] [PubMed]
Kientz, B.; Luke, S.; Vukusic, P.; Péteri, R.; Beaudry, C.; Renault, T.; Simon, D.; Mignot, T.; Rosenfeld, E. A unique self-organization of bacterial sub-communities creates iridescence in *Cellulophaga lytica* colony biofilms. *Sci. Rep.* **2016**, *6*, 19906. [CrossRef] [PubMed]
Hamidjaja, R.; Capoulade, J.; Catón, L.; Ingham, C.J. The cell organization underlying structural colour is involved in *Flavobacterium* IR1 predation. *ISME J.* **2020**, *14*, 2890–2900. [CrossRef] [PubMed]
Schertel, L.; van de Kerkhof, G.T.; Jacucci, G.; Catón, L.; Ogawa, Y.; Wilts, B.D.; Ingham, C.J.; Vignolini, S.; Johansen, V.E. Complex photonic response reveals three-dimensional self-organization of structural coloured bacterial colonies. *J. Roy. Soc. Interface* **2020**, *17*, 20200196. [CrossRef] [PubMed]
Zhang, L.; Mazo-Vargas, A.; Reed, R.D. Single master regulatory gene coordinates the evolution and development of butterfly color and iridescence. *Proc. Natl. Acad. Sci. USA* **2017**, *114*, 10707–10712. [CrossRef]
Jaiswal, S.K.; Gupta, A.; Saxena, R.; Prasoodanan, V.P.; Sharma, A.K.; Mittal, P.; Roy, A.; Shafer, A.; Vijay, N.; Sharma, V.K. Genome sequence of peacock reveals the peculiar case of a glittering bird. *Front. Gen.* **2018**, *9*, 392. [CrossRef] [PubMed]
Vignolini, S.; Moyroud, E.; Glover, B.J.; Steiner, U. Analysing photonic structures in plants. *J. R. Soc. Interface* **2013**, *10*, 20130394. [CrossRef] [PubMed]

Journal of
Fungi

icle

ngal Biomarkers Stability in Mars Regolith Analogues after imulated Space and Mars-like Conditions

ssia Cassaro [1], Claudia Pacelli [1,2,*], Mickael Baqué [3], Jean-Pierre Paul de Vera [4], Ute Böttger [5], enzo Botta [1], Raffaele Saladino [1], Elke Rabbow [6] and Silvano Onofri [1]

1 Department of Ecological and Biological Sciences, University of Tuscia, Largo Dell'Università snc, 01100 Viterbo, Italy; cassaro@unitus.it (A.C.); lorenzo.botta@unitus.it (L.B.); saladino@unitus.it (R.S.); onofri@unitus.it (S.O.)
2 Italian Space Agency, Via del Politecnico snc, 00133 Rome, Italy
3 German Aerospace Center (DLR), Planetary Laboratories Department, Institute of Planetary Research, Ruthefordstraße 2, 12489 Berlin, Germany; mickael.Baque@dlr.de
4 MUSC, German Aerospace Center (DLR), Space Operations and Astronaut Training, 51147 Köln, Germany; jean-pierre.devera@dlr.de
5 German Aerospace Center (DLR), Institute of Optical Sensor Systems, 12489 Berlin, Germany; ute.Boettger@dlr.de
6 Radiation Biology Division, Institute of Aerospace Medicine, DLR, Linder Höhe, 51147 Köln, Germany; elke.rabbow@dlr.de
* Correspondence: claudia.pacelli@asi.it; Tel.: +39-068567466

tion: Cassaro, A.; Pacelli, C.; é, M.; de Vera, J.-P.P.; Böttger, U.; , L.; Saladino, R.; Rabbow, E.; ri, S. Fungal Biomarkers Stability ars Regolith Analogues after lated Space and Mars-like litions. *J. Fungi* **2021**, *7*, 859. s://doi.org/10.3390/jof7100859

emic Editor: Laurent Dufossé

ived: 30 August 2021
pted: 9 October 2021
ished: 14 October 2021

isher's Note: MDPI stays neutral regard to jurisdictional claims in ished maps and institutional affil- ns.

Abstract: The discovery of life on other planets and moons in our solar system is one of the most important challenges of this era. The second ExoMars mission will look for traces of extant or extinct life on Mars. The instruments on board the rover will be able to reach samples with eventual biomarkers until 2 m of depth under the planet's surface. This exploration capacity offers the best chance to detect biomarkers which would be mainly preserved compared to samples on the surface which are directly exposed to harmful environmental conditions. Starting with the studies of the endolithic meristematic black fungus *Cryomyces antarcticus*, which has proved its high resistance under extreme conditions, we analyzed the stability and the resistance of fungal biomarkers after exposure to simulated space and Mars-like conditions, with Raman and Gas Chromatography–Mass Spectrometry, two of the scientific payload instruments on board the rover.

Keywords: spectroscopy; Mars exploration; life-detection; pigments; nucleic acids

1. Introduction

In the next few years, in situ space exploration missions will be devoted to the detection of biogenic signatures of extinct or extant life on Mars. The driver for searching for life on Mars is the findings supporting that ancient environments on Mars could have supported microbial life [1]. One of the primary issues in the search for life is that the present-day Martian surface presents a very inhospitable habitat for life as we know it because of the intense radiation, highly oxidizing conditions, concentrated evaporative salts, and extremely low water activity [2]. Despite the Martian surface having been cold and predominantly dry for at least the last three billion years (i.e., the Amazonian Period, immediately following the Hesperian), the subsurface could have sustained stable reservoirs of geothermally heated liquid water for the majority of this time. These conditions could represent a long-lived habitat that maintained hypothetical living cells [3–6]. For these reasons, the next planetary mission, ESA-Roscosmos ExoMars Rosalind Franklin, will collect and analyze samples in the subsurface, up to 2 m depth [7–9], to access places where organic molecules may be well preserved even after billions of years [10].

In this context, our best chance to find traces of extant or recently extinct life on Mars is to look for biomarkers [11]. A molecular biomarker is defined as a pattern, distribution of molecules or molecular structures that in nature derives uniquely from past or pres biological processes [12,13]. Indeed, we consider as a possible biomarker a range molecules indicative for life as we know it, such as DNA, amino acids, lipids, carbohydra pigments, intermediary metabolites [13], their degradation products and their gene characteristics as preservation potential, specificity and extractability [14]. Assuming t hypothetical Martian life is similar to that on Earth, the identification of these compour would be diagnostic of extant or recently extinct life because they are the indispensal biological components of the organisms as we know it. Any extinct or existing for of microbial life on Mars may have produced biomolecules that may still be preserv and detectable in Martian rocks. In fact, minerals and organic molecules are stric linked: minerals provide surfaces to support, concentrate, and preserve organic molecu In support of this, organic matter was recently found within the three-billion-year- mudstones of Gale Crater [15].

Due to the capability to detect biomolecules or their alteration products within mine grains, Raman spectroscopy is part of the analytical instrumentation in the payloads of rovers. This technique, on top of its main goal to provide mineralogical identification the samples, can also detect a wide range of potential biomarkers in the rock substrate has been chosen for its non-destructive properties, for its sensibility during the detectior microbial life closed in their niches, and for its efficiency in in situ analyses also in prese of mineral rocks [16].

The Mars Organic Molecule Analyzer (MOMA), among various instruments, w operate with Pyrolysis–Gas Chromatography–Mass Spectrometry (Pyr-GCMS) in order analyze volatile compounds in the Martian subsurface. In this context, it is important understand the interaction effect of high temperatures with regolith and organic mat In extreme environments, life could find refuge inside rocks, as in terrestrial analog en ronments, where in harsh conditions, extremophilic microorganisms dwelling inside subsurface produce characteristic compounds protecting themselves from stressed cor tions [17,18], among which are carotenoids and melanin pigments. Organic, metabolic a morphological structure or compounds of the organisms could persist in mineral structu and are considered good biomarkers of microorganisms' presence. On Earth, biologi matter can be preserved in sedimentary rocks as carbonaceous macromolecules [12] a maybe, if life exists or existed on Mars, its rocks could preserve organic deposits: orga components may be preserved either as degraded molecules within mineral structu or as disseminated molecules chemically bonded to mineral particles of rocks, such phyllosilicates [19].

In this context, the BIOMEX (BIOlogy and Mars EXperiment) project, aimed at vestigating the endurance of extremophiles and the stability of biomolecules under spa and Mars-like conditions in the presence of Martian regolith analogs [20]. This experim involved 16 months of real space and a close approach to Mars-like conditions expos outside the International Space Station (ISS). The experiment was placed externally abo the EXPOSE-R2 exposure payload and comprised a series of ground-based simulation te including Experiment Verification Test (EVT) and Science Verification Tests (SVTs), carr out before the flight. In the frame of ground-based tests, the stability/degradation biomolecules of the extremotolerant microorganism *Cryomyces antarcticus* was investiga the fungus, isolated from the McMurdo Dry Valleys (South Victoria Land, Antarctica), w chosen for its widely proved ability to withstand stressors similar to the ones encounte in space and Mars-like environments (e.g., ionizing and no ionizing radiation, vacuum Martian atmosphere, temperature cycles) [21,22].

In total, fungal colony samples were investigated after exposure to simulated spa and Mars-like conditions during the ground-based experiments before flight. Fun colonies were spread on three different cultivation media consisting of Malt Extract A (MEA) and three different regoliths: the Original Substrate (i.e., Antarctic sandsto OS), the Phyllosilicatic Mars Regolith Simulant (P-MRS) analogue and the Sulfatic M Regolith Simulant (S-MRS) analogue. The P-MRS simulates igneous rocks altered hydrous fluids (neutral to basic) while the S-MRS mimics a more acidic environment with sulfate deposits [23]. The objectives of this study were to identify potential fungal biomolecules to be accounted as biomarkers and to understand if and how simulated space and Mars environment could modify them. As the primary goal of this study was the detection of microbial signatures within the Martian regolith analogues, the same methods, Raman spectroscopy and Gas Chromatography–Mass Spectrometry (GC-MS), planned for the ExoMars Rosalind Franklin mission, were used. In addition, extracted melanin pigments and nucleic acids were analyzed by using UV–VIS spectrophotometry and by quantitative Polymerase Chain Reaction (qPCR) technique, respectively, to detect any changes in structure after simulated space and Mars exposure.

2. Materials and Methods

2.1. Ground-Based Simulations

2.1.1. Science Verification Tests (SVTs)

The SVTs were performed at the Planetary and Space Simulation facilities (PSI) at the Institute of Aerospace Medicine (German Aerospace Center, DLR, Koln, Germany). SVTs are designed to ensure that all samples are appropriately prepared to successfully withstand hardware integration, conditions experienced during the mission, and post flight de-integration. The application of mission-equivalent space parameters allows for testing the resilience of samples toward the extreme environmental conditions of space and simulated Mars exposure on the ISS. Samples were accommodated in wells with a diameter of 12 mm, within square aluminum alloy carriers, with a side of 76 mm. Following the accommodation plan scheduled for the EXPOSE-R2 mission, SVTs allowed only one replicate per sample [20]. To simulate space-like test conditions, the sample that was grown on OS analogue was exposed to vacuum (10^{-5} Pa) and cycling temperatures between −25 °C (16 h in the dark) and +10 °C (8 h during irradiation), alone (Bottom samples) or in combination with polychromatic UV (200–400 nm) radiation produced by the solar simulator SOL2000 (Top samples). The dose of 570 MJ/m^2 was reached by running the solar simulator (SOL2000; Dr. Hönle GmbH, Germany, Rabbow et al., 2017) for 125 h at 1271.2 W/m^2. In parallel, Mars test parameters were simulated by low temperature (−25 °C), Mars-like atmosphere (95.55% CO_2, 2.70% N_2, 1.60% Ar, 0.15% O_2, *370 ppm H_2O; Praxair Deutschland GmbH), and Mars-like pressure of 10^3 Pa for S-MRS and P-MRS analogues, alone or in combination (Top samples) with the same radiation as described above. Neutral density filters (0.1%) were used to attenuate radiation in all tests performed; all conditions were simulated for a period of 28 days. The applied fluency corresponds to the long-term space experiment of 1 year of exposure outside the ISS, as estimated from previous EXPOSE data and simulations [24,25]. It should be noted that space parameters cannot be fully mimicked in the laboratory (space vacuum and complex radiation environment), for example, deep UV, that is, solar UV radiation below 200 nm. Below the irradiated samples, an identical set of samples (space dark samples and Mars dark samples/bottom samples) was kept in the dark and experienced all simulation parameters except UV radiation exposure. Controls (Ctr) were kept at DLR in the dark at room temperature. The exposure conditions are summarized in Table 1.

Table 1. Exposure conditions during the Scientific Verification Tests (SVTs).

Test Parameters	Duration
Vacuum (2×10^{-4}) + polychromatic UV irradiation (200–400 nm), with SOL2000 at 1271.2 W/m^2, attenuated with 0.1% neutral density filter.	28 days SOL2000 125 h
Simulated CO_2 Mars atmosphere 10^3 Pa+ polychromatic UV irradiation (200–400 nm), with SOL2000 at 1271.2 W/m^2, attenuated with 0.1% neutral density filter.	28 days SOL2000 125 h
Control experiment, 1 atm air, dark, room temperature	28 days

SOL = Solar simulator.

2.1.2. Fungal Melanin Extraction

Melanin was extracted from dried colonies of *C. antarcticus* grown on different su strata, optimizing the protocol reported by [26]. Fungal cells were collected by centrifu tion at 16,100 rcf for 20 min, washed with phosphate-buffered saline (PBS) (pH 7.4). Ce were suspended in 20 μL of a buffer composed by 0.25 μL of enzyme from *Trichoderr harzarium* (Sigma #L1412, St. Louis, MO, USA), 50 μL of 0.2 M sodium citrate (pH 5. 15 μL of 0.2 M citric acid and 18.2 g of sorbitol, up to the final volume of 500 μL.

Samples were incubated overnight at 30 °C, shaken at 50 rpm, washed with P two times and collected by centrifugation (10 min at 16,100 rcf). The supernatant w removed by adding 1 mL of 4 M guanidine thiocyanate (VWR International srl, Radn PA, USA), and samples were incubated overnight on stirrer. Samples were washed wi PBS and centrifuged for 15 min at 16,100 rcf for two times. Then, 1 mg/mL of Proteina K, previously dissolved in the reaction buffer (4 μL of TRIS HCl, 1 μL of $CaCl_2$ and 1 n of SDS, up the final volume of 20 mL) was added to the samples; they were incubated 37 °C for 4 h and after centrifuged for 5 min at 16,100 rcf. Samples were washed two tim by adding 1 mL of PBS 4×, and three times by adding 1 mL of chloroform.

A volume of 2 mL of a 6 M HCl solution was added to the samples, and they we boiled for 1 h. Samples were transferred in dialysis membrane (SnakeSkin Dialysis Tubi 3.5 K MWCO-Thermo Scientific, Waltham, MA, USA) in sterile water for 3 days, changi the water every day. Finally, samples were lyophilized overnight with the Lyophiliz FreeZone 2.5 L (Freeze Dry Systems, LabConco, Kansas city, MO, USA) and then used f the analyses.

2.1.3. Spectrophotometric Analysis

After the extraction, purified pigments were dissolved in 500 μL of NaOH 1 M and UV-Visible spectrum was measured in a UV spectrophotometer (UV 1600 PC Spectroph tometer, VWR International) by using M.Wave professional 2.0. A standard graph f estimation was used and was made using synthetic melanin. NaOH 1 M was used as blank and the instrument was set in a range of 200–800 nm for the analysis. The correlatio between absorbance and wavelength was defined. To determine the concentration extracted melanin, synthetic DHN (1,8-DiHydroxyNaphthalene) melanin (Thermo-Fish Scientific, Waltham, MA, USA) was prepared in 1 M NaOH at concentrations of 500 mg/n and as reported in [27], a standard curve at 650 nm was obtained.

2.1.4. Confocal Raman Spectroscopy Analyses

Raman spectroscopy was performed at German Aerospace Center in Berlin, usi a 532 nm excitation laser, with a Confocal Raman microscope (WITec alpha300), at roo temperature, under ambient atmospheric conditions. The spectral resolution of the spe trometer is 4–5 cm^{-1}. Before the analyses, the spectrometer was calibrated with pure silico and paracetamol test samples. A 10× Nikon objective, with a 0.25 numerical apertu was used to focus the laser on a 1.5 μm spot. For single spectra, all measurements we performed at 0.1 mW laser power with 10 s integration and 50 accumulations. Image sca were done at 0.7 mW with 1 s and 1 accumulation (to avoid signal saturation or damagi effects) on three distinct areas up to 100 μm × 100 μm and up to 500 image points, th collecting a minimum of 1000 measurements per sample.

All data analyses were performed with the WITec Project FIVE software. Paramete used for analysis were the value of signal coverage on random zones on the samples, show the presence/absence of the signal defined as spectra having Signal to Noise Rat (SNR) values superior to 5 and the value of SNR. To analyze the presence of the signal, t region of interest between 200 and 2000 cm^{-1} was cropped, then a fifth-order polynomi function was applied for background subtraction; and finally, SNR masks were applied. previously described in [28] for carotenoid pigments, the SNR was defined as the heig of the 1600 cm^{-1} peak divided by the noise represented by the standard deviation o spectral region without features (1750 to 1950 cm^{-1}). The position of the 1600 cm^{-1} peak was also derived from masks available in the WITec Project FIVE software.

2.2. Nucleic Acid Analysis

2.2.1. Nucleic Acid Extractions from Synthetic Mars and Terrestrial Soils

DNA was extracted from colonies, using the Nucleospin Plant kit (Macherey-Nagel, Düren, Germany) following the protocol optimized for black fungi as reported in [29]. Before amplification, DNA was quantified using QUBIT system and diluted at the concentration of 0.1 ng/µL for the following analyses.

2.2.2. Acid Nucleic Detection through Quantitative Real-Time PCR (qPCR)

Three different target genes were amplified with qPCR approach to investigate the differences in acid nucleic detection: a long-repeated fragment (Large Sub-Units, LSU gene of 939 bp), a short-repeated fragment of the same gene (LSU gene of 330 bp) and a small but non-repeated fragment in the genome (β-actin gene of 330 bp). qPCR was performed with a BioRad CFX96 real time PCR detection system (BioRad, Hercules, CA, USA) using primers targeting the fungal LSU rRNA gene and the β-actin gene: LR0R (ACCCGCTGAACTTAAGC, [30]) and LR5 (TCCTGAGGGAAACTTC, [31]), and ACT512-F (ATGTGCAAGGCCGGTTTCGC3) and ACT783-R (TACGAGTCCTTCTGGCCCAT) [32], respectively, each at 5 pmol final concentration.

The primers LR0R-LR5 and LR0R-LR3 were used to amplify a 939 bp and 300 bp products, respectively, spanning the LSU region of rRNA encoding genes. The standard qPCR cycling protocol for both products, consisting of a denaturation step at 94 °C for 5 min, followed by 35 cycles of denaturing at 94 °C for 45 s, annealing at 52 °C for 30 s, and elongation at 72 °C for 2 min, was performed. The primers ACT512-F and ACT783-R are used to amplify a 330 bp product spanning the β-actin gene. The standard qPCR cycling protocol, consisting of a denaturation step at 95 °C for 10 min, followed by 35 cycles of denaturation at 95 °C for 15 s, annealing at 61 °C for 20 s, and elongation at 72 °C for 15 s, was performed. Fluorescence measurements were recorded at the end of each annealing step. After 35 cycles, a melt curve analysis was performed by recording changes in fluorescence as a function of raising the temperature from 60–90 °C in 0.5 °C per increments. All tests were performed in triplicate.

2.2.3. Statistical Analyses

For multiple data points, mean and standard deviation were calculated. Statistical analyses were performed by one-way analysis of variance (Anova) and pair wise multiple comparison procedure (*t* test), carried out using the statistical software SigmaStat 2.0 (Jandel Scientific, San Jose, CA, USA).

2.2.4. Organic Compounds Detection by Gas Chromatography–Mass Spectrometry

Gas chromatography associated to mass spectroscopy has been selected for the analysis of samples onboard of the rover of the ESA-ExoMars mission [33]. Each sample (pellet) was grinded in agate mortar and then suspended in 2 mL of ethyl acetate. The mixture was left 4 h under magnetic stirring at room temperature. After this time, the suspension was filtered to remove the solid, and the solution obtained was concentrated under reduced pressure. After the extraction and fractionation of the samples with *N*,*N*-bis-trimethylsilyl trifluoroacetamide in pyridine (620 µL) at 60 °C for 4 h in the presence of betulinic acid [3β -hydroxy-20(29)-lupaene-oic acid] as the internal standard (0.2 mg), the Gas Chromatography–Mass Spectrometry was performed. Mass spectrometry was carried out through the following program: injection temperature 280 °C, detector temperature 280 °C, gradient 100 °C for 2 min and 10 °C for 60 min. To identify the structure of the products, two strategies were followed. First, the spectra were compared with commercially available electron mass spectra libraries such as NIST (Fison, Manchester, UK). Second, GC-MS analysis was repeated with standard compounds. All products have

been recognized with a similarity index (SI) greater than 98% compared with that of
reference standards. The analysis was limited to products of ≥1 ng/mL quantity, and
yield was calculated as micrograms of isolated product.

3. Results

3.1. Detection of Pigments by Spectrophotometric Analyses

Spectrophotometric analyses were performed on fungal melanin extracted fror *antarcticus* colonies after exposure to simulated space conditions (OS) and Mars-like cor tions (P-MRS and S-MRS). The wavelength of maximum absorbance was scanned at a ra of 200 to 800 nm. The UV-visible absorbance spectrum of the purified pigments showe strong absorbance in the UV region, and a characteristic absorption peak was observec 230 nm corresponding to typical melanin UV absorption. The strong absorption in the region with a progressive decrease at high wavelength is due to the presence of comp conjugated structures in the melanin molecule [34,35]. The decrease in the absorption w increasing wavelength is linear in the case of melanin.

Figure 1 shows the melanin spectra extracted from samples exposed to simula space UV radiation (OS Top; red line), vacuum (OS Bottom; blue line), to Mars-like radiation (P-MRS and S-MRS Top; red line) and Mars-like atmosphere (P-MRS and S-M Bottom; blue line) compared with relative control (black line). No change in melar absorbance at 230 nm has been reported in all the experimental conditions (Top, Bott and Control), compared with those reported in [36] for extracted melanin pigments fror *antarcticus*. In addition, UV–VIS analyses of melanin pigments extracted for P-MRS a S-MRS analogues revealed a bulge at ~300 nm, not shown in OS spectra, probably due the presence of regolith during the detection process.

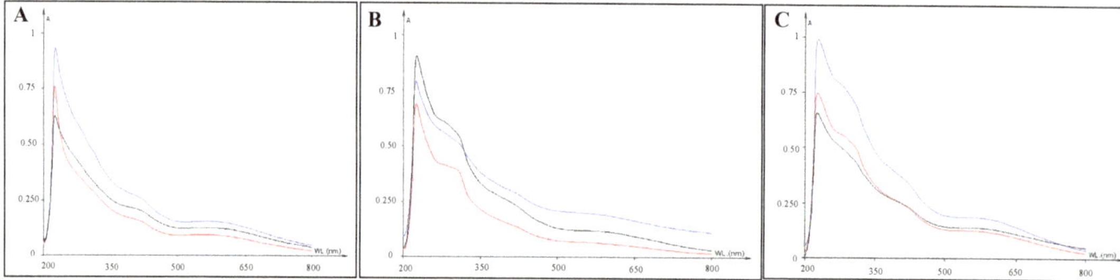

Figure 1. UV–VIS-spectra of melanin pigment extracted from *C. antarcticus* colonies grown on (**A**) OS exposed to simulated space conditions; (**B**) P-MRS and (**C**) S-MRS exposed to Mars-like conditions on different layers: Top (exposed to sun light with 0.1% Neutral Density filters) = red spectrum, Bottom (dark control in space, not exposed to space radiation) = blue spectrum and CTR (sample kept in the lab, in the dark at room temperature) = black spectrum. OS = Original Substrate, P-MRS = Phyllosilicatic Mars Regolith Simulant, S-MRS = Sulfatic Mars Regolith Simulant.

3.2. Detection of Pigments by Confocal Raman Spectroscopy

Colonies of *C. antarcticus* grown on different analogues were analyzed by Confocal man spectroscopy after exposure to simulated space and Mars-like conditions. The resu of Raman analyses identified the melanin spectra with two main peaks, according to [an intense and broad peak at 1590–1605 cm^{-1} and a second peak at lower wavenuml at 1340 cm^{-1} (Figure 2). The presence of these peaks is probably due to the aromatic C bonds for the first peak [38] and to the stretching of the C-C bonds within the rings of aromatic melanin monomers for the second peak [39]. The presence of other peaks a shoulders in all spectra, with the most prominent one around 1425 cm^{-1}, served as a pr that the samples did not undergo thermal degradation and that the signal acquired is melanin and not burnt organic matter (or amorphous carbon). Figure S1 shows the sig coverage for the image scan analyses, calculated from the application of a SNR superior

5; results were normalized to the corresponding SNR of the non-irradiated sample (Ctr) for each analogue.

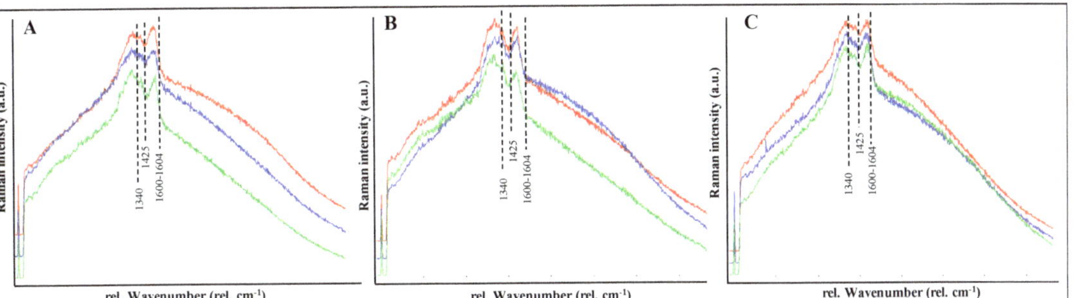

Figure 2. Raman spectra of melanin pigments of *C. antarcticus* grown on (**A**) OS exposed to space simulated conditions; (**B**) P-MRS and (**C**) S-MRS exposed Mars-like conditions on different layers: Top (exposed to sun light with 0.1% Neutral Density filters) = red spectrum, Bottom (dark control in space, not exposed to space radiation) = blue spectrum and Control (sample kept in the lab, in the dark at room temperature) = green spectrum. OS = Original Substrate, P-MRS = Phyllosilicatic Mars Regolith Simulant, S-MRS = Sulfatic Mars Regolith Simulant.

No significant melanin changes are reported in all the collected spectra; similar peak positions (1602–1604 cm^{-1}) were detected in samples exposed to simulated space conditions (OS) and in samples exposed to Mars-like conditions (P-MRS and S-MRS), both in Bottom and Top conditions, in comparison with the respective Control samples (Figure S2A,B).

3.3. Detection of Nucleic Acids through Amplification Method

Although a thermocycler instrument is not one of the pieces of equipment foreseen for the imminent exploration missions to Mars, it is a good candidate instrument to search for Earth-like life beyond Earth [40]. It is specific and sensitive, detecting even a single DNA molecule in a sample. The persistence of intact DNA has been tested on samples after a ground-based experiment simulating 16 months of exposure to space and Mars-like conditions outside the ISS. First, two different gene lengths of 939 bp and 330 bp, spanning the ribosomal LSU, have been targeted to be quantitatively amplified by qPCR based on the principle that damaged DNA is not amplifiable. Then, since the ribosomal genes occur in multiple copies in the genomes, we decided to amplify a region of 330 bp of the housekeeping gene β-actin, which is present in a single copy in the genome, to compare the level of detection based on the different gene length and number of copies in the genome.

In Figure 3, we reported the amplification of nucleic acids amplifying 939 bp and 330 bp of LSU gene and 330 bp of β-actin; the results showed a high amount of amplified DNA in all the experimental conditions tested and despite the gene type amplified: 13,170 DNA copies on average and never less than 10^2 copies were amplified. The test highlighted a common trend for all the samples, i.e., a lower amount of amplified DNA for samples exposed to UV radiation and vacuum (Top), and a higher copy number for samples exposed to vacuum but no radiation (Bottom) and in the control samples.

To conclude, all the gene amplifications worked out, and the number of amplified DNA copies was never under the amplification limit (one copy of a target sequence in genomic DNA), even when we decided to use the single copy gene β-actin.

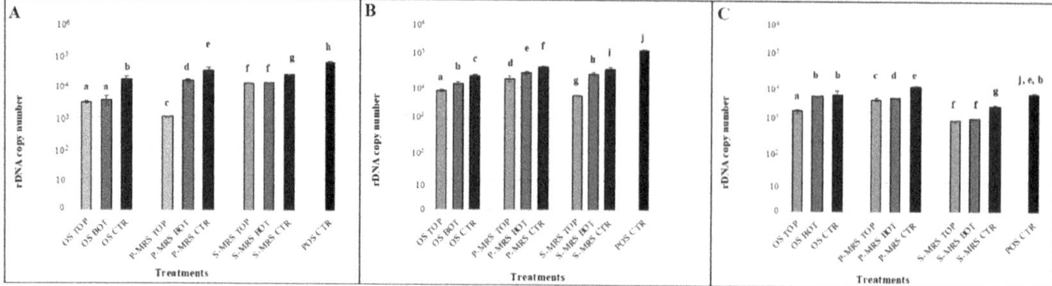

Figure 3. (**A**) Quantitative PCR (qPCR) of a 939 bp target LSU (Large SubUnit) gene; (**B**) a 330 bp target gene (LSU) and (**C**) a 330 bp target gene (actin) of *C. antarcticus* DNA, after exposure to SVT treatments. On the axis of the ordinates the number of amplified copies on a logarithmic scale is shown; on the abscissa axis, the treatments are as follows: DNA from samples exposed to space simulated conditions (OS) and samples exposed to Mars-like conditions (P-MRS and S-MRS). OS = Original Substrate (sandstone); P-MRS = Phyllosilicatic Mars Regolith Stimulant; S-MRS = Sulfatic Mars Regolith Stimulant. Top (exposed to sun light with 0.1% Neutral Density filters), Bottom (dark control in space, not exposed to space radiation) and CTR (sample kept in the lab, in the dark at room temperature); POS CTR = DNA of *C. antarcticus* colony growth in physiological conditions (cultivated on MEA and incubated at 15 °C). The same letters above the bars indicate that the values are not statistically significant according to the *t* test ($p \leq 0.05$). OS = Original Substrate, P-MRS = Phyllosilicatic Mars Regolith Simulant, S-MRS = Sulfatic Mars Regolith Simulant.

3.4. Organic Compounds Detection by Gas Chromatography Associated to Mass Spectrometry

The Mars Organic Molecule Analyzer instrument onboard the ExoMars rover w employ thermal extraction protocols [41]. The characterization of thermal extracts will carried out by GC-MS detection [42]. This analysis was performed only on a selection samples, control and samples exposed to simulated space and Mars-like conditions (T condition).

The GC-MS analysis revealed the presence of several compounds using mass- charge ratios (*m/z*): azelaic acid (1,7-epta-didecanoic acid), myristic acid (1-tetradecanc acid), pyruvic acid (2-Oxopropanoic acid), glucose, fructose, glucitol, glycerol (1,2,3-t hydroxy propane) and ethylene glycol (1,2-dihydroxy ethane). The identified compoun are shown in Table 2; the *m/z* fragmentation spectra of the identified compounds a reported in Table 3. Particularly, the results of pyrolysis analyses showed the presen of low-molecular-weight saturated carboxylic acids. Dicarboxylic acids, such as azela acid, were detected in Control and Top samples in OS, P-MRS, and S-MRS analogu (Table 2). In particular, azelaic acid was found in all examined samples, allowing consider this molecule as a distinctive biomarker for the presence of *C. antarcticus*. Myris acid was detected only in S-MRS samples exposed to Top conditions and in the relati Control (Table 2). Surprisingly, the amount of azelaic and myristic acids increased aft the treatments. A low amount of pyruvic acid was found only in S-MRS control sampl no amount was shown in exposed samples, probably due to the effect of radiation. The results are similar to those reported for ethylene glycol, detected only in the case of S-M and P-MRS Top samples.

Table 2. Abundance of main compounds detected with Gas Chromatography–Mass Spectrometry analysis after ground-based simulation.

ompounds [a]	OS Top	OS Control	P-MRS Top	P-MRS Control	S-MRS Top	S-MRS Control
Azelaic acid	0.99	0.13	0.69	0.45	0.30	0.02
Myristic acid	nd	nd	nd	nd	0.48	0.42
Pyruvic acid	nd	nd	nd	nd	nd	0.17
Glucose	3.14	2.78	nd	nd	0.68	13.87
Fructose	0.64	7.53	nd	nd	nd	4.50
Glucitol	nd	3.58	nd	nd	0.88	2.50
Glycerol	nd	nd	nd	nd	nd	nd
thylene glycol	nd	nd	1.73	nd	0.41	nd

[a] The analyses were performed after silylation with *N,N*-bis-trimethylsilyl trifluoroacetamide in pyridine (620 μL) at 60 °C for 4 h in the presence of betulinic acid [3b-hydroxy-20(29)-lupaene-oic acid] as the internal standard (0.2 mg). All quantities are expressed in μg. GC-MS analysis of OS, P-MRS, and S-MRS samples. Original substrate (OS), Phyllosilicatic Mars Regolith Simulant (P-MRS) and the Sulfatic Mars Regolith Simulant Phyllosilicate (S-MRS). Top samples exposed on the Top of the payload in comparison with Control samples, not exposed to treatments. nd: not determined.

Table 3. Mass-to-charge ratio (*m/z*) value and the abundance of peaks of identified compounds.

Products [a]	*m/z* (%)
Azelaic Acid [c]	317 (25) [M-CH_3], 302 (3) [M-2xCH_3], 243 (2) [M-OSi$(CH_3)_3$], 201 (15) [M-Si$(CH_3)_3$-CO_2-CH_3], 186 (3) [M-2xSi$(CH_3)_3$], 170 (4) [M-OSi$(CH_3)_3$- Si$(CH_3)_3$], 73 (100).
Myristic Acid [b]	300 (10) [M], 285 (95) [M-CH_3], 257 (3) [M-2xCH_3], 73 (100).
Palmitic Acid [b]	328 (20) [M], 313 (100) [M-CH_3], 73 (100).
Stearic Acid [b]	356 (20) [M], 341 (90) [M-CH_3], 327 (2) [M-CH_3- CH_2], 313 (50) [M-CH_3-2xCH_2].
Lactic Acid [b]	219 (6) [M-CH_3], 190 (14) [M-CO_2], 147 (71) [M-Si$(CH_3)_3$-CH_3], 133 (7), 117 (76) [M-Si$(CH_3)_3$-$(CH_3)_3$].
Pyruvic acid [b]	160 (10) [M], 145 (7) [M-CH_3], 88 (14) [M-Si$(CH_3)_3$], 71 (12) [M-Si$(CH_3)_3$-OH], 43 (100) [M-HSi$(CH_3)_3$- CO_2].
Glucose [e]	437 (5) [M-Si$(CH_3)_3$-2xCH_3], 394 (4) [M-2xSi$(CH_3)_3$], 305 (5) [M-OSi$(CH_3)_3$-2xSi$(CH_3)_3$], 217 [g] (30), 204 [g] (100), 191 [g] (75).
Fructose [e]	437 (5) [M-Si$(CH_3)_3$-2xCH_3], 217 [g] (30), 204 [g] (100), 146 (75).
Glucitol [f]	319 (60) [M], 297 (94) [M-CH_3], 282 (30) [M-2xCH_3], 267 (40) [M-3xCH_3]; 217 [g] (90), 204 [g] (80); 147 (40).
Glycerol [d]	293 (3) [M-CH_3], 263 (2) [M-3xCH_3], 218 (20) [M-OSi$(CH_3)_3$], 205 (60) [M-OSi(CH3)$_3$-CH_3], 191 (3) [M-OSi$(CH_3)_3$-2xCH_3], 171 (4) [M-OSi$(CH_3)_3$-3xCH_3].
thylene Glycol [c]	191 (25) [M-CH_3], 147 (100) [M-4xCH_3], 133 (5) [M-Si$(CH_3)_3$], 103 (20) [M-Si$(CH_3)_3$-2xCH_3].

[a] Mass spectroscopy was performed by using a GC-MS Varian 410 GC-320 MS. The peak abundance is reported in parenthesis [b] Product analyzed as the monosilyl derivative; [c] Product analyzed as the bis-silyl derivative; [d] Product analyzed as the tris-silyl derivative; [e] Product analysed as the penta-silyl derivative; [f] Product analysed as the hexa-silyl derivative; [g] Ions characteristic for EI/MS sugar degradation: *m/z* 217 [$(CH_3)_3$SiOCH=CH-CH=OSi$(CH_3)_3$]$^+$, *m/z* 204 [$(CH_3)_3$SiOCH=CHOSi$(CH_3)_3$]$^{+\bullet}$, *m/z* 191 [$(CH_3)_3$SiOCH=OSi$(CH_3)_3$]$^+$.

4. Discussion

The detection of biomolecular markers in the subsurface regolith of Mars is a primary goal for astrobiology. Investigation on the stability or degradation-rate of any biomarker after exposure to simulated or real space conditions is of high interest for space exploration. The stability of any biomarker is dependent on its initial form, the matrix in which it is hosted, and the chemical and physical processes to which it is subjected over time [19].

In this paper, we aimed at investigating the biomarkers associated with fungal colonies in Martian regolith analogues under simulated space and Mars-like conditions, tested during the Scientific Verification Test of the BIOMEX project, with the same techniques planned for the robotic exploration of Mars. Assuming that life on Mars evolved with the same characteristics than that on Earth or that there was an exchange of material inside meteorites between Earth and Mars [43], it is reasonable to search for traces of Earth-like life on Mars. Among biomarkers, we first focused on pigments. The black fungus *C. antarcticus* can withstand the hyper-arid and extremely cold conditions of McMurdo Dry

Valleys in Antarctica, thanks to its strongly melanized cell wall that helps to protect t fungus from the environmental stressors of these remote areas. It is known that the fung is able to survive in different stressor conditions, such as ionizing radiation [36], heavy io in hydrated and de-hydrated conditions [44,45].

Melanins are known to be involved in cellular resistance against a multitude factors, such as toxic metals, hyperosmotic conditions and pH variations [46], but al extreme temperatures, desiccation and radiation, such as ultraviolet (UV) light, oxidizi agents and even ionizing radiation [35]. We aimed at detecting the presence of melar pigments in samples subjected to simulated space and Mars-like exposure with differe methods. Spectral analyses were used to confirm the presence and the possible alterati of melanin pigments after treatments. The UV spectrum (Figure 1) was characterized the unaltered typical absorption profile of *C. antarcticus* melanin pigments [36]. It show strong absorption in the UV region (230 nm) with a progressive reduction as the waveleng increased, due to the presence of many complex conjugated structures in the melar molecule [34,35]. This kind of analyses is relevant for the upcoming NASA Mars 20 exploration rover Perseverance, which has onboard the SHERLOC (Scanning Habital Environments with Raman and Luminescence for Organics and Chemicals) instrument deep UV laser with emission wavelengths below 250 nm that will provide an integrat analysis to avoid the fluorescence background during Raman measurements [47]. T importance of using a multiple technique approach to characterize melanic pigments due to its absorption wavelength, which is similar to other compounds (e.g., amorpho carbon), making its detection more challenging.

Indeed, to further investigate the state of preservation and characterize the fung melanin, we performed Raman spectroscopy analysis. The Raman spectra were dominat by two main peaks around 1340 cm^{-1} and 1600 cm^{-1} and one smaller at 1425 cm (Figure 2). According to the literature, the presence of the main peak at 1600 cm^{-1} is d to the stretching vibration of the aromatic C = C stretching modes and the 1340 cm^{-1} pe to the C-N stretching band of indole [48]: these peaks together with the smaller peak 1425 cm^{-1} are attributed to the presence of melanin pigments and, hence, to the fung colonies.

It is well known that melanin pigments contribute to protecting microorganis against the effect of radiation, which are abundant in space. Melanin pigments may ha helped potential Earth-like life to survive in the harsh and highly irradiated environme of Mars, as it occurs in some extreme environments on Earth, where microorganisms ha adopted this survival strategy. This assumption is supported by previous studies on t black fungus *C. antarcticus*; the melanized strain survived better the exposure to increasi doses of ionizing radiation compared to the non-melanized counterpart [49], confirmi the protective role of melanin against ionizing radiation.

Furthermore, the ExoMars rover is equipped with a pyrolysis gas chromatograph mass spectrometry device with the aim of detecting organic molecules such as carboxy acids in Martian subsurface materials [35,41–52]. Assessing the possible thermal destr tion and transformation of organic matter induced by pyrolysis is of utmost importance, the next rover missions. Indeed, pyrolysis, may alter and transform organic matter [53,5 hiding the presence of organic signatures from samples. For example, organosulfur co pounds that were detected by the Sample Analysis at Mars (SAM) instrument onboard t Curiosity rover may derive from artificial secondary reactions of sulfur (from decom sition of sulfates) with organic radicals [15], due to the high temperature of the pyroly processes. In this context, the main question is: do potential fungal biomarkers rem unaltered during pyrolysis processes? Our study reports that pyrolysis GC-MS is able detect fungal signals, even when the biomarkers are present in low amounts. Indeed, c data highlighted the presence of low molecular weight carboxylic acids, such as pyru acid (only in S-MRS Control samples) and myristic acid (only in S-MRS Top and Cont samples). Although pyruvic acid is a key intermediate in several metabolic pathwa throughout the cell, it cannot be accounted as a biomarker since it has been found in c bonaceous meteorites, and it can be produced through the chemistry of interstellar nitriles, HCN/CN and ketene [55]. Conversely, myristic acid, which can be part of the phospholipid bilayer of the plasma membrane of the eukaryotic cell, can be considered as biomarker. Nevertheless, in our study, it has been found only in Top and Control S-MRS samples. Interestingly, azelaic acid, which is a saturated dicarboxylic acid involved in stress resistance, was found in all tested samples, without any alteration in GC-MS profiles (Table 2). In this framework, we can define this dicarboxylic acid directly related to the presence of fungal colonies, and we can define it as a good biomarker for the search of Earth-like life elsewhere. Although they can be generated by both biotic and abiotic synthesis (Fischer–Tropsch reactions), in the latter, carbons are added one at a time. In most microorganisms, fatty acids are generated biochemically by the addition of two carbon atoms at a time [56]. This can help discriminating biotic and abiotic origins of the molecules, and therefore, it allows to consider fatty acids with an odd C number as a potential biomarker. The other detected compounds in our experiment, such as carbohydrates components (glucose and fructose) may derive from the malt extract used for the cultivation of *C. antarcticus*. The low amount of ethylene glycol detected only in S-MRS and P-MRS irradiated samples most probably derived from the degradation of more complex substances. Even if we can correlate the presence of the colony with the detection of azelaic acid through the GC-MS analyses, the compound was detected in a very small amount; this has to be taken into account in future space exploration missions, in order to avoid negative outcomes. One of the restrictions of this technique is the undetectability of high molecular weight compounds. Indeed, one of the main features of the black fungus, the melanin pigments, was not detected using GC-MS instrumentation, showing the importance of using multiple and connected techniques to avoid false negative results.

Then, we investigated the possibility to detect fungal DNA after the exposure through quantitative PCR. Nucleic acids are an ideal target biomarker due to their unambiguity, non-specificity and mainly the impossibility of generation in the absence of life [57,58]. DNA potentially preserved in Martian soil may be destroyed by the effects of UV, cosmic ray exposure and radioactive decay on nucleic acids [59,60]. Ancient DNA is prone to damage (e.g., hydrolysis, depurination) and fragmentation [9,61] once it is no longer actively repaired in a biological system. However, as previously reported in [62] and in [63], DNA showed to be still detectable in environmental samples by PCR and stable in Mars-like conditions [64]. In addition, the bound between DNA and minerals may have facilitated the preservation on different geological timescales, even more in cold and dry environments, such as the Gale Crater on Mars with an average temperature of −48 °C [65]. In the framework of planetary exploration, the implementation of techniques, such as the qPCR that allows to amplify fragments of a DNA molecule, would detect even minimal traces of DNA, even after partial degradation. Such kind of instruments are also very easy to miniaturize. It is already used in field campaigns with the MinION device [66]. This miniaturized technology was used for the first time in 2016 in the frame of Biomolecule Sequencer project [67,68] with the aim to detect unambiguous signs of life through nucleic acids sequencing, starting from a small amount of sample. The advantage of the in situ DNA detection with a miniaturized tool also allows minimizing any eventual contamination from terrestrial microorganisms [69]. Accordingly, our results suggested that, in spite of the treatments, the nucleic acids are still detectable by qPCR technique. In particular, we found no differences in the amplified region: an average of ~13,000 DNA copy number were detected. The advantage of the amplification of long-repeated LSU gene allows detecting any modification at the genomic level in an extended region. On the other hand, these results may give a false outcome, disguising possible radiation damage on DNA, due to the repetition of the gene in the whole genome. Indeed, the choice of β-actin gene amplification permits to amplify a small non-repeated gene region, with the objective to identify nucleic acid damage in a single copy gene. Therefore, we recommend that space agencies should investigate the utility of establishing and promoting a miniaturized PCR instrument (e.g., MinION) to promote the detection of nucleic acids beyond Earth. Specific primers should be used for each of the three domains: bacteria, archaea, and eukarya. order to identify microorganisms by their sequence, an automated sequencer needs included in addition to the thermocycler apparatus. Our results showed that, even wh using instrumentation already planned for future missions, each analytical technique h constraints that limit the set of information produced. Further improvements in techniq and instrumentation, as well as cross comparisons in varying approaches, will be requir to optimize data interpretation for the upcoming missions to Mars.

5. Conclusions

The final aim of this work was to find a good approach to detect fungal biomarkers a to evaluate their detectability after exposure to simulated space and Mars-like conditio as well as to test instrumentation applicability for in situ analyses. Additionally, this wo is useful to outline a database of biomarker of Earth microbial life, to improve the detecti of biomarkers at Mars and to eliminate false-positives or negative detections [70]. In th context, we may consider melanin pigments as a fungal biomarker, owing to their hi stability when detected in Martian regolith analogues after simulated space and Mars-li exposure. In addition, the detection of fatty acids through a GC-MS approach is of utmo importance since these molecules are found in the membranes of all living organism and then, considered as potential biomarkers for life. Finally, although nucleic aci represent a controversial issue in the context of biomarkers, due to their lower stability a preservation over the time, our results demonstrate a good amplification and stability aft the exposure to simulated space and Martian conditions. The results of this work are ve appropriate given that the ExoMars Rosalind Franklin rover includes within its payloa Raman Laser Spectrometer and a Gas Chromatography–Mass Spectrometry instrument

Supplementary Materials: The following are available online at https://www.mdpi.com/article/ .3390/jof7100859/s1, Figure S1: Raman signal coverage, Figure S2: Raman peak position.

Author Contributions: Conceptualization: J.-P.P.d.V., E.R. and S.O.; methodology, A.C., C.P., M. and L.B.; formal analysis, A.C., C.P., M.B. and L.B.; investigation, C.P., A.C., M.B., L.B., R.S. a U.B.; resources, S.O.; data curation, A.C., C.P., M.B. and L.B.; writing—original draft preparati A.C. and C.P., writing—review and editing, A.C., C.P., M.B., L.B., R.S., U.B., E.R., J.-P.P.d.V. a S.O.; supervision, S.O.; funding acquisition, S.O. All authors have read and agreed to the publish version of the manuscript.

Funding: This work was supported by the Italian Space Agency (BIOMEX MicroColonial Fun Experiment on ISS for tracking biomarkers on Martian and lunar rock analogues—ASI N.2013-063-F and BioSigN MicroFossils—ASI N. 2018-6-U.0).

Institutional Review Board Statement: Not applicable.

Informed Consent Statement: Not applicable.

Data Availability Statement: Not applicable.

Acknowledgments: The Italian National Program of Antarctic Researches (PNRA) and the Itali National Antarctic Museum "Felice Ippolito" (MNA) are also acknowledged for funding the collecti of Antarctic samples CCFEE. Special acknowledgments also to Timm Roegler and its DLR-departme Programmatics in Space Research and Technology and to the Ministry of Economics and Energy f supporting these studies within BIOMEX and BioSigN.

Conflicts of Interest: The authors declare no conflict of interest.

References

1. Grotzinger, J.P.; Sumner, D.Y.; Kah, L.C.; Stack, K.; Gupta, S.; Edgar, L.; Rubin, D.; Lewis, K.; Schieber, J.; Mangold, N.; et al. Habitable Fluvio-Lacustrine Environment at Yellowknife Bay, Gale Crater, Mars. *Science* **2013**, *343*, 1242777. [CrossRef] [PubMe
2. Davila, A.F.; Duport, L.G.; Melchiorri, R.; Jaenchen, J.; Valea, S.; de Los Rios, A.; Wierzchos, J. Hygroscopic salts and the po-tent for life on Mars. *Astrobiology* **2010**, *10*, 617–628. [CrossRef] [PubMed]

Clifford, S.M.; Lasue, J.; Heggy, E.; Boisson, J.; McGovern, P.; Max, M.D. Depth of the Martian cryosphere: Revised estimates and implications for the existence and detection of subpermafrost groundwater. *J. Geophys. Res. Space Phys.* **2010**, *115*, E07001. [CrossRef]
Ehlmann, B.L.; Mustard, J.F.; Murchie, S.; Bibring, J.-P.; Meunier, A.; Fraeman, A.; Langevin, Y. Subsurface water and clay mineral formation during the early history of Mars. *Nature* **2011**, *479*, 53–60. [CrossRef]
Thomas, R.J.; Hynek, B.M.; Osterloo, M.M.; Kierein-Young, K.S. Widespread exposure of Noachian phyllosilicates in the Margaritifer region of Mars: Implications for paleohydrology and astrobiological detection. *J. Geophys. Res. Planets* **2017**, *122*, 483–500. [CrossRef]
Travis, B.J.; Rosenberg, N.D.; Cuzzi, J.N. On the role of widespread subsurface convection in bringing liquid water close to Mars' surface. *J. Geophys. Res. Space Phys.* **2003**, *108*, 8040. [CrossRef]
Dartnell, L.R.; Desorgher, L.; Ward, J.; Coates, A.J. Modelling the surface and subsurface Martian radiation environment: Implications for astrobiology. *Geophys. Res. Lett.* **2007**, *34*, L02207. [CrossRef]
Dartnell, L.R.; Page, K.; Jorge-Villar, S.E.; Wright, G.; Munshi, T.; Scowen, I.J.; Edwards, H.G. Destruction of Raman biosig-natures by ionising radiation and the implications for life detection on Mars. *Anal. Bioanal. Chem.* **2012**, *403*, 131–144. [CrossRef]
Hassler, D.M.; Zeitlin, C.; Wimmer-Schweingruber, R.F.; Ehresmann, B.; Rafkin, S.; Eigenbrode, J.L.; Brinza, D.E.; Weigle, G.; Böttcher, S.; Böhm, E.; et al. Mars' Surface Radiation Environment Measured with the Mars Science Laboratory's Curiosity Rover. *Science* **2013**, *343*, 1244797. [CrossRef]
Vago, J.; Witasse, O.; Svedhem, H.; Baglioni, P.; Haldemann, A.; Gianfiglio, G.; Blancquaert, T.; McCoy, D.; De Groot, R. ESA ExoMars program: The next step in exploring Mars. *Sol. Syst. Res.* **2015**, *49*, 518–528. [CrossRef]
Westall, F.; Foucher, F.; Bost, N.; Bertrand, M.; Loizeau, D.; Vago, J.L.; Kminek, G.; Gaboyer, F.; Campbell, K.; Bréhéret, J.-G.; et al. Biosignatures on Mars: What, Where, and How? Implications for the Search for Martian Life. *Astrobiology* **2015**, *15*, 998–1029. [CrossRef] [PubMed]
Summons, R.E.; Amend, J.P.; Bish, D.L.; Buick, R.; Cody, G.D.; Marais, D.J.D.; Dromart, G.; Eigenbrode, J.L.; Knoll, A.H.; Sumner, D. Preservation of Martian Organic and Environmental Records: Final Report of the Mars Biosignature Working Group. *Astrobiology* **2011**, *11*, 157–181. [CrossRef]
Vago, J.L.; Westall, F.; Teams, L.S.P.I.; Coates, A.; Jaumann, R.; Korablev, O.; Ciarletti, V.; Mitrofanov, I.; Josset, J.-L.; De Sanctis, M.C.; et al. Habitability on Early Mars and the Search for Biosignatures with the ExoMars Rover. *Astrobiology* **2017**, *17*, 471–510. [CrossRef]
Aerts, J.W.; Röling, W.F.; Elsaesser, A.; Ehrenfreund, P. Biota and biomolecules in extreme environments on Earth: Implica-tions for life detection on Mars. *Life* **2014**, *4*, 535–565. [CrossRef] [PubMed]
Eigenbrode, J.L.; Summons, R.E.; Steele, A.; Freissinet, C.; Millan, M.; Navarro-González, R.; Sutter, B.; McAdam, A.C.; Franz, H.B.; Glavin, D.P.; et al. Organic matter preserved in 3-billion-year-old mudstones at Gale crater, Mars. *Science* **2018**, *360*, 1096–1101. [CrossRef]
Ellery, A.; Wynnwilliams, D.D. Why Raman Spectroscopy on Mars?—A Case of the Right Tool for the Right Job. *Astrobiology* **2003**, *3*, 565–579. [CrossRef]
Wynn-Williams, D.; Edwards, H. Antarctic ecosystems as models for extraterrestrial surface habitats. *Planet. Space Sci.* **2000**, *48*, 1065–1075. [CrossRef]
Onofri, S.; Selbmann, L.; Zucconi, L.; Pagano, S. Antarctic microfungi as models for exobiology. *Planet. Space Sci.* **2004**, *52*, 229–237. [CrossRef]
Preston, L.J.; Barcenilla, R.; Dartnell, L.R.; Kucukkilic-Stephens, E.; Olsson-Francis, K. Infrared Spectroscopic Detection of Biosignatures at Lake Tírez, Spain: Implications for Mars. *Astrobiology* **2019**, *20*, 15–25. [CrossRef]
de Vera, J.P.; Boettger, U.; de la Torre Noetzel, R.; Sánchez, F.J.; Grunow, D.; Schmitz, N.; Rettberg, P. Supporting Mars ex-ploration: BIOMEX in Low Earth Orbit and further astrobiological studies on the Moon using Raman and PanCam technology. *Planet. Space Sci.* **2012**, *74*, 103–110. [CrossRef]
Onofri, S.; De La Torre, R.; De Vera, J.-P.; Ott, S.; Zucconi, L.; Selbmann, L.; Scalzi, G.; Venkateswaran, K.J.; Rabbow, E.; Iñigo, F.J.S.; et al. Survival of Rock-Colonizing Organisms After 1.5 Years in Outer Space. *Astrobiology* **2012**, *12*, 508–516. [CrossRef] [PubMed]
Onofri, S.; de Vera, J.P.; Zucconi, L.; Selbmann, L.; Scalzi, G.; Venkateswaran, K.J.; Horneck, G. Survival of Antarctic crypto-endolithic fungi in simulated Martian conditions on board the International Space Station. *Astrobiology* **2015**, *15*, 1052–1059. [CrossRef] [PubMed]
Böttger, U.; de Vera, J.P.; Fritz, J.; Weber, I.; Hübers, H.W.; Schulze-Makuch, D. Optimizing the detection of carotene in cya-nobacteria in a martian regolith analogue with a Raman spectrometer for the ExoMars mission. *Planet. Space Sci.* **2012**, *60*, 356–362. [CrossRef]
Rabbow, E.; Rettberg, P.; Barczyk, S.; Bohmeier, M.; Parpart, A.; Panitz, C.; Horneck, G.; Von Heise-Rotenburg, R.; Hoppenbrouwers, T.; Willnecker, R.; et al. EXPOSE-E: An ESA Astrobiology Mission 1.5 Years in Space. *Astrobiology* **2012**, *12*, 374–386. [CrossRef] [PubMed]
Rabbow, E.; Rettberg, P.; Barczyk, S.; Bohmeier, M.; Parpart, A.; Panitz, C.; Horneck, G.; Burfeindt, J.; Molter, F.; Jaramillo, E.; et al. The astrobiological mission EXPOSE-R on board of the International Space Station. *Int. J. Astrobiol.* **2014**, *14*, 3–16. [CrossRef]
Rosas, A.L.; Nosanchuk, J.D.; Feldmesser, M.; Cox, G.M.; McDade, H.C.; Casadevall, A. Synthesis of Polymerized Melanin by Cryptococcus neoformans in Infected Rodents. *Infect. Immun.* **2000**, *68*, 2845–2853. [CrossRef] [PubMed]

27. Raman, N.M.; Ramasamy, S. Genetic validation and spectroscopic detailing of DHN-melanin extracted from an environmer fungus. *Biochem. Biophys. Rep.* **2017**, *12*, 98–107. [CrossRef]
28. Baqué, M.; Hanke, F.; Böttger, U.; Leya, T.; Moeller, R.; De Vera, J.-P. Protection of cyanobacterial carotenoids' Raman signatu by Martian mineral analogues after high-dose gamma irradiation. *J. Raman Spectrosc.* **2018**, *49*, 1617–1627. [CrossRef]
29. Selbmann, L.; De Hoog, G.S.; Mazzaglia, A.; Friedmann, E.I.; Onofri, S. Fungi at the edge of life: Cryptoendolithic black fu from Antarctic desert. *Stud. Mycol.* **2005**, *51*, 1–32.
30. Cubeta, M.A.; Echandi, E.; Abernethy, T.; Vilgalys, R. Characterization of anastomosis groups of binucleate Rhizoctonia spec using restriction analysis of an amplified ribosomal RNA gene. *Phytopathology* **1991**, *81*, 1395–1400. [CrossRef]
31. Vilgalys, R.; Hester, M. Rapid genetic identification and mapping of enzymatically amplified ribosomal DNA from seve Cryptococcus species. *J. Bacteriol.* **1990**, *172*, 4238–4246. [CrossRef] [PubMed]
32. Carbone, I.; Kohn, L.M. A method for designing primer sets for speciation studies in filamentous ascomycetes. *Mycologia* **1999**, 553–556. [CrossRef]
33. Holland, P.M.; Chutjian, A.; Darrach, M.R.; Orient, O.J. Miniaturized GC/MS instrumentation for in situ measurements: Mi gas chromatography coupled with miniature quadrupole array and Paul ion trap mass spectrometers. *Proc. SPIE* **2003**, *4878*, [CrossRef]
34. Bell, A.A.; Wheeler, M.H. Biosynthesis and functions of fungal melanins. *Annu. Rev. Phytopathol.* **1986**, *24*, 411–451. [CrossRe
35. Cockell, C.S.; Knowland, J. Ultraviolet radiation screening compounds. *Biol. Rev.* **1999**, *74*, 311–345. [CrossRef]
36. Pacelli, C.; Cassaro, A.; Baqué, M.; Selbmann, L.; Zucconi, L.; Maturilli, A.; Botta, L.; Saladino, R.; de Vera, J.-P.P.; Onofri, S.; e Fungal biomarkers are detectable in Martian rock-analogues after space exposure: Implications for the search of life on Mars. *J. Astrobiol.* **2020**, in press.
37. Culka, A.; Jehlička, J.; Ascaso, C.; Artieda, O.; Casero, M.C.; Wierzchos, J. Raman microspectrometric study of pigments melanized fungi from the hyperarid Atacama desert gypsum crust. *J. Raman Spectrosc.* **2017**, *48*, 1487–1493. [CrossRef]
38. Samokhvalov, A.; Liu, Y.; Simon, J.D. Characterization of the Fe(III)-binding Site in Sepia Eumelanin by Resonance Ram Confocal Microspectroscopy. *Photochem. Photobiol.* **2007**, *80*, 84–88. [CrossRef]
39. Galván, I.; Jorge, A.; Ito, K.; Tabuchi, K.; Solano, F.; Wakamatsu, K. Raman spectroscopy as a non-invasive technique for quantification of melanins in feathers and hairs. *Pigment. Cell Melanoma Res.* **2013**, *26*, 917–923. [CrossRef]
40. Thiel, C.S.; Ehrenfreund, P.; Foing, B.; Pletser, V.; Ullrich, O. PCR-based analysis of microbial communities during the roGeoMars campaign at Mars Desert Research Station, Utah. *Int. J. Astrobiol.* **2012**, *10*, 177–190. [CrossRef]
41. Goesmann, F.; Brinckerhoff, W.B.; Raulin, F.; Goetz, W.; Danell, R.; Getty, S.A.; Siljeström, S.; Mißbach, H.; Steininger, Arevalo, R.D.; et al. The Mars Organic Molecule Analyzer (MOMA) Instrument: Characterization of Organic Material in Mart Sediments. *Astrobiology* **2017**, *17*, 655–685. [CrossRef]
42. Mahaffy, P.R.; Webster, C.R.; Cabane, M.; Conrad, P.G.; Coll, P.; Atreya, S.K.; Arvey, R.; Barciniak, M.; Benna, M.; Bleacher, L.; e The Sample Analysis at Mars Investigation and Instrument Suite. *Space Sci. Rev.* **2012**, *170*, 401–478. [CrossRef]
43. Melosh, H.J. Exchange of Meteorites (and Life?) Between Stellar Systems. *Astrobiology* **2003**, *3*, 207–215. [CrossRef]
44. Claudia, P.; Alessia, C.; Siong, L.M.; Lorenzo, A.; Ralf, M.; Akira, F.; Silvano, O. Insights into the Survival Capabilities Cryomyces antarcticus Hydrated Colonies after Exposure to Fe Particle Radiation. *J. Fungi* **2021**, *7*, 495.
45. Aureli, L.; Pacelli, C.; Cassaro, A.; Fujimori, A.; Moeller, R.; Onofri, S. Iron Ion Particle Radiation Resistance of Dried Colonies *Cryomyces antarcticus* Embedded in Martian Regolith Analogues. *Life* **2020**, *10*, 306. [CrossRef] [PubMed]
46. Eisenman, H.C.; Casadevall, A. Synthesis and assembly of fungal melanin. *Appl. Microbiol. Biotechnol.* **2011**, *93*, 931– [CrossRef] [PubMed]
47. Bhartia, R.; Hug, W.F.; Reid, R.D.; Beegle, L.W. Explosives detection and analysis by fusing deep ultraviolet native fluoresce and resonance Raman spectroscopy. In *Laser-Based Optical Detection of Explosives*; CRC Press: Boca Raton, FL, USA, 2015.
48. Samokhvalov, A.; Garguilo, J.; Yang, W.-C.; Edwards, G.S.; Nemanich, R.J.; Simon, J.D. Photoionization Threshold of melanosomes Determined Using UV Free Electron Laser−Photoelectron Emission Microscopy. *J. Phys. Chem. B* **2004**, 16334–16338. [CrossRef]
49. Pacelli, C.; Bryan, R.A.; Onofri, S.; Selbmann, L.; Zucconi, L.; Shuryak, I.; Dadachova, E. The effect of protracted X-ray expos on cell survival and metabolic activity of fast and slow growing fungi capable of melanogenesis. *Environ. Microbiol. Rep.* **2018**, 255–263. [CrossRef] [PubMed]
50. Van de Meent, D.; Brown, S.C.; Philp, R.; Simoneit, B.R. Pyrolysis-high resolution gas chromatography and pyrolysis chromatography-mass spectrometry of kerogens and kerogen precursors. *Geochim. Cosmochim. Acta* **1980**, *44*, 999–1013. [CrossR
51. Horsfield, B. Practical criteria for classifying kerogens: Some observations from pyrolysis-gas chromatography. *Geochim Cosmochim. Acta* **1989**, *53*, 891–901. [CrossRef]
52. Eglinton, T.; Damste, J.S.; Pool, W.; de Leeuw, J.W.; Eijkel, G.; Boon, J.J. Organic sulphur in macromolecular sediment organic matter. II. Analysis of distributions of sulphur-containing pyrolysis products using multivariate techniques. *Geochim Cosmochim. Acta* **1992**, *56*, 1545–1560. [CrossRef]
53. Reinhardt, M.; Goetz, W.; Thiel, V. Testing Flight-like Pyrolysis Gas Chromatography–Mass Spectrometry as Performed by Mars Organic Molecule Analyzer Onboard the ExoMars 2020 Rover on Oxia Planum Analog Samples. *Astrobiology* **2020**, 415–428. [CrossRef]

Faure, P.; Jeanneau, L.; Lannuzel, F. Analysis of organic matter by flash pyrolysis-gas chromatography–mass spectrometry in the presence of Na-smectite: When clay minerals lead to identical molecular signature. *Org. Geochem.* **2006**, *37*, 1900–1912. [CrossRef]
Moldoveanu, S. Chapter 17 Pyrolysis of Carboxylic Acids. *Tech. Instrum. Anal. Chem.* **2010**, *28*, 471–526. [CrossRef]
Dorn, E.D.; Nealson, K.H.; Adami, C. Monomer Abundance Distribution Patterns as a Universal Biosignature: Examples from Terrestrial and Digital Life. *J. Mol. Evol.* **2011**, *72*, 283–295. [CrossRef] [PubMed]
McKay, C.P. What is life—And how do we search for it in other worlds? *PLoS Biol.* **2004**, *2*, e302. [CrossRef] [PubMed]
Mustard, J.F.; Adler, M.; Allwood, A.; Bass, D.S.; Beaty, D.W.; Bell, J.F.; Edgett, K.S. Report of the Mars 2020 science definition team. *Mars Explor. Progr. Anal. Gr.* **2013**, *150*, 155–205.
Neveu, M.; Hays, L.E.; Voytek, M.A.; New, M.H.; Schulte, M.D. The Ladder of Life Detection. *Astrobiology* **2018**, *18*, 1375–1402. [CrossRef]
Kminek, G.; Bada, J. The effect of ionizing radiation on the preservation of amino acids on Mars. *Earth Planet. Sci. Lett.* **2006**, *245*, 1–5. [CrossRef]
Dabney, J.; Meyer, M.; Pääbo, S. Ancient DNA damage. *Csh. Perspect. Biol.* **2013**, *5*, a012567. [CrossRef]
Millar, C.D.; Lambert, D.M. Ancient DNA: Towards a million-year-old genome. *Nature* **2013**, *499*, 34–35. [CrossRef] [PubMed]
Alvarez, M.; Zeelen, J.P.; Mainfroid, V.; Rentier-Delrue, F.; Martial, J.A.; Wyns, L.; Maes, D. Triose-phosphate Isomerase (TIM) of the Psychrophilic Bacterium Vibrio marinus Kinetic and Structural Properties. *J. Biol. Chem.* **1998**, *273*, 2199–2206. [CrossRef] [PubMed]
Isenbarger, T.A.; Carr, C.E.; Johnson, S.S.; Finney, M.; Church, G.M.; Gilbert, W.; Zuber, M.T.; Ruvkun, G. The Most Conserved Genome Segments for Life Detection on Earth and Other Planets. *Orig. Life Evol. Biosphere* **2008**, *38*, 517–533. [CrossRef] [PubMed]
Fajardo-Cavazos, P.; Schuerger, A.C.; Nicholson, W.L. Exposure of DNA and Bacillus subtilis Spores to Simulated Martian Environments: Use of Quantitative PCR (qPCR) to Measure Inactivation Rates of DNA to Function as a Template Molecule. *Astrobiology* **2010**, *10*, 403–411. [CrossRef]
Haberle, R.M.; Juarez, M.D.L.T.; Kahre, M.A.; Kass, D.M.; Barnes, J.R.; Hollingsworth, J.L.; Harri, A.-M.; Kahanpää, H. Detection of Northern Hemisphere transient eddies at Gale Crater Mars. *Icarus* **2018**, *307*, 150–160. [CrossRef]
John, K.K.; Botkin, D.S.; Burton, A.S.; Castro-Wallace, S.L.; Chaput, J.D.; Dworkin, J.P.; Stahl, S. The Biomolecule Sequencer Project: Nanopore Sequencing as a Dual-Use Tool for Crew Health and Astrobiology Investigations. 2016. Available online: https://pdxscholar.library.pdx.edu/chem_fac/123/ (accessed on 8 October 2021).
Castro-Wallace, S.L.; Chiu, C.Y.; John, K.K.; Stahl, S.E.; Rubins, K.H.; McIntyre, A.B.; Stephenson, T.A. Nanopore DNA se-quencing and genome assembly on the International Space Station. *Sci. Rep.* **2017**, *7*, 1–12. [CrossRef]
Maggiori, C.; Stromberg, J.; Blanco, Y.; Goordial, J.; Cloutis, E.; García-Villadangos, M.; Parro, V.; Whyte, L. The Limits, Capabilities, and Potential for Life Detection with MinION Sequencing in a Paleochannel Mars Analog. *Astrobiology* **2020**, *20*, 375–393. [CrossRef]
Javaux, E.J. Extreme life on Earth—past, present and possibly beyond. *Res. Microbiol.* **2006**, *157*, 37–48. [CrossRef]

MDPI
St. Alban-Anlage 66
4052 Basel
Switzerland
Tel. +41 61 683 77 34
Fax +41 61 302 89 18
www.mdpi.com

Journal of Fungi Editorial Office
E-mail: jof@mdpi.com
www.mdpi.com/journal/jof

www.ingramcontent.com/pod-product-compliance
Lightning Source LLC
LaVergne TN
LVHW070923170726
843515LV00005B/853

* 9 7 8 3 0 3 6 5 5 8 1 2 7 *